FIRST INTERNATIONAL CONFERENCE ON
PHYSICAL COASTAL PROCESSES, MANAGEMENT
AND ENGINEERING

COASTAL PROCESSES

CONFERENCE CHAIRMEN

C.A. Brebbia
Wessex Institute of Technology, UK

G. Benassai
University of Pharthenope, Italy

G. Rodriguez
University of Las Palmas, Spain

INTERNATIONAL SCIENTIFIC ADVISORY COMMITTEE

Organised by

Wessex Institute of Technology, UK
University of Pharthenope, Italy
University of Las Palmas (Canary Islands), Spain

Sponsored by

WIT Transactions on Ecology and the Environment

WIT Transactions

Transactions Editor

Carlos Brebbia
Wessex Institute of Technology
Ashurst Lodge, Ashurst
Southampton SO40 7AA, UK
Email: carlos@wessex.ac.uk

Editorial Board

Coastal Processes

Editors

C.A. Brebbia
Wessex Institute of Technology, UK

G. Benassai
University of Naples Parthenope, Italy

G.R. Rodriguez
Universidad de Las Palmas de Gran Canaria, Spain

WIT*PRESS* Southampton, Boston

Editors

C.A. Brebbia
Wessex Institute of Technology, UK

G. Benassai
University of Naples Parthenope, Italy

G.R. Rodriguez
Universidad de Las Palmas de Gran Canaria, Spain

Published by

WIT Press
Ashurst Lodge, Ashurst, Southampton, SO40 7AA, UK
Tel: 44 (0) 238 029 3223; Fax: 44 (0) 238 029 2853
E-Mail: witpress@witpress.com
http://www.witpress.com

For USA, Canada and Mexico

Computational Mechanics Inc
25 Bridge Street, Billerica, MA 01821, USA
Tel: 978 667 5841; Fax: 978 667 7582
E-Mail: infousa@witpress.com
http://www.witpress.com

British Library Cataloguing-in-Publication Data

A Catalogue record for this book is available
from the British Library

ISBN: 978-1-84564-200-6
ISSN: 1746-448X (print)
ISSN: 1743-3541(online)

The texts of the papers in this volume were set individually by the authors or under their supervision. Only minor corrections to the text may have been carried out by the publisher.

Printed in Great Britain by Athenaeum Press Ltd.

Preface

Coastal regions present a complex dynamic web of natural and human related processes. Although coastal zones are narrow areas extending a few kilometres on either side of the shoreline, and occupying small strip of ocean and land, they play a very important role as they account for nearly a quarter of all oceanic biological production, which in turn supplies approximately 80% of the world's fish. About 60% of the human population live in the coastal zone, and around 70% of big cities are placed in this narrow area. Concomitantly, more than 90% of the pollutants generated by human economic activities end up in the coastal zone.

The unstoppable demand of the coast for recreational and tourism activities has increased the need for shore and beach protection as well as the construction of artificial beaches, ports and harbours. Most of the coastlines are subjected to the direct impact of wind waves, swell and storm wave activity. As a result, wind waves and wave driven currents are the dominant mechanisms controlling littoral sand transport and determining the nearshore morphology. In addition, many other physical phenomena, such as tides and associated currents, long waves and storm surges, among others, can play a significant role in the dynamic behaviour of the coastal zone.

Coastal zones represent potential sources of renewable energy generated from winds, waves, tides, currents and thermal gradients. However, the coastal zone is also exposed to risks related to energy generation. Thus, for instance, extraction and transportation of hydrocarbons can give rise to major ecological disasters. Furthermore, thermal and nuclear power plants are often located in the coastal zone and use large volumes of cooling water which are discharged into the marine environment. If this occurs in shallow water, the physical properties of sea water and the local hydrodynamics are affected.

It is well known that distinctive features of the coastal zone dynamics are not only due to the nearshore hydrodynamics, but also to the complex local behaviour of the atmospheric dynamics. Thus, understanding the meteorology of the coastal zone is complicated by the inherent heterogeneity of its atmospheric boundary layer, due to the irregularity of the coastal topography, the different land-sea surface roughness and thermal properties. As a result, complex interactions occur between the atmosphere, ocean and land, inducing large temporal and spatial variations in air-sea exchange processes and in the strength and direction of the wind.

Due to its great socio-economic importance, the physical aspects of coastal processes have been of concern for decades, but recent advances in a number of areas, including satellite remote sensing, are giving rise to significant progress in this field. In particular, the use of satellite and imaging systems has significantly enhanced the monitoring and understanding of coastal processes.

Accordingly, it has become clear that the ocean side of the coastal zone represents a very sensitive and particularly vulnerable sector of the ocean to any kind of man-made action or natural extreme events. Consequently, the problem of environmental protection and conservation takes special relevance in this zone, and any decision concerning its viability must be preceded by a forecast of its consequences. Their adequate prediction is only possible on the basis of a clear understanding and careful analysis of the fundamental dynamic processes occurring in such areas.

In order to reach satisfactory solutions for the demands imposed on the coastal areas and the protection of its environment, one needs to understand very different aspects and their interaction. The problems are essentially interdisciplinary and scientists need to be able to exchange ideas with colleagues from other disciplines with a variety of different experiences.

The application of the principles of sustainable development on coastal zones, together with the need to protect the environment and control the physical mechanisms acting on them is the reason why this book provides an interdisciplinary approach.

The book comprises the edited papers of the first International Conference on Coastal Processes held in Malta in 2009, and grouped into the following topics:

- Wave modelling
- Wave transformation hydrodynamics
- Extreme events and sea level rise
- Sea defence and energy recovery
- Hydrodynamic forces and sediment transport
- Pollution and dispersion
- Planning and beach design

The Editors are grateful to all the authors for their excellent contributions as well as to the members of the International Scientific Advisory Committee for the review of both the abstracts and the papers included in this book. The quality of the material makes this volume a most valuable and up-to-date tool for professionals, scientists and managers to appreciate the state-of-the-art in this important field of knowledge.

The Editors
Malta 2009

Contents

Section 1
Wave modelling

Modelling mean wave direction distribution with the von Mises model

J. L. Vega & G. Rodríguez
Departamento de Física, Universidad de Las Palmas de Gran Canaria, Spain

Abstract

This paper presents the probabilistic modelling of the mean wave direction derived from directional spectral analysis of waves recorded by directional buoys. The analysis is performed on the mean wave direction in terms of the climatic season, the sea state severity and the period of the dominant waves. The usefulness of the von Mises theoretical models to describe the empirical kernel density estimates is examined. It is observed that the single von Mises theoretical model results are useful to fit the observed distribution only for moderate and severe sea states while the mixture of two von Mises distributions enhances significantly the degree of fitness.
Keywords: wave modelling, mean wave direction, kernel density estimation, circular variables, von Mises distribution, von Mises mixtures.

1 Introduction

Probabilistic design and assessment of marine structures interacting with sea waves requires a reliable knowledge of the long-term wave climate. In this context, it is generally assumed that directional wave spectra provide a complete description of a given sea state. However, it is common practice to accept that a sea state in simpler terms is reasonably well characterized by means of three parameters derived from it. These are the significant wave height H_{m0}, the spectral peak period T_p and the mean direction θ_m. Accordingly, it is generally assumed that long time series of these parameters allows one to obtain a convenient description of the long-term wave climate by estimating the joint and

WIT Transactions on Ecology and the Environment, Vol 126, © 2009 WIT Press
www.witpress.com, ISSN 1743-3541 (on-line)
doi:10.2495/CP090011

marginal probability density functions of these three parameters. Nonetheless, the sea state severity is commonly presented in the form of univariate and bivariate histograms of significant wave height and the peak period, or the mean zero-upcrossing wave period.

The above commented procedure does not give any consideration to the directions of waves approaching at a site. It implicitly assumes that all wave directions are equally likely to occur, or in other words, there are no preferred directions for the sea states approaching the point of measurement. However, directional wave information is often required for a variety of applications, including coastal engineering and nearshore dynamics, waste dispersal and pollution studies, sediment transport and beach erosion. Thus, for example, wave-driven currents are the principal mechanism for sand transport on most of the world coasts. The direction of these currents is governed by the direction of the deepwater ocean waves and their subsequent refraction over the shoaling zone. As a consequence, the long-term wave climate is properly described by the joint probability density function of the significant wave height, spectral peak period and average wave direction. Nevertheless, in practical applications using directional information, it is common to use the conditional joint distributions of wave height and period given the mean wave direction, for a certain number of directional sectors.

This study focuses on the long-term scale mean wave direction variability, which is necessary to identify the wave climate in a given area. In particular, the purpose of the current paper is to establish the main properties of the mean wave directional regime in the area of the directional buoy located off Estaca de Bares, a location in the Galician coast, at the Spanish North-Atlantic coast, and to assess the usefulness of the von Mises probability model to fit the observed probability distribution, considering the single symmetric and unimodal von Mises and the two components mixture of von Mises theoretical probability models.

To reach this goal, long time series of H_{m0}, T_p, and θ_m, are analysed. The marginal probabilistic structure of the mean wave direction, as well as its variability as a function of climatic seasons is examined, and contrasted with the single von Mises and a mixture of two von Mises models for circular random variables. Furthermore, the conditional distributions of mean wave direction for various wave height and period thresholds are estimated and the usefulness of these theoretical models to fit the observed empirical distributions is assessed.

The presentation of the study is structured as follows. Section 2 describes the instrumental wave measurement data set examined in the study and indicates the location of the buoy used for recording the data. This section includes a brief summary of the kernel density estimation procedure with emphasis on its use for circular data. Next, in section 3, the main properties of the von Mises probability distribution family, used as theoretical model to fit the observed density functions are summarised. The discussion of results obtained by examining the observed time series is presented in section 4. Concluding remarks are given in section 5.

2 The field site and data analysis

2.1 Field site and recorded data

Wave observations used in this study were performed by a buoy of the offshore network implemented by Puertos del Estado (Ministerio de Fomento). This is a Seawatch directional buoy deployed in the Galician coast (Northwest of Spain), offshore from A Coruña, at a point of latitude 44º 03,94' N and longitude 07º 37,27' W. This location is shown in Figure 1 for helping the comments of results. The water depth at the measuring point was 387 meters.

The Seawatch buoy measured short-term records at three hour intervals during 1997 and hourly till 2003. Measurements used in this study span over a relatively long period, starting on January 1997 and lasting on December 2003.

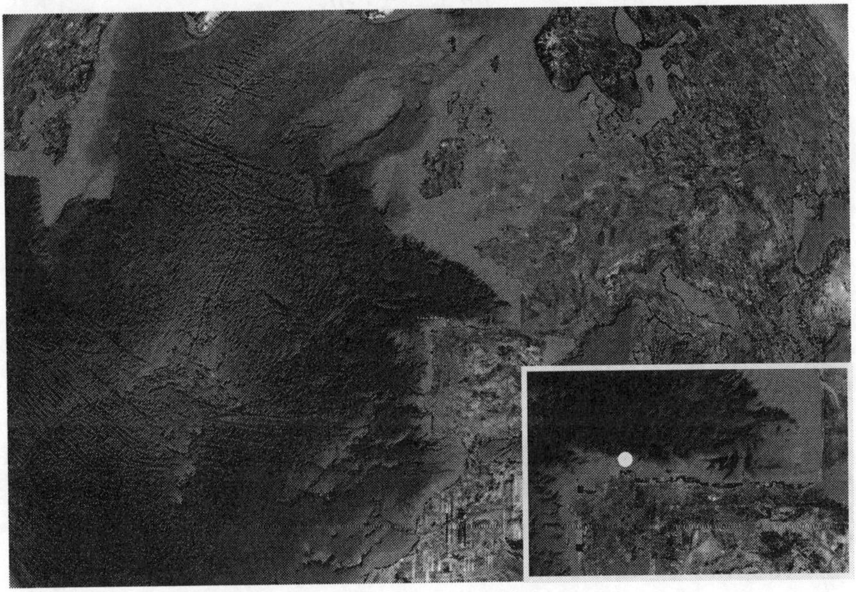

Figure 1: Location of the directional buoy offshore of Estaca de Bares, (Galician Coast) Northwest Spain.

Due to problems with power supply, change of the internal battery of the buoy, failure in remote data transmission via satellite, retrieval for cleaning biofouling growth on the outer surface, and other logistical mishaps caused loss of data during some periods. Hence 100% data could not be collected. Also the collected time series were subjected to error checks and only the records which were found suitable were included for posterior analysis. A total of 43227 usable records were obtained over this period of seven years.

Short-term analysis of directional wind-wave spectra permits to derive various characteristic parameters, such as the significant wave height H_{m0}, the peak period T_p and the mean wave direction θ_m, among others.

Thus, the data base used in this study consisted of an hourly (except during 1997) valued trivariate time series $\{H_{m0}, T_p, \theta_m\}$.

2.2 Kernel density estimation

Kernel density estimation is a nonparametric method of estimating a pdf from data that is related to the histogram [1]. Given a set of n observations $x_1; \ldots; x_n$, the kernel density estimate (*kde*) is defined as.

$$f(x) = \frac{1}{nh}\sum_{i=1}^{n} K\left(\frac{x - x_i}{h}\right) \tag{1}$$

where K is a function named as the kernel and given by some smooth density. In practice however, the choice of kernel appears to have very little effect on the performance of the kernel estimator, and in most cases the Gaussian kernel is used for simplicity, such as in this study. In contrast, the choice of bandwidth, h, is of crucial importance for the performance of the kde. The bandwidth is a positive number. The value of h basically decides how many observations are included in the estimation of f(x) at the point x. So a small choice of bandwidth means that only observations very close to x are used in the estimation, while a large bandwidth includes most of the observations in the sample. Since the observations close to x are more likely to carry information about the density's behaviour at that point, we would expect precision of the density estimator to increase, and thereby the bias to decrease, as we decrease h. On the other hand, as we decrease h, fewer observations are used to estimate f(x), so we would expect the variance of our estimator to increase as we decrease h. So, there is a tradeoff between choosing a small vs. a large bandwidth. Nevertheless, some practical rules which permit to obtain reasonable values have been proposed. In the present paper that suggested in Fisher [2] has been used. That is,

$$h = \sqrt{7}\frac{\sigma}{n^{-1/5}} = \sqrt{\frac{7}{\kappa}}n^{-1/5} \tag{2}$$

In the particular case of directional data, the natural domain of definition of random variables is limited to an interval bounded on both sides. A useful approach to deal with data on the finite interval $[0, 2\pi)$ is to impose periodic boundary conditions. That is, to wrap the kernel round the circle. Nevertheless, from a computational point of view, a simpler procedure consists in augmenting the data set by replicating it twice on the intervals $[-2\pi, 0]$ and $[2\pi, 4\pi]$. The estimated probability density is then drawn in the range $[0, 2\pi)$, such as suggested by Silverman [1]. This procedure has been used by Vega and Rodriguez [3].

3 The von Mises probability distribution

A circular random variable, θ, is defined as a random variable with support on the unit circle, i.e. the angle is in the range (0, 2π) radians or (0º, 360º). The probability density function of a circular variable, f(θ), is always positive

$$f(\theta)\geq 0 \quad \forall \quad 0\leq\theta\leq 2\pi \tag{3}$$

satisfies the normalization condition

$$\int_0^{2\pi} f(\theta)d\theta = 1 \tag{4}$$

and the periodicity condition

$$f(\theta)=f(\theta+n2\pi) \quad ;n=0,\pm 1,\pm 2,\cdots \tag{5}$$

This condition is peculiar to circular random variables. A linear random variable, R, is defined as a continuous random variable with support on the whole real line or an interval on it. Consequently, circular random variables must be analysed by using techniques differing from those appropriate for the usual Euclidean type variables because the circumference is a bounded closed space, for which the concept of origin is arbitrary or undefined.

The probability distribution model most frequently used in applications involving circular random variables is the von Mises family. A circular random variable, θ, is said to follow a von Mises distribution with parameters μ and κ, VM(μ, κ), if its probability density function is given by

$$f(\theta)=\frac{1}{2\pi I_0(\kappa)}\exp[\kappa\cos(\theta-\mu)] \quad ;0\leq\mu\leq 2\pi \quad ,\kappa>0 \tag{6}$$

where μ is the mean direction and κ is called the concentration parameter, and $I_0(\kappa)$ is the modified Bessel function of the first kind and order zero. Details on the estimation of these parameters can be found in Mardia and Jupp [4].

Unfortunately, the von Mises density function of a circular random variable is unimodal and symmetric. These facts make the von Mises model unsuitable for analysing circular data with more complicated features such as multimodality and/or skewness. One possible alternative in these situations is to use a mixture of von Mises distributions, given by

$$f(\theta)=\sum_{i=1}^{N}\lambda_i\frac{1}{2\pi I_0(\kappa_i)}\exp[\kappa_i\cos(\theta-\mu_i)] \tag{7}$$

where, for the *i*th component, p_i is the mixing proportion, μ_I is the mean direction and κ_i is the concentration parameter. The finite mixtures of von Mises distributions in both mean direction and concentration parameters are widely used in many disciplines, including astronomy, ecology, geology and medicine. However, it is worthwhile to mention that the use of mixtures of von Mises distributions present two main drawbacks. One is that testing the order or the number of components necessary in a circular mixture is a challenging problem, and the other one is the complexity of numerical routines necessary to fit these models with the maximum likelihood estimation [5]. For these reasons, the study limits to the use of second order von Mises mixtures and its comparison with a single von Mises distribution. Various methods for estimating the parameters in a von Mises mixture have been suggested in the literature. A comparison of several of these procedures has been presented in Spurr and Koutbeiy [6]. In this study, the estimation of the unknown parameters has been carried out by means of a non-linear least squares method, based on the Levenberg-Marquardt algorithm.

4 Results and discussion

The mean wave direction of the complete data set is represented, as a rose diagram in Figure 2 (left). It is observed that the largest part of sea states arrives at the zone coming from the WNW-NW sector. This is a clear indication of the fetch restrictions due to the geographical location of the buoy. Sea states approaching the buoy with South component are strongly restricted by the orientation of the North Spanish coast. Furthermore, the Gulf of Vizcaya conforms a relatively short basin in the sector N-E, in comparison with the sector N-W, open to Northwest Atlantic, where longer fetches can be delineated.

Detailed examination of the wave direction rose of Figure 2 reveals the presence of a secondary peak around the NE-ENE sector. The bimodality precludes the possibility that the single von Mises distribution fits properly the empirical distribution, such as observed in the right part of Figure 2. In this case the agreement between the mixture of two VM and the empirical distribution seems to be adequate, even though it probably could be improved by adding one more component to the von Mises mixture.

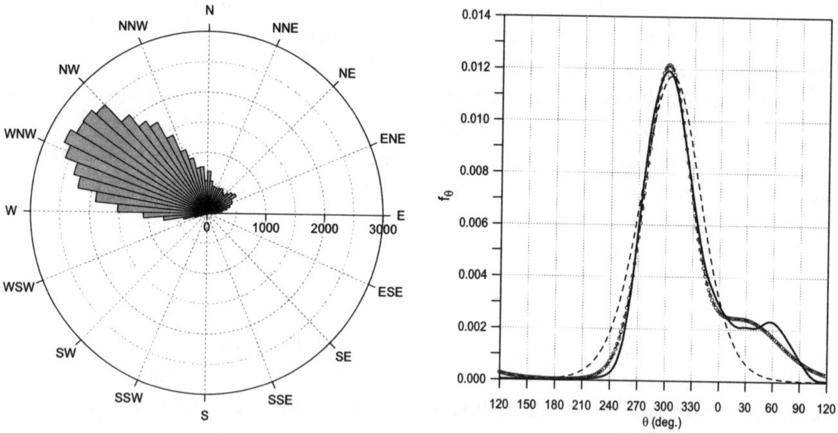

Figure 2: Mean wave direction rose (left) and empirical probability distribution of mean wave direction (solid line), and fitted VM (dashed line) and two VM mixture (dotted line) models (right), for the whole analysed period.

In spite of the bimodal character of the empirical distribution of mean wave direction, it is considerably clustered around a value close to 317 degrees, mainly in the sector between 240° and 360°. Thus, the contribution to the kde of the secondary peak could be due to some atmospheric conditions prevailing during some period during the year, or could be associated to some weather conditions occurring during a non specific time period. With this in mind, sea states have been classified by climatic seasons, sea state severity and the length or period of the dominant waves present in wave trains.

Results in terms of climatic seasons are shown in Figure 3. It can be observed that the distinction between the principal and the secondary peaks is smaller during summer. During this period the frequency of sea states approaching from the NE-ENE sector is lower and the bimodality reduces with an improvement in the fit with the two VM mixtures. The distinction between both peaks enhances progressively toward winter and accordingly the agreement between the mixture model and the *kde* get worse. Hence, separation of sea states by climatic seasons reveals that the bimodal character persists during the year and that the single VM model is not able to fit the observed probability distribution. In this context, the mixture of two VM improves the degree of fitness in all the seasons especially during summer, when the distribution is clearly asymmetric but bimodality reduces. This fact is reflected in the maximum likelihood estimated value of κ, the concentration parameter of the VM model, which is a measurement of the concentration around the mean direction, such as revealed by values given in Table 1, which gives the number of sea states, the average mean wave direction and κ for the full data set and for each season. It is observed that κ increases in summer enhancing the distribution peakedness.

Taking into account that the bimodal character of the observed mean wave direction distribution is not dependent of the climatic season, the variability of the observed distribution has been examined in terms of the sea state severity and by considering whether sea states approaching the measurement site have been locally or remotely generated, that is, filtering the data set for different significant wave height and spectral peak period thresholds.

The directional relationships during calmer periods are less relevant in the design of offshore and coastal systems; the greater interest is in the behaviour at higher sea states. The conditional mean wave direction distribution for six significant wave height thresholds, from 1 to 6 meters, is shown in Figure 4. It is evidenced that empirical distributions narrows and enhance around the modal value of θ_m, which shifts westward, as the H_{m0} threshold increases. This evolution is reflected by the statistical values given in Table 2. It is remarkable that for low thresholds the observed distribution present a similar structure to that associated to the full data set, but as the significant wave height threshold increases the von Mises, model, and naturally the two components mixture, fits more and more properly to the empirical density estimate because of the distribution becomes more concentrated about the mean direction, the symmetry increases and the bimodal character tends to disappear. These results reveal that the secondary peak is associated with sea states of low or moderate severity.

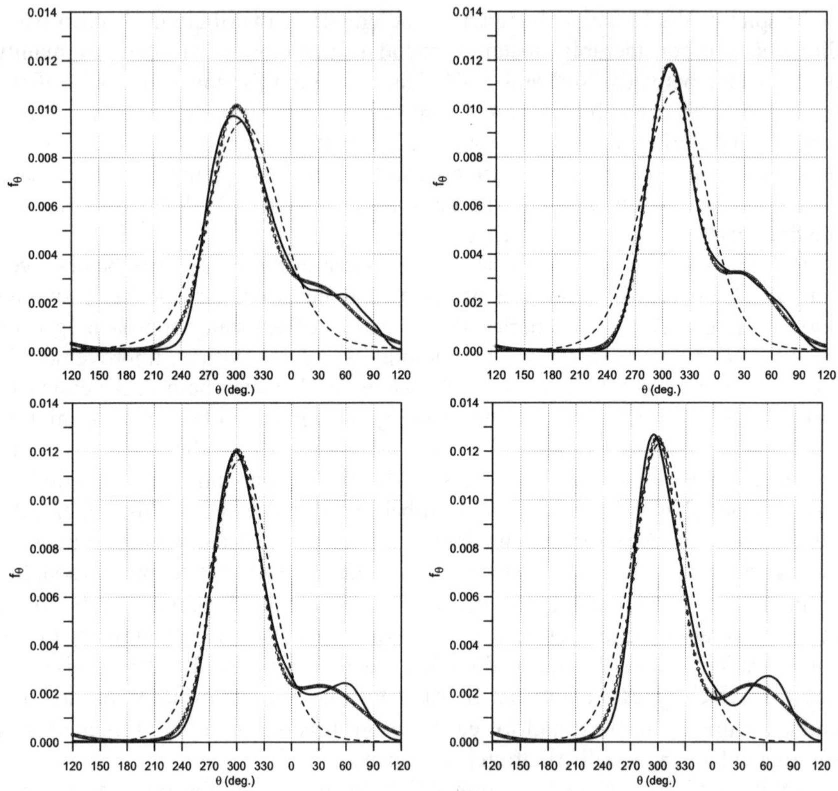

Figure 3: Empirical probability distribution of mean wave direction (solid line), the VM (dashed line) and two VM mixture models fitted (dotted line) for spring (up-left), summer (up-right), autumn (down-left), and winter (down-right) seasons.

The evolution of the empirical density function of the mean wave direction and the fit to a single VM and a two VM mixture for four threshold spectral peak period is shown in Figure 5. The corresponding values of the thresholds, as well as the basic circular statistics are given in Table 3. Results show that empirical distributions narrows and enhance around θ_m, which shifts westward, as the T_P threshold increases. Furthermore, it can be observed that for large values of T_P, that is, by removing low period sea states, the empirical distribution fits better to the von Mises models. This fact is particularly true for the mixture of two VM because the bimodal character disappear by filtering the sea states with low wave spectral period but even for large thresholds the distribution remains skewed, with a smoothly decaying plateau in the NE quadrant, which makes the single VM fail to fit the empirical *kde*.

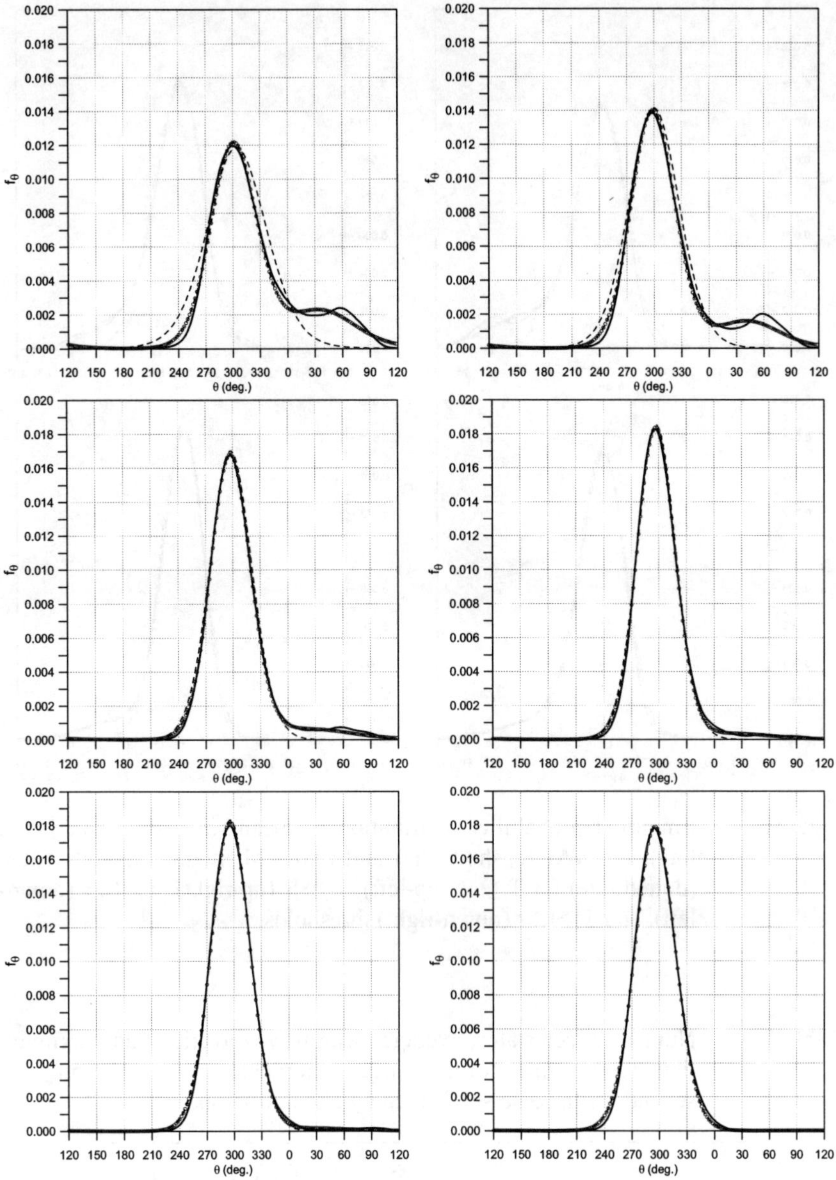

Figure 4: Empirical probability distribution of mean wave direction (solid line), the VM (dashed line) and two VM mixture models fitted (dotted line) for $H_{m0} \geq 1m$ (up-left), $H_{m0} \geq 2m$ (up-right), $H_{m0} \geq 3m$ (middle-left), $H_{m0} \geq 4m$ (middle-right), $H_{m0} \geq 5m$ (down-left), and $H_{m0} \geq 6m$ (down-right) thresholds.

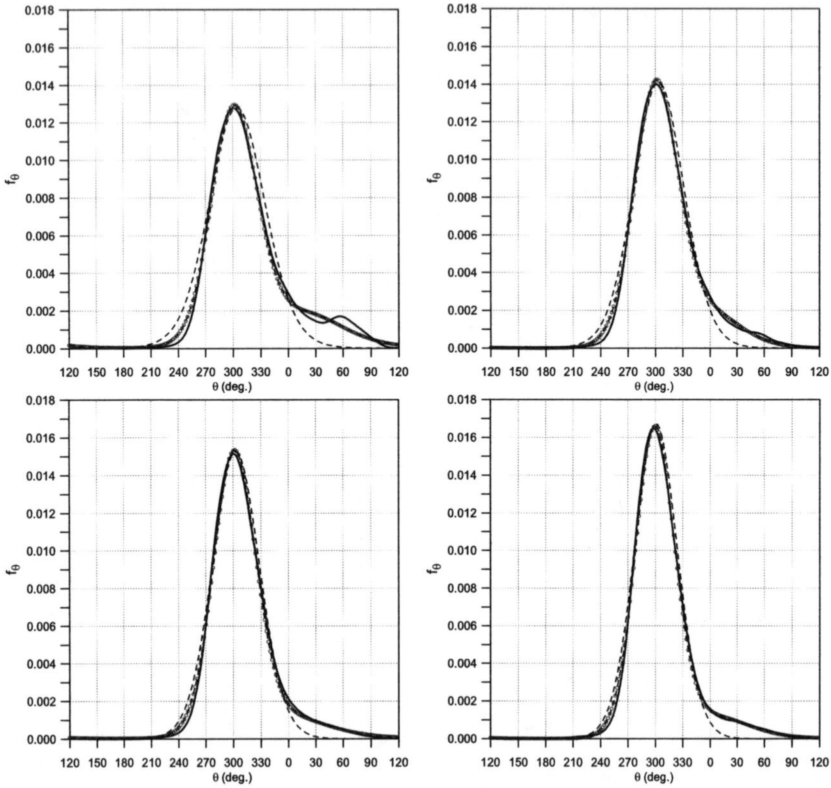

Figure 5: Empirical probability distribution of mean wave direction (solid line), the VM (dashed line) and two VM mixture models fitted (dotted line) for $T_z \geq 6$ s (up-left), $T_z \geq 8$ s (up-right), $T_z \geq 10$ s (down-left), and $T_z \geq 12$ s (down-right) thresholds.

Table 1: Number of sea states, average mean wave direction and maximum likelihood estimate of the concentration parameter for the full data set and for the data set filtered by climatic season.

	N	θ_m (°)	κ
Complete set	43 227	317.5	2.18
Spring	10 663	319.8	1.91
Summer	10 727	326.0	2.20
Autumn	11 726	316.3	2.08
Winter	10 111	315.8	2.00

Table 2: Number of sea states, average mean wave direction and maximum likelihood estimate of the concentration parameter for the data set filtered with different significant wave height thresholds.

	N	θ_m (°)	κ
H > 1 m.	33 782	325.5	3.85
H > 2 m.	15 840	319.6	5.73
H > 3 m.	6 219	316.1	10.93
H > 4 m.	2 264	316.7	12.83
H > 5 m.	837	315.5	14.37
H > 6 m.	218	314.6	16.46

Table 3: Number of sea states, average mean wave direction, and maximum likelihood estimate of the concentration parameter for the data set filtered with different spectral peak period thresholds.

	N	θ_m (°)	κ
Ts > 6 s.	39 854	313.7	2.58
Ts > 8 s.	33 004	309.5	3.50
Ts > 10 s.	20 351	307.7	3.98
Ts > 12 s.	9 646	305.4	4.29

5 Concluding remarks

The study reveals that a single VM distribution is not adequate in general to characterize the probabilistic structure of the mean wave direction in the study site, even when considering the various climatic seasons independently. However, this model becomes useful when examining mean wave directions associated to sea states with moderate and large periods, removing low period sea states, or for the more severe wave conditions, in terms of significant wave heights.

The use of a mixture of two VM models significantly improves the degree of fitness in all the cases, especially when the empirical distribution presents a bimodal character, or when is unimodal but significantly skewed, such as in the case of the full data set, the seasonal distributions, or when severe sea states are considered.

The present analysis is site specific and no attempt has been made to draw general conclusions for wider sea areas. However, the used methodology is of wide application. Furthermore, results derived from this study should help the development of joint distributions of the three parameters considered to characterise long-term wave climate.

Acknowledgement

The authors are grateful to Puertos del Estado, Ministerio de Fomento, Spain, for providing the data used in this study.

References

[1] Silverman, B.W. 1986. *Density Estimation for Statistics and Data Analysis*. London: Chapman and Hall.
[2] Fisher, N.I. 1993. *Statistical Analysis of Circular Data*. New York: Cambridge University Press.
[3] Vega J.L and G. Rodriguez, 2007. Modelling long term distribution of mean wave direction, *Proc. of the 12th Int. Conf. of the International Maritime Association of the Mediterranean*, Eds. Guedes Soares and Kolev, Balkema, 839-846.
[4] Mardia, K.V. and Jupp, P.E., 1999. *Directional Statistics*. Wiley, Chichester.
[5] Mooney, J. A., P.J. Helms, and I.T. Jolliffe, 2003. Fitting mixtures of von Mises distributions: a case study involving sudden infant death syndrome. *Computational Statistics and Data Analysis*, **41**: 505-513.
[6] Spurr, B.D. and M. A. Koutbeiy, 1991. A comparison of various methods for estimating the parameters in mixtures of von Mises distributions. *Communications in Statistics*, **20**: 725-741.

An analysis of measurement from a 3D oceanic wave field

P. C. Liu[1], C. H. Wu[2], K. R. MacHutchon[3] & D. J. Schwab[1]
[1]NOAA Great Lakes Environmental Research Laboratory, USA
[2]Department of Civil and Environmental Engineering, University of Wisconsin-Madison, USA
[3]Liebenberg & Stander International (Pty) Ltd, South Africa

Abstract

We present here a preliminary examination and analysis of a small suite of 3D wave data to explore what new insight or inference we can garner – particularly toward the realm where conventional approaches have not yet been. While we caught a few glimpses that might indicate a need for new conceptualizations, it by no means negates the vast positive contributions that the conventional approaches have allowed us to make in the past century. We feel it is timely to encourage further 3D ocean wave measurement and thereby facilitate fresh new states of study and to enhance our understanding of ocean waves.
Keywords: wind waves, 3D wave measurements, ocean waves, wave data analysis.

1 Introduction

We consider here the configuration of ocean waves to be a three-dimensional phenomenon in the sense that it is a function of (x, y, z, t) or more specifically, z = f(x, y, t). Note that conventional time series wave measurement at a fixed single point, $z = f(x_0, t)$, is basically a function of one-dimensional surface fluctuation at a given single point, x_0, with respect to time t. Incidentally data of the case $z = f(x, t_0)$, which is in essence a snapshot of an ocean segment at a single time point, t_0, with respect to an one-dimensional direction x, is also a single point measurement.

So for over six decades, the ocean wave research community has been content with a general perception of ocean waves that was predominantly based on

WIT Transactions on Ecology and the Environment, Vol 126, © 2009 WIT Press
www.witpress.com, ISSN 1743-3541 (on-line)
doi:10.2495/CP090021

single point in-situ wave measurement, (x_0, t), or (x, t_0), either Eulerian from fixed probes or Lagrangian from floating instruments. As a result, the present day conventional conceptualization of ocean wave studies has been strictly (x_0, t) oriented, with seemingly three-dimensional dynamics and models built around it. It has thus forged a kind of subjective reality for which whole ocean wave processes are described through this one-dimensional single-point realization of conventional wave measurements.

But the need for more realistic ocean wave measurements is gradually being recognized. At the recent OMAE 2008 Conference, there were at least two separate presentations, by Liu *et al.* [3] and Gallego *et al.* [1], independently advocating non-intrusive stereo imaging measurement with three and two digital video cameras systems respectively. The technology of digital cameras has advanced by leaps and bounds in recent years. And at the same time, the study of wind waves and wave modelling, actuated through five decades of single-point wave measurements, may have been "reaching a cul-de-sac, yielding 'no more great revelations or revolutions, but only incremental, diminishing returns'" as Horgan [2] regarded as the predicament of general science over a decade ago.

The ultimate goal of the development of these new stereo measurement systems is undoubtedly to provide three-dimensional wave surface fluctuations with respect to time, $z = f(x, y, t)$. Hopefully this new approach will become the mainstay of wave measurement and analysis studies and replace the traditional one-dimensional single-point time-series data analysis method. Provided, of course, we can manage to rise above our deeply seated, familiar comfort zone of one-dimensional mentality.

2 Before single point wave measurement

Nearly two decades before the advent of conventional single-point wave measurements in the mid-1940's, and prior to the start of the present-day use of pressure cell, step-resistance staff, or buoy accelerometer wave measurements that are all confined at a single location, early 20th century attempts at measuring ocean waves were focused mainly on a broader area of the actual ocean such as using stereophotogrammetry as described in Sverdrup *et al.* [4]. An example given in the book is shown in Figure 1.

Considering that the contour results were made long before conventional wave spectrum conceptualization and single point wave measurements, it is certainly a remarkable accomplishment in the early part of the last century. What is also interesting in this topographic contour plot is the 9 m high point in the centre as compared with the 1 m low point in the upper right corner, a rising of 8 m in elevation and 57 m in horizontal distance. While we may not now be surprised by the appearance of the contour picture, Sverdrup *et al.* [4] did comment about the striking irregularity in the topography. Nevertheless it is an extraordinary snapshot of a portion of the ocean surface for its time.

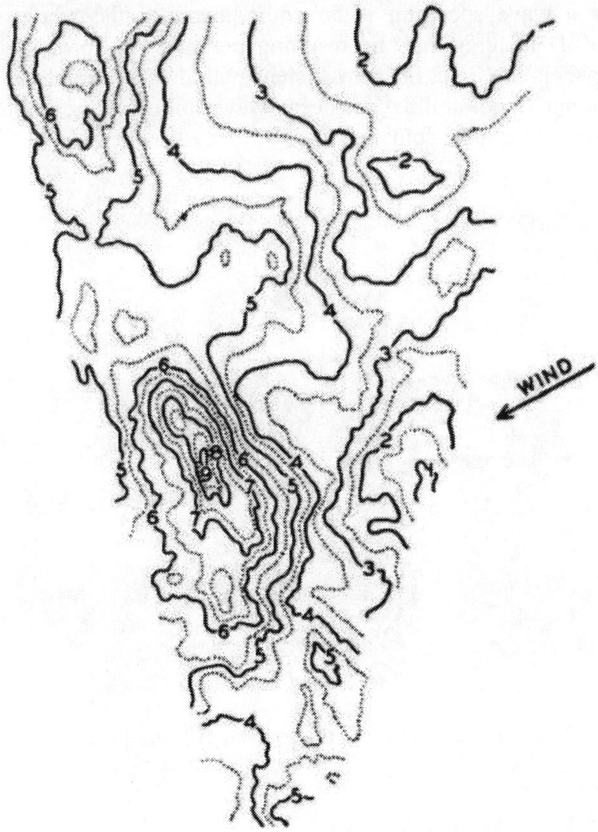

Figure 1: Topography of the sea surface derived from stereophotogrammetry of the sea surface taken onboard Meteor on January 23, 1926. From Sverdrup *et al.* [4].

3 The ATSIS measurement

Now fast forward some eight decades, where technology advances have made possible the use of digital video cameras for ocean wave measurement. Warnek and Wu [5] developed the Automated Trinocular Stereo Imaging System (ATSIS) for non-intrusively measuring the temporal evolution of three-dimensional wave characteristics. The ATSIS system can provide a contour picture of the ocean surface every fraction of a second or more, depending on whatever resolution is required. So while the idea of stereo 3D imaging of the ocean surface is not exactly new, the emerging state of the art ATSIS system could provide measurement of 3D ocean wave fields with respect to time. It is certainly a new arena we are only starting to explore.

For those of us who were brought up during the era where single point wave measurement was the only practice available and have been conditioned to believe that a wave spectrum is an embodiment of the whole ocean wave process, the 3D imaging and the resulting pertinent and profuse data will be clearly overwhelming. But the new system will also be capable of unlocking a whole new range of possibilities for ocean wave studies, invigorating a field that has been stagnant for quite some time.

Figure 2: The three camera set up for the ATSIS system deployed in the field.

4 Data from a 3D wave field

So the data available is in general,

$$\eta = f(x_0, t) \quad (1)$$

for the conventional single point wave measurement; and

$$\eta = f(x, y, t) \quad (2)$$

for the new 3D measurement. As the measurement of eqn. (1) typifies the familiar time series, an example of the measurement of eqn. (2) at $t = t_0$, on the other hand, is shown in Figure 3 as the topography of the ocean surface.

As the actual wave measurement from the ATSIS system is a video recording not a single image, the full extent cannot be shown in print form. But in the age of internet, a portion of the measured video used in this study has been posted and can be seen on the internet at http://www.youtube.com/watch?v=pMYENsrSLN4.

Now the digital data used in this study, as derived from the video recording, is a three dimensional (x, y, η) data set of [441x251x151] coordinates, that in essence represents 15 sec of data with x and y coordinates that remain the same for each time step, while the η coordinate varies with respect to (x, y, t), sampled at a frequency of 10 Hz.

Figure 3: 3D topography of a portion of the ocean surface at a given instant of time.

5 A grappling with 3D wave data analysis

To some extent, viewing the afore-mentioned video recording is no different than watching the ocean surface from a cruise ship out there in the deep ocean (e.g. http://www.youtube.com/watch?v=MnoTj7Jx4L4.) Only now we are capable of making more realistic wave measurements. The key question to ask is really: what would be the pertinent course to follow to analyze this new wealth of data?

Undoubtedly new approaches will continue to evolve as more data become available. In the mean time, we are confronted with the availability of 15 seconds of (x, y, η, t) data, that was recorded in a small lake, specifically Lake Mendota, near Madison, Wisconsin. To proceed, we shall start with an exploratory approach by performing conventional analysis for each individual single data point. For a pixel grid of 441x 251, the data is in fact equivalent to having 110,691 single points with each point providing time series data for 15 seconds with 10 Hz resolution. To effectively visualize all these volumetric data, we simply calculate the total energy represented by each individual standard deviation wave height, i.e., 4*standard deviation, to demonstrate their essential character and then plot them in 3D space as shown in Figure 4. It is rather surprising and pleasant to see this well formed result from basically a statistical estimate for each pixel point. Furthermore the distribution of the standard deviation of the wave heights is shown by the histogram in Figure 5. One might be tempted to try fitting a distribution function to this clearly skewed case. We do not feel that really serves much meaningful purpose. Suffice it to say that in the 3D framework, we can readily obtain useful information using only 15 sec data. This is inconceivable in the conventional framework.

Now with the variations exhibited, an immediate question is: what is the wave height representative in this 3D wave field?

Being completely nurtured in the conventional conceptualization, we are conditioned to regard a wave height as the distance between the trough and crest in a single location. This is very perceptive and straightforward when we look at a customary plot of time series data at a single location. But in the open ocean, how do we sift through a distance between a crest and a trough? So what is the wave height for a given region of the ocean?

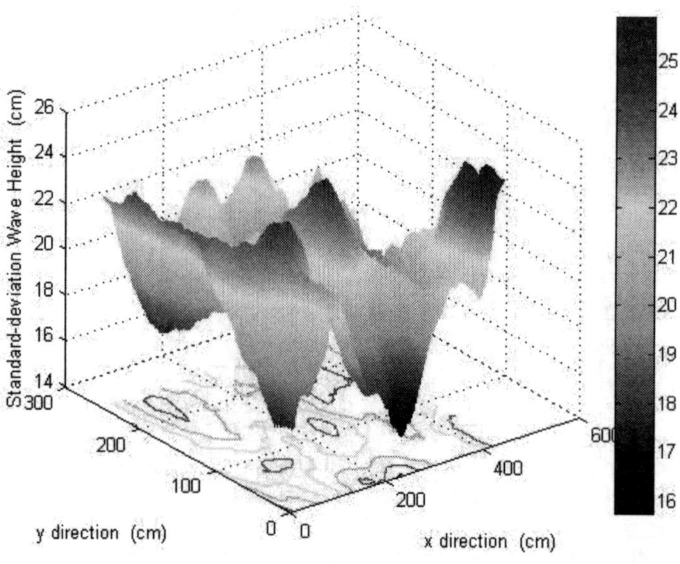

Figure 4: A 3D plot of the standard deviation wave heights for the pixels, each represents a single point data set.

The conventional practice of using one single-point significant wave height to represent the wave height of a region of the ocean surface is clearly no longer valid in a 3D wave field. Conceivably when a seafarer in the open ocean talks about a wave height, it is most likely the height of a visible crest rather than something between trough and crest. So we choose to first examine the highest crest and lowest trough at each instance of the data set.

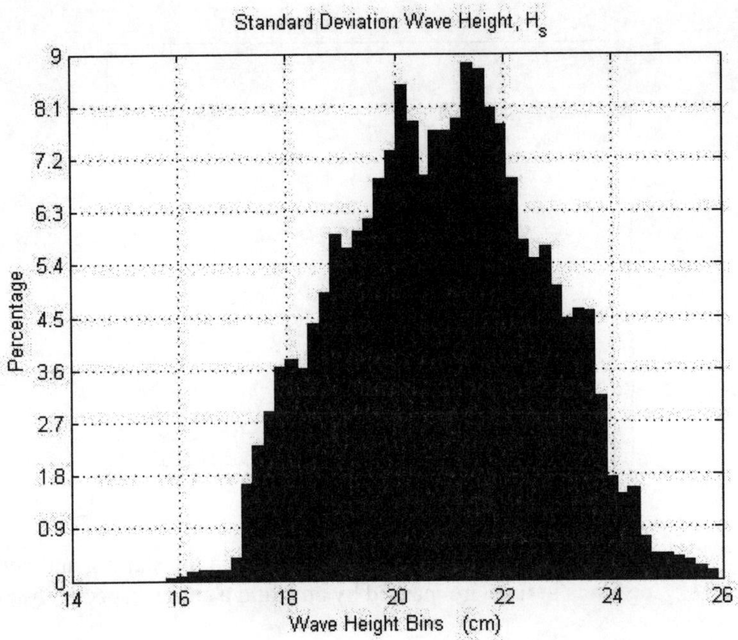

Figure 5: Histogram of the standard deviation wave height given in Figure 4.

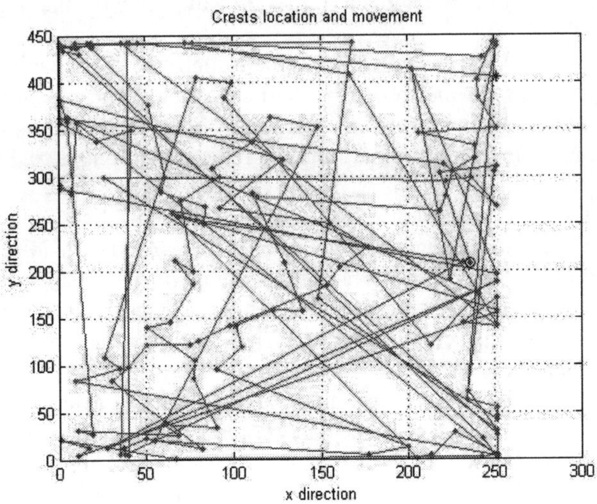

Figure 6: The crest locations of the data set. The starting and ending locations in time are marked by an * and a circle respectively.

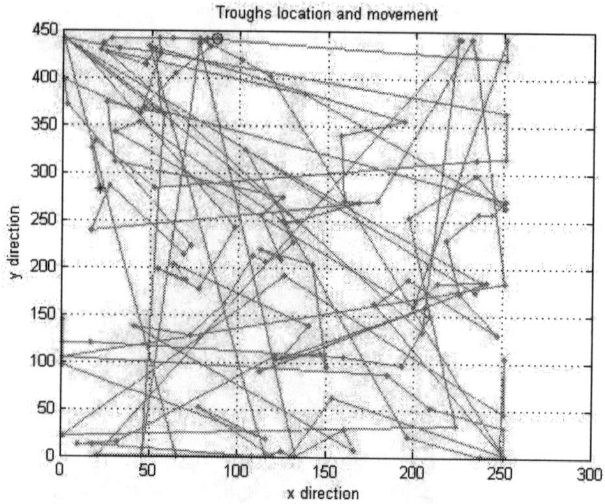

Figure 7: The trough locations of the data set. Again the starting and ending locations in time are marked by an * and a circle respectively.

As shown in Figures 6 and 7, the crest and the trough, being the highest and lowest points at each instance, move around constantly, but they never occur at the same location at the same instance. Thus it is understandable that the familiar notion of wave height evolved from elementary trigonometry and time series analysis cannot be generalized to the 3D wave field as one might wish to bring it into play.

As a matter of fact, if we connect from the trough to crest at each instant, and then connect the crest to the trough of the next instant, and repeat the process throughout the whole data set, the result is Figure 8. It is interesting to note that the points are fairly evenly spread around the region, but none is really on top of each other.

Alternatively we also plotted the crests, troughs, and the sums of corresponding crest and trough, with respect to time as shown in Figure 9. It gives us some indication of the surface fluctuations of the ocean surface in that region. This is aimed at practical reference, which may or may not be meaningful. But the question regarding what is the wave height in a 3D wave field remains unanswered.

Finally, we have also tried to examine the possibility of calculating the wave number spectra for each instant of the data set. For the instant surfaces shown in Figure 10, their corresponding wave number spectra are given in Figure 11. Because the data only covered 15 sec, it may not have sufficient oscillations to provide transient processing and consequential interpretations. At any rate, it is only an indication of what can be done with the data set. Under the conventional approach, a 15 sec measurement will certainly not provide any information about the underlying wave process.

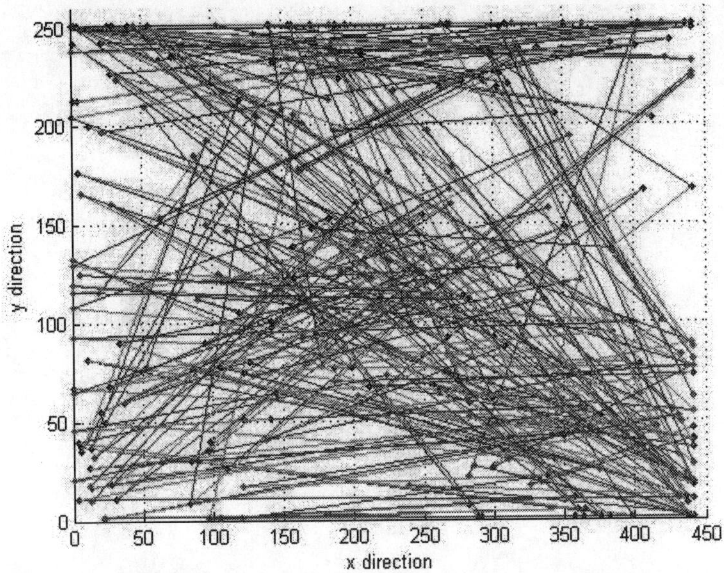

Figure 8: Connecting the trough of each instant to the crest, in dark grey, and connecting the crest to the trough of the next instant, in light grey, and repeating the process throughout the data set.

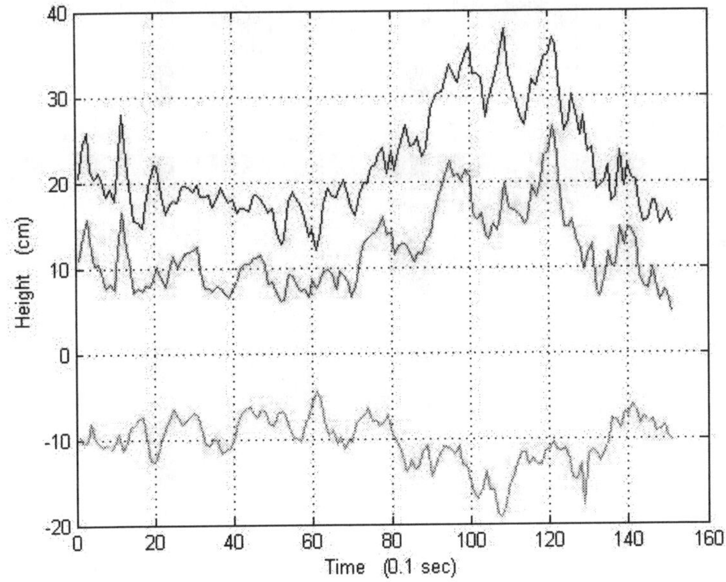

Figure 9: Plot of the height of crest (middle), trough (bottom), and the sum of crest + trough (top) with respect to time.

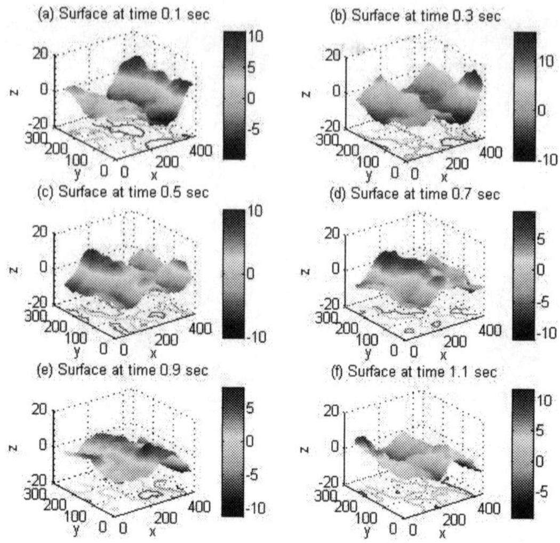

Figure 10: Examples of recorded instant wave surface during the first second of measurement.

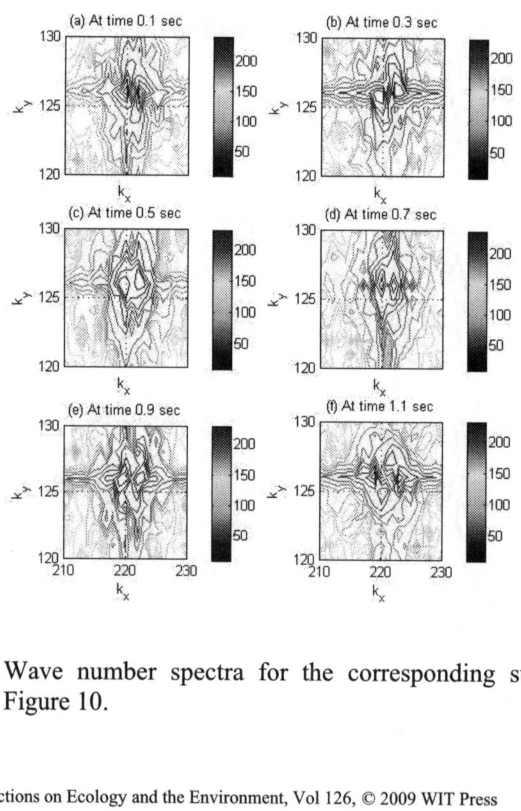

Figure 11: Wave number spectra for the corresponding surfaces given in Figure 10.

All of these analyses presented here are tentative. One of the basic premises of the conventional conviction on the long standing single location ocean wave measurement is that the data from this single location measurement represents the wave condition of a general region. So if waves are measured from another point of this region, it should be generally the same. We found that is not the case. Even for a small pixel region the data are vary from one pixel to the next. Herein lies the answer to why Komen *et al.* [6] were not able to make wave predictions that always fall within the error bands of the observations as expected. Single-point in-situ wave measurement is simply incapable of embodying the realistic ocean waves for more than just that one single-point location. The theoretically-refined wave prediction model cannot be made in accord with the observations at a single point because the single point observation does not represent the reality that the theoretical model is trying to portray. It cannot be over stressed that we need the more comprehensive ATSIS 3D ocean wave measurement.

6 Concluding remarks

After six decades of dominating ocean wave conceptualization as the innate reality, single-point wave measurements have served well the general wave studies. However the progress and model refinement has been in stagnate during the last dozen years or more, it is timely that a new system of ocean wave measurement should be initiated and implemented. Along with this new frontier of ocean wave data measurement, there will be a whole new realm of wave data analysis. New paradigms and new conceptualizations that have not yet been contemplated should be further explored. For instance, instead of the distance between a trough and an adjacent crest at a single point, what should a wave height in the three-dimensional wave field be? There should never be any shortage of incentive or excitement in this new ocean wave measurement frontier. It is time to open up to fresh new perspectives and ideas.

Acknowledgement

This is GLERL Contribution No. 1512.

References

[1] Gallego, G., A. Benetazzo, A. Yezzi, and F. Fedele, 2008: Wave statistics and spectra via a variational wave acquisition stereo system, Proceedings, OMAE2008, Portugal.

[2] Horgan, J., 1996: The End of Science, Addison-Wesley.

[3] Liu, P. C., D. J. Schwab, C. H. Wu, and K. MacHutchon, 2008: Wave heights in a 4D ocean wave field, Proceedings, OMAE2008, Portugal.

[4] Sverdrup, H. U., M. W. Johnson, and R. H. Fleming, 1942: The Oceans, their physics, chemistry, and general biology. Prentice-Hall, Inc., 1087p.

[5] Wanek, J. M. and C. H. Wu, 2006: Automated trinocular stereo imaging system for three-dimensional surface wave measurements. Ocean Engineering, vol. 33, 723-747.
[6] Komen, G.J., L. Cavaleri, M. Donelan, K. Hasselmann, S. Hasselmann, and P.A.E.M. Janssen, 1994: Dynamics and Modelling of Ocean Waves, Cambridge, 532p.

Tidal effect on chemical spills in San Diego Bay

P. C. Chu[1], K. Kyriakidis[1], S. D. Haeger[2] & M. Ward[3]
[1]*Naval Postgraduate School, USA*
[2]*Naval Oceanographic Office, Stennis Space Center, USA*
[3]*Applied Science Associates, Inc., USA*

Abstract

A coupled hydrodynamic-chemical spill model is used to investigate chemical spills in San Diego Bay. The hydrodynamic model shows that San Diego Bay is tidally dominated. Two different patterns of chemical spill were found with pollutants (methanol, benzene, liquefied ammonia, etc.) released at 0.5 m depth in the northern bay (32°43'N, 117°13.05'W) and in the southern bay (32°39'N, 117°07.92'W). For the north-bay release, the chemical pollutants spread in the whole basin with a fast speed of spill in the northern part (12 hours) and a slow speed of spill in the southern part (20 days) with very low concentration. For the south-bay release, the chemical pollutants are kept in the southern part. Very few pollutants reach 32°41'N parallel (the boundary between the north and south bays).
Keywords: San Diego Bay, water pollution, water quality management, chemical fate model, tidal basin, chemical spill, hydrodynamic model.

1 Introduction

San Diego Bay (Figure 1(a)) is located near the west coast of southern California. It is a relatively small basin (43-57 km^2) about 25 km long and 1-4 km wide. It is a flipped Γ-type shape and extends to the north to the city of San Diego and to the south to Coronado Island and Silver Strand, with a northwest to southeast orientation. The topography is not homogeneous (Figure 1(b)), and the average depth is of 6.5 m (measured from the mean sea level). The northern/outer part of the bay is narrower (1-2 km wide) and deeper (reaching a depth of 15 m) and the southern/inner part is wider (2-4 km wide) and shallower (depth less than 5 m). Near the mouth of the bay, the north-south channel is about 1.2 km wide, bounded by Point Loma to the west and Zuniga jetty to the

WIT Transactions on Ecology and the Environment, Vol 126, © 2009 WIT Press
www.witpress.com, ISSN 1743-3541 (on-line)
doi:10.2495/CP090031

east with depths between 7 and 15 m [1]. The western side of the channel is shallower than the east side.

The shoreline landscape of San Diego Bay is spotted with highly polluting shipbuilding and ship repair facilities. Ship operations including recreational boating and Navy operations are other sources of pollution in San Diego Bay. These toxins threaten public health and the environment. Investigation of the chemical dispersion of floating chemicals such as methanol, benzene and ammonia is very important for water quality control.

(a) (b)

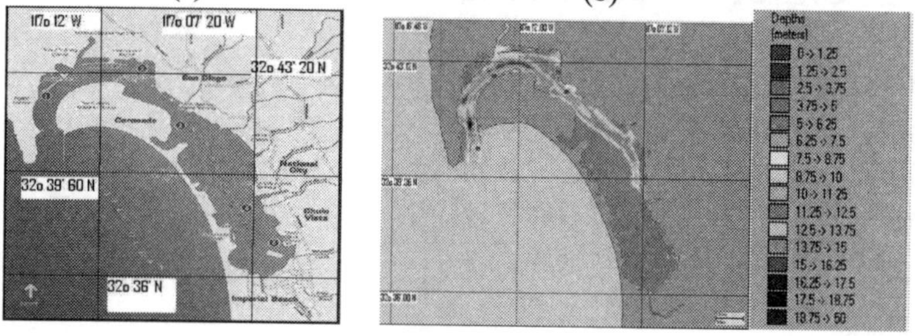

Figure 1: San Diego Bay: (a) geographical locations, and (b) bathymetry.

2 Tidal basin

San Diego Bay is a tidal basin connected to the ocean by an inlet with an artificial jetty (Zuniga) built to control beach erosion. The Zuniga jetty extends almost one mile offshore of Zuniga Point and most of it is not clearly visible at high water. Obviously, the bay has been intensively engineered to accommodate shipping activities. Ninety percent of all available marsh lands and fifty percent of all available inter-tidal lands have been reclaimed and dredging activities within the bay have been equally extensive [2, 3]. Kelp forests extend approximately 2 km south of Point Loma (Figure 1(a)) and along its western side. They are quite thick and they create seasonal dumping of currents to about one-third their values outside [4].

The currents in San Diego Bay are predominately produced by tides [3]. This tidal exchange between the ocean and the bay is a result of a phenomenon called "tidal pumping" [5]. The "pumping" of water is due to the flow difference between the ebb and the flood flows. Being located at mid-latitude, tides and currents within San Diego Bay are dominated by a mixed diurnal-semidiurnal component [2]. The tidal range from mean lower-level water (MLLW) to mean higher-high water (MHHW) is 1.7 m with extreme tidal ranges close to 3 m [1]. Typical tidal current speeds range between 0.3-0.5 m/s near the inlet and 0.1-0.2 m/s in the southern region of the bay. The phase propagation suggests that the tides behave almost as standing waves with typical lags between the mouth and the back portion of the bay of 10 min and a slight increase in tidal amplitude in the inner bay compared to the outer bay. The overall tidal prism for the bay is

5.5×10^7 m^3 and the tidal excursion is larger than the mouth with a value of 4.4 km [6].

3 Water quality monitoring

In 1960, an earthquake with a Richter scale of 9 in Chile caused the biggest sudden rise in sea level ever recorded in the San Diego area of 1.07 m at the Scripps pier. There is a natural protection due to the 160 km wide continental shelf of San Diego. There is a fault off San Diego Bay, but it is inactive. These are the reasons why from the 15 locally generated tsunamis in California since 1812, only two have occurred in Southern California, and only one in San Diego, dating back to 1862.

There is widespread toxicity in San Diego Bay sediments attributable to copper, zinc, mercury, polycyclic aromatic hydrocarbons, polychlorinated biphenyls (PCBs) and chlordane. No single chemical or chemical group has a dominant role in contributing to the identified toxicity. Contributions of trace metals from vessel activities have long been suspected as a potentially large source to San Diego Bay. Actually, Shelter Island Yacht Basin, a semi-enclosed boat harbor, has been added to the State's list of impaired water bodies. These contributions arise from specially formulated paints, impregnated with biocides, and applied to boat hulls to retard the growth of fouling organisms such as barnacles.

4 Hydrodynamic model

The numerical hydrodynamic model implemented for San Diego Bay is a boundary fitted tidal and residual circulation model known as the Water Quality Management and Analysis Package (WQMAP) [7, 8] developed at the Applied Science Associates Inc. WQMAP consists of three basic components: a boundary-fitted coordinate grid creation module, a three-dimensional hydrodynamics model, and a water quality or pollutant transport model. These models are executed on a boundary fitted grid system. They can also be operated on any orthogonal curvilinear grid or a rectangular grid, which are special cases of the boundary fitted grid. The model is configured to run in a vertically averaged (barotropic) mode or as a fully three-dimensional (baroclinic) mode. Several assumptions are made in the model formulation, including the hydrostatic (shallow water) approximation, the Boussinesq approximation, and incompressibility. In this study, the 2D version is used. WQMAP for San Diego Bay covers an area of 43 km^2. The computational mesh has 150×200 (30,000) grid nodes with an average horizontal resolution of 40 m. Model bathymetry is determined from depth sounding data provided by NOAA and supplemented by data from published navigation charts. Recently Navy conducted bathymetry surveys show that the water depths in regions near the bay entrance are significantly deeper than the water depths shown on the NOAA navigation chart [3]. The most up-to-date bathymetry data are used in the model.

Surface elevation and velocity are set to zero, and temperature and salinity are assigned as the characteristic values for San Diego Bay (16°C, 34 ppt) at all grid points. The model is allowed to "spin up" from quiescent initial condition for one day before any model results are used for analysis. A six-minute time step is chosen for time step. At this time step the CFL condition is satisfied. Temporally varying sea surface elevation (or tidal harmonic constituents) along the open boundary (entrance of San Diego Bay) is taken as the model forcing function. Such data are available at the NOAA Centre for Operational Oceanographic Products and Services website. The elevation data with six-minute interval are archived from time 0000 on 22 June 1993 to 2354 on 27 August 1993 for San Diego Bay entrance, in accordance with NOAA San Diego Station number 9410170, located at (32°42'48"N, 117°10'24"W).

High correlation (>90%) between prediction and observation exists in phase and amplitude. For nb1, the u speed between the data and the model has a correlation coefficient of 91.87% and can be verified. The observational u-velocity ranges between -51.8 and 44.5 cm/s and the modeled u-velocity changes between -46.9 and 40.8 cm/s. The difference between the observational and modeled mean u-velocity is 0.49 cm/s (Figure 2).

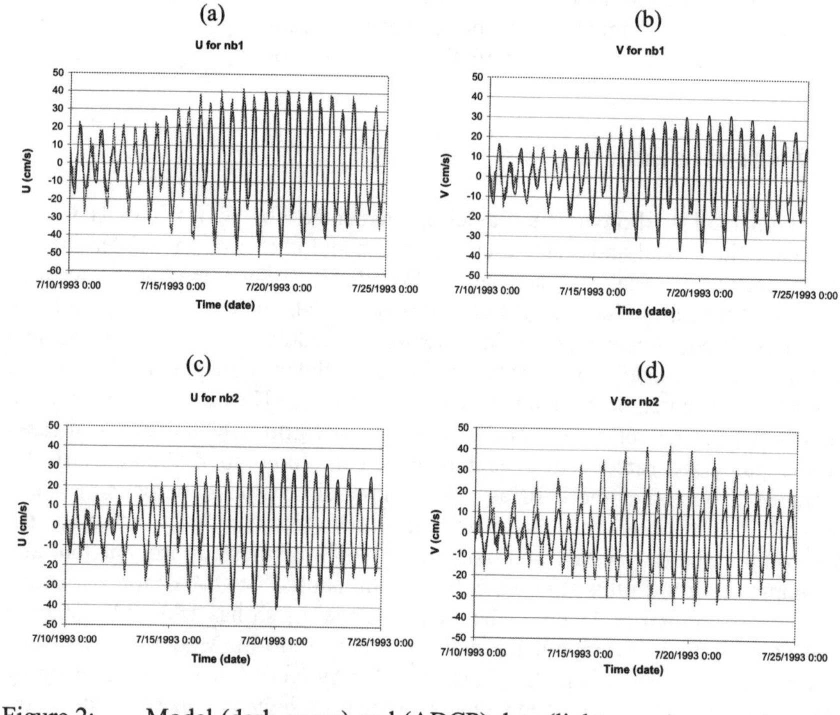

Figure 2: Model (dark curve) and (ADCP) data (light curve) comparison for station-nb1 (upper panels) and nb2 (lower panels): (a) *u*-component, and (b) *v*-component.

Overall, the model results are reasonably good, especially taking into account that the comparison between data and model is not at exactly the same position and the proximity of the ADCPs to the shore. If finer grid and more accurate bathymetry are used, the model results may be further improved.

5 Chemical spill model

A chemical spill model (CHEMMAP, developed at the Applied Science Associates Inc.) is used to predict the trajectory and fate of floating, sinking, evaporating, soluble and insoluble chemicals and product mixtures. It estimates the distribution of chemical elements (as mass and concentrations) on the surface, in the water column and in the sediments. The model is initialized for the spilled mass at the location and depth of the release. The state and solubility are the primary determining factors for the initialization algorithm. If the chemical is highly soluble in water and is either a pure chemical (e.g., the benzene scenario) or dissolved in water (e.g., the methanol scenario), the chemical mass is initialized in the water column in the dissolved state and in a user-defined initial volume. For insoluble or semi-soluble gases released underwater (e.g., the naphthalene gas scenario), the spilled mass is initialized in the water column at the release depth in a user-defined plume volume, as bubbles. The median particle size is characterized by a user-defined diameter [9].

The model simulates adsorption onto suspended sediment, resulting in sedimentation of material. The Stokes Law is used to compute the vertical velocity of pure chemical particles or suspended sediment with adsorbed chemical. If rise or settling velocity overcomes turbulent mixing, the particles are assumed to float or settle to the bottom. Settled particles may later re-suspend (assumed to occur above 20 cm/s current speed). Wind-driven current (drift) in the surface water layer (down to 5m) is calculated within the fates model, based on hourly wind speed and direction data. Surface wind drift of oil has been observed in the field to be 1-6% of wind speed in the direction of 0-30 degrees to the right (in the northern hemisphere) of the down-wind direction (Youssef and Spaulding 1993 [10]). The user may also specify the wind drift speed and angle [9].

6 Chemical spill patterns

The coupled hydrodynamical-chemical model (WQMAP-CHEMMAP) is used to investigate the chemical spill patterns for floating, sinking, gaseous chemicals. Since the WQMAP is integrated for the period from 0000 on 22 June 1993 to 2354 on 27 August 1993 for San Diego Bay, the following scenarios were suggested: A small boat drops one barrel of chemical (e.g., methanol) in less than 12 minutes on midnight July 4, 1993 (Independence Day) at (1) northern San Diego Bay (32°43'N, 117°13.05' W) (Point 2 in Figure 1(a)), and (2) southern San Diego Bay (32°39'N, 117°07.92' W) (Point 4 in Figure 1(a)). The release depth is 1 m and the initial plum thickness is 0.5 m. Two distinct spill patterns are found for all the chemicals. Here, spill of methanol is presented for illustration.

6.1 Pollutants released at North San Diego Bay

The chemical spill pattern is described as follows. In 3 hours, the methanol is in San Diego port (Figure 3(a)) and in 10 hours it is spread all over the North San Diego Bay. However, the south part of the Bay is contaminated much later. After two days, there are no pollutant particles south of 32°40'N (Figure 3(b)). After 3 days the northern part is heavily impacted but after 9 days, there are still no pollutant particles south of 32°39'N. The methanol reaches the south end of the Bay only after 20 days (Figure 3(c)), but its concentration in the water column can be neglected. Figure 4 shows the swept area after 2 days and 32 days. In such a case, it can be concluded that there is plenty of time to take protective measures for the southern part of the Bay where the results of such an incident would be minimal.

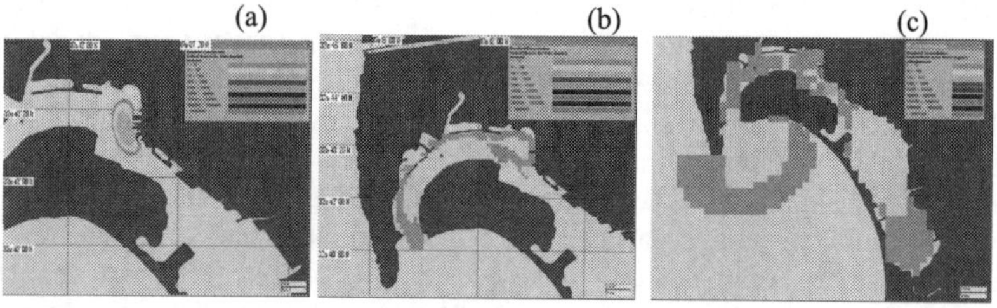

Figure 3: Dissolved concentration in San Diego after (a) 3 hours, (b) 2 days, and (c) 20 days after methanol dropped in North San Diego Bay.

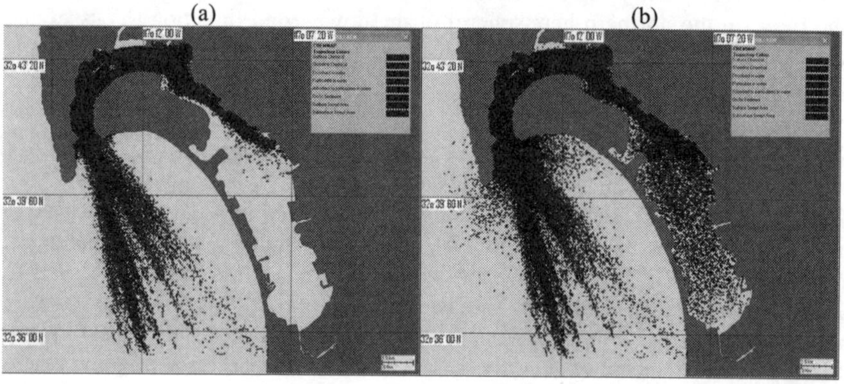

Figure 4: Swept area after (a) 2 days and (b) 32 days for methanol dropped in North San Diego Bay.

Furthermore, after five entire days, one third of the methanol is still in the water column (Figure 5). Note that it takes almost 12 days for the concentration in the water column to reach 10% and 15 days for the decayed methanol to reach a level of 80%. Moreover, the end-state is the contamination not only of the San Diego Bay but also a considerable part of the sea outside the Bay. The scenario is repeated by increasing the amount of methanol, but nothing changes fundamentally. The mass balance curves and the area contaminated remain the same.

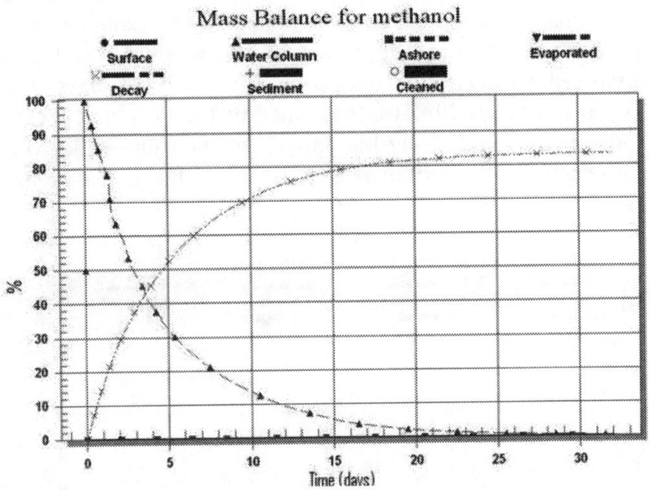

Figure 5: Mass balance for methanol dropped in North San Diego Bay.

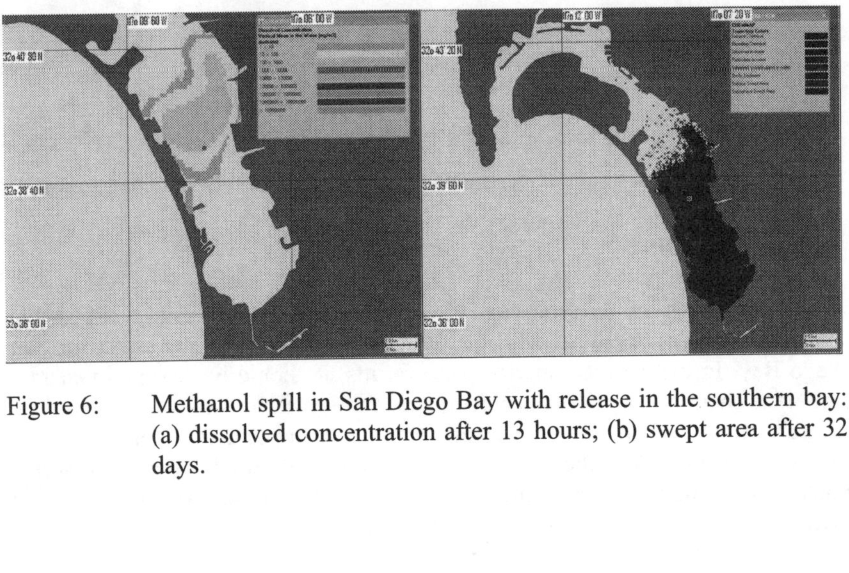

Figure 6: Methanol spill in San Diego Bay with release in the southern bay: (a) dissolved concentration after 13 hours; (b) swept area after 32 days.

6.2 Pollutants released at South San Diego Bay

The chemical spill pattern is described as follows. In 13 hours, the methanol reaches the central San Diego Bay (Figure 6(a)). But, very few pollutants reach 32°41'N parallel. Figure 6(b) shows the swept area after 32 days. It is crucial for protective measures to highlight this fact because a chemical attack in the South San Diego Bay will have minimal effects, or at least much less considerable than an attack (or accident) in the north part of the bay. Figure 7 shows a similar but different result as regards the mass balance curves. Thus, the decayed methanol reaches 80% in only nine days, mainly due to the inert nature of methanol in combination to the shallow bathymetry of the southern part of the Bay. It is important to single out that in the first case (methanol spill over in the north), the dissolved concentration disappears after only 15 days, but in the second case (south), it needs 29 days. It is noted that the ecological catastrophe that can be caused with a relatively big amount of methanol spill over is very considerable, especially if the spill over is in the north. It can also be harmful to humans.

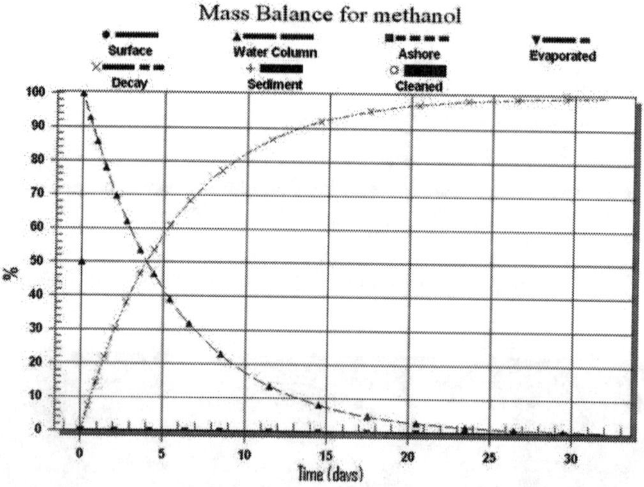

Figure 7: Mass balance for methanol dropped in South San Diego Bay.

7 Conclusions

This study shows the vulnerability of a semi-enclosed tidal basin in a possible chemical attack or accident, with the aforementioned particular results for San Diego Bay. In order to summarize these results, it should be repeated that in a case of a chemical attack or accident, first the sensitive eco-system would be severely damaged, no matter the nature of the event and the location. If the chemical were a sinker, the results would be more catastrophic than if it were a floater. Since the water exchange with the Pacific Ocean occurs only through a narrow entrance, the water would be contaminated for long time.

Two regimes of the chemical dispersion were found in this thesis. The first was the case of an attack/accident in the North San Diego Bay. In that case the entire Bay would be contaminated. In 3 hours the chemical would reach San Diego port and city, in 12 hours the entire northern part of the Bay would be affected and in 2-5 days the south part of the bay would be contaminated as well. The rest of the Bay would be reached much later. The second regime was an attack/accident in the South San Diego Bay. In such case, the incident would have minimal effects on the city and the shores of Coronado Island (located in the north part of the bay) and none outside the Bay. On the other hand, when the spill occurs in the southern part of the Bay, a larger percentage of the chemical remains in the water column and for longer period of time, which makes it more "effective", which in a case of a chemical attack means lethal.

For the aforementioned reasons, the propagation model shows that the northern part of the Bay is more likely to be a target because it would affect the city, and it would reach, even slightly, the South San Diego Bay and would spread outside the Bay as well. In general, results concerning San Diego Bay can also be applied to studies in other semi-closed, barotropic, no-wind driven circulation basins.

As regards recommendations for future research, it should be mentioned that the use of more accurate bathymetry and of a finer grid would give better results in a similar case. Moreover, the use of more recent ADCP measurements, during a longer period of time would further improve the results and verify the overall conclusions. It would be helpful if the ADCPs used in the future were located in a bigger distance from the shore.

A more detailed comparison of 3D vs. 2D model is encouraged, as well as its application for drift and for instantaneous current prediction. Last but not least, as regards chemical propagation, a classified research with data unavailable to foreigners about real chemical threats (e.g. anthrax) should be conducted.

Acknowledgements

This work was funded by the Naval Oceanographic Office, the Office of Naval Research, and the Naval Postgraduate School.

References

[1] Chadwick, D. B. & Largier, J. L., Tidal exchange at the bay-ocean boundary. *Journal of Geophysical Research*, **104 (C12)**, 29901-29924, 1999a.

[2] Peeling, T. J., A Proximate Biological Survey of San Diego Bay, California. Naval Undersea R&D Centre, San Diego, California, Technical Report No. TP389, 1975.

[3] Wang, P. F., Cheng, R. T., Richter, K., Gross, E. S., Sutton, D., & Gartner, J. W., Modeling tidal hydrodynamics of San Diego Bay, California. *Journal of American Water Resources Association*, **34 (5)**, 1123-1140, 1998.

[4] Jackson, J. A. & Winant, C. D., Effect of a kelp forest on coastal currents. *Continental Shelf Research*, **2 (1)**, 75-80, 1983.

[5] Fischer, H. B., List, E. J., Koh, R. C. Y., Imberger, J. & Brooks, N. H., *Mixing in Inland and Coastal Waters*. Academic Press, pp. 483, 1979.

[6] Chadwick, D. B. & Largier, J. L., The influence of tidal range on the exchange between San Diego Bay and the ocean. Journal of Geophysical Research, 104 **(C12)**, 29885-29899, 1999b.

[7] Muin, M. & Spaulding, M. L. Two-dimensional boundary fitted circulation model in spherical coordinates. *Journal Hydraulic Engineering*, **122 (9)**, 512-520, 1996.

[8] Muin, M. & Spaulding, M. L., Three-dimensional boundary fitted circulation model. *Journal Hydraulic Engineering*, **123 (1)**, 2-12, 1997.

[9] Fench-McCay, D.P. & Isaji, T., Evaluation of the consequences of chemical spills using modeling: chemicals used in deepwater oil and gas operations. *Environmental Modeling & Software*, **19(7-8)**, 629-644, 2004.

[10] Youssef, M. & Spaulding, M. L., Drift current under the action of wind and waves. *Proceedings of 16th Arctic and Marine Oil Spill Program Technical Seminar,* Calgary, Alberta, Canada, pp. 587-615, 1993.

Section 2
Wave transformation hydrodynamics

Use of video imagery to test model predictions of surf heights

D. Huntley[1], A. Saulter[2], K. Kingston[1] & R. Holman[3]
[1]*Marine Institute and School of Earth, Ocean and Environmental Sciences, University of Plymouth, UK*
[2]*National Centre for Ocean Forecasting, Met Office, UK*
[3]*College of Oceanic and Atmospheric Sciences, Oregon State University, Oregon, USA*

Abstract

This paper describes a new method of estimating breaking wave heights from video images of the surf zone, and uses the method to test real-time numerical model predictions based on global and regional winds. The test site is an exposed beach on the southwest coast of the United Kingdom (Perranporth, Cornwall). Breaking wave height estimates based on the video technique are found to be accurate to at least ±30%. The model predictions show a linear correlation with video-derived wave heights with a regression coefficient of 0.82 and slope of 0.93 However individual comparisons can differ by up to a factor of 2 for wave heights around 1m, reducing to around 1.5 for a wave height of 3m. The primary causes of error are likely to be inadequate bathymetry near the coast and wind speed errors offshore.

Keywords: breaking wave heights, coastal video imagery, wave modelling.

1 Introduction

Wave breaking at the shoreline creates one of the most energetic natural environments and is responsible for the generation of strong currents and for moulding the constantly changing morphology of the coastal environment. The processes of wave breaking and its consequences are however complex and still far from being fully understood.

Modelling and hindcasting breaking wave heights and associated currents and evolving nearshore morphologies over timescales of weeks and years appear to

WIT Transactions on Ecology and the Environment, Vol 126, © 2009 WIT Press
www.witpress.com, ISSN 1743-3541 (on-line)
doi:10.2495/CP090041

be gaining some skill [1, 2], though parameter values can be physically unrealistic. In contrast, real time prediction of breaking wave heights has received relatively little attention, despite its obvious importance for the safe day-to-day management of shoreline and beach activities.

The UK Met Office has, since 2004, been providing real time forecasts of breaker heights at three locations around the UK for the Royal Navy in support of beach-based training activity. However to date there has been no independent, quantitative assessment of the accuracy of these breaker height predictions. The purpose of this paper is therefore to describe and assess a method of testing the predictions using video measurements of the surf zone at an exposed site on the southwest coast of the United Kingdom.

2 The Met Office breaking wave height model

The UK Met Office has for over 15 years provided forecasts of sea state on a global and regional scale using a suite wave models [3, 4]. The models are forced with hourly wind fields generated by Met Office Numerical Weather Prediction (NWP) models, which include observational data from ships and data buoys in their assimilation schemes. Wave energy (swell as well as wind-sea) is advected through the model domain at the group velocity, with bottom friction and refraction included for depths less than 200m. Output from the models is regularly updated and provides forecasts with lead times of up to 5 days.

This study used configurations based on the (then operational) Met Office second generation wave model. The Global Wave Model operates on a 5/9° latitude by 5/6° longitude grid (approximate 60 km square at UK latitude). This provides input to the boundaries of the UK Waters Wave Model, which uses a 1/9° latitude, 1/6° longitude grid (approximately 12 km square). This finer resolution model includes the effects of time-varying currents on the UK continental shelf, taking hourly currents from a 12km resolution Storm Surge Model.

In order to predict depth-limited breaker heights at the coast, the nearest grid point of the UK Waters Wave Model is used as input to an implementation of the SWAN model (Simulating WAves Nearshore [5, 6]). For the site to be discussed later, this grid point was 7.5 km from the shore in 30 m water depth. The SWAN model includes two shallow water effects not present in the offshore models; depth-limited wave breaking and triad (as opposed to quadruplet) wave-wave interactions. The Met Office SWAN configuration typically uses a grid size of 100-250 m, limited largely by the resolution of available local bathymetry and run time, and brings the waves to a minimum depth of 5m chart datum.

Two wave shoaling and breaking algorithms are then used to define the extent of the surf zone (assumed shoreward of the SWAN boundary point). The model of Goda [7, 8], uses empirical formulae for wave height transformation through shoaling and breaking based upon a large number of laboratory measurements and checked against a limited number of field measurements. The model divides the propagation of waves into a shoaling zone, a transition zone and a depth-limited breaking zone, with coefficients which depend upon the offshore wave steepness (ratio of wave height to wavelength) and the local beach slope. The Met Office prediction uses measured intertidal beach profiles with interpolation to offshore bathymetry, combined with modelled tide and surge levels, to provide the depth profile over which the waves propagate. The predicted quantity, the wave height at the outer edge of the surf zone, is taken as the height at the junction of the depth-limited and transition zones.

The alternative model uses the method of Battjes and Janssen [9]. The maximum depth-limited wave height is defined by an equation after Battjes and Stive [10], resulting in a ratio of wave height to water depth which is a function of the offshore wave steepness and the local wavelength of the waves (itself a function of the wave period and local depth).

The output of these wave breaking prediction schemes consists of hourly predictions of significant breaking wave heights, maximum breaking wave heights (defined as the highest of a run of 250 waves) and the associated breaker depths, celerities and wavelengths.

3 The field site and video monitoring system

3.1 Field site

The site chosen for this hindcasting intercomparison was Perranporth, on the northern coast of Cornwall, UK (Figure 1). The beach at Perranporth is long and essentially straight, backed by a large dune system and facing Atlantic swell from the west. Typical offshore wave heights range between 1 and 3m, rising to more than 4m during major storm conditions. Perranporth is macrotidal, with a typical spring tide range of 7m and neap range of 3m. The beach slope is approximately 0.012 below the low water line but rises to approximately 0.04 at the spring high water line.

3.2 The Argus video system

Breaking wave heights were estimated using an Argus video camera system (Holman and Stanley [11] provide a recent review). Two video cameras were installed at a Youth Hostel on the crest of a headland to the south of the beach at Perranporth, at a height of 48m above the mean water line. Data from the cameras are controlled by a computer housed in the Hostel. The usual sampling regime involves capture every hour of images in the form of a snapshot, a 10-minute time exposure, a variance image (a measure of the variance of intensity at each image pixel over a 10 minute interval) and an image giving the brightest intensity measured at each pixel over a 10 minutes interval. These images are archived at the controlling computer and then downloaded to a server at the University of Plymouth each night when the video system is not operating. From Plymouth the data is placed on the Argus web site (http://cil-www.coas.oregonstate.edu:8080). This site has been in operation since August 1996 and has therefore to date built up over 13 years of almost continuous imagery of the beach at Perranporth.

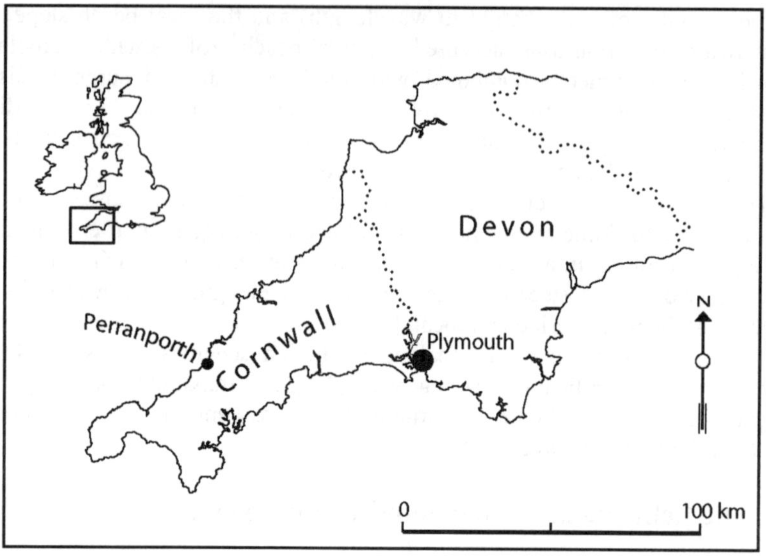

Figure 1: Location map.

Algorithms developed through the Argus program can be used to project the position of each image pixel onto the equivalent map location on either the UK national grid (OSGB36) or a local Argus grid chosen so that the x and y directions are cross-shore and alongshore respectively. These algorithms use GPS-surveyed ground control points in the field of view of the cameras and projects pixel locations onto the horizontal map plane at the known tide and surge water level. For the Perranporth data described in this paper, the pixel resolution is typically better than 5m in the cross-shore direction and 12m in the alongshore direction.

4 Estimation of breaker height using video images

4.1 Imaging breaking waves

Breaking waves create a clear signature in video images due to the high reflection from aerated water at the wave crest. The seaward edge of the surf zone in principle shows the location of the highest breaking waves and therefore can be used to estimate breaker heights. The 'brightest' images, showing the highest intensities over a 10 minute interval, provide particularly clear images of the outer edge of the surf zone. Figure 2 is an example from Perranporth which shows how clearly the surf zone is revealed. The 'brightest' images have not to date been used for breaker height estimation but they have a number of advantages over alternative methods. In particular they give a sharp and unambiguous outer edge (Figure 3 shows an example), in contrast to other images where the intensity profile often approximates a Gaussian shape. They thus provide a direct video-based measure of the location where the largest wave

Figure 2: A 'brightest' image from camera 1 (landward-facing), showing the brightest intensity for each pixel over a 10 minute interval. The dark line shows the transect line used in this study.

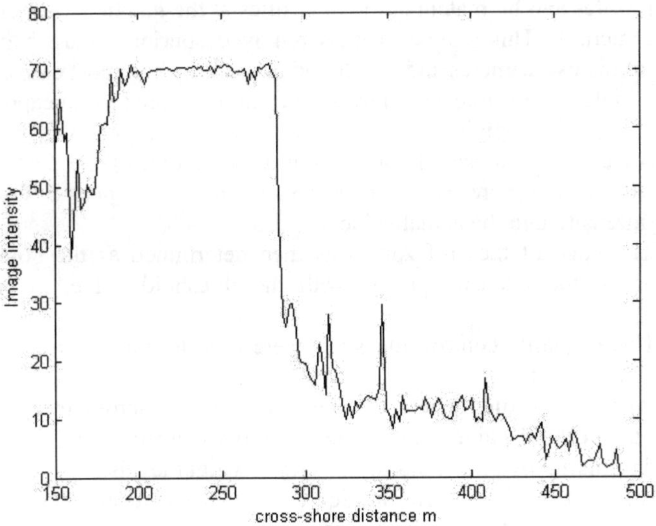

Figure 3: An intensity profile along the line shown in Figure 2 from a 'brightest' image from camera 1. The surf zone is characterised by a high and almost constant intensity, with sharp cut-offs, particularly at the seaward edge at approximately 280m offshore.

over the 10 minute interval started to break. One potential disadvantage, of course, is the statistical uncertainty in the height of the largest wave over a 10 minute period; in contrast to other images the brightest images do not involve averaging over many waves. We shall see later however that this uncertainty is limited and does not overwhelm the advantages. We have therefore chosen to use brightest images for this study.

4.2 Use of brightest images to estimate breaker height

The process of estimating breaker heights at Perranporth involved a number of steps.

1. Intensity profiles along a cross-shore line were extracted from the images by rectifying each image based on the mean water level at the time the image was captured. For this purpose tide level was determined by interpolation of coastal tide gauge records in the region using a 2DH numerical model (K. George, University of Plymouth – personal communication), and surge levels were taken from a Proudman Oceanographic Laboratory Storm Surge model run operationally at the Met Office. Figure 3 is an example profile from camera 1.

2. The seaward-facing camera, camera 2, was found generally to have a lower intensity than camera 1 and a systematic variation of intensity across the field of view, with the highest intensity in the centre of the image. In order to merge images from the two cameras, these effects in camera 2 were removed by fitting a simple parabolic curve, as a function of offshore distance, to the intensity profiles and by matching the intensities at the junction with the field of view of camera 1. This simple process removed spurious sharp edges at the boundary of the two cameras and was found to work well in most cases.

3. Single sharp intensity peaks due to offshore white-capping were reduced in height by applying a two-point low pass filter to the intensity profiles.

4. An intensity threshold was chosen as a fraction of the peak intensity. Values of 0.7 or 0.5 were used but the results were, as expected, found to be relatively insensitive to the actual value.

5. The edge of the surf zone was then determined as the point of first intersection of the intensity profile with the threshold value, moving from offshore.

6. Three 'quality control' measures were used to omit values which were deemed to be suspect:

   a) Omit profiles where the intensity contrast across the profile falls below a threshold value. This process removes profiles from poor images caused primarily by poor visibility or very low light levels.

   b) Omit profiles where the tide level is below mean water level. This ensures that most edges are within or close to the surveyed beach profile line and thus minimises the influence of seaward extrapolation of the profile.

   c) Remove estimated breaker locations which are near the seaward boundary of the profile. This also removes points well seaward of the measured beach profile line as well as points where the camera distortion is largest.

7. Finally a visual check was made on outliers. In fact very few points were found to be in error in this process, though correction was made for one point where the surf zone edge coincided with the junction between the two cameras.

The resulting values for the location of the edge of the surf zone were then converted to local breaker depths using a measured beach profile and known mean water level, as shown schematically in Figure 4. As for the image rectification, the water level for each image time was determined from the interpolated coastal tides and the surge level from the Storm Surge model. Pixel resolution, described in section 3.2, has a negligible effect on the accuracy of these height estimates, giving an uncertainty of only around 0.05m in the worst case scenario of waves over the steepest part of the beach profile.

In principle the video system itself can provide regular beach profiles by imaging the shoreline at different tide and surge water levels [12], but for this study we used data from a single total station (Electronic Distance Meter, EDM) survey carried out on 4th July 2000. Some error is introduced by assuming that this survey profile remains appropriate for the periods used in this study, but there is evidence that for this wide macrotidal beach the error is likely to be small. The maximum difference in beach elevations along the profile line for surveys on 7 May 1997 and on 4 July 2000 was found to be 0.4m, with a standard deviation value of 0.2m, for offshore distances between 80 and 550m, and these changes are equivalent to maximum and standard deviations of breaker height of only 0.12m and 0.06m respectively.

The final step is to convert the resulting breaker depths to an equivalent significant breaker height. There are various ways in which this can be done, including the use of a simple constant breaker index. However a more satisfactory method is to convert the depth into the height of the highest breaking wave, H_{max}, and then relate that height to the equivalent significant wave height, H_s. Of the available equations relating individual maximum wave heights and water depths, we have used the equations of either Goda ([7, 8]):

$$H_{max} = A\, L_0 \,[\, 1 - \exp\,(-\,1.5\, \pi\, (h_b/ L_0)\, (1+15 \tan^{4/3}\theta))] \qquad (1)$$

or Battjes and Janssen [9] and Battjes and Stive [10]:

$$H_{max} = (0.88/ k)\, \tanh\, (\gamma\, k\, h_b\, /\, 0.88) \qquad (2)$$

$$\gamma = 0.5 + 0.4 \tanh\, (33\, H_0\, /L_0) \qquad (3)$$

where A is a constant, taken to be 0.17, L_0 is the deep water wavelength, h_b is the depth of breaking, θ is the angle of beach slope, k is the local wavenumber (2π/wavelength) and H_0 is the deep water wave height.

Calculation of L_o in equations 1 and 3, and k in equation 2 requires input of the incident wave period. In principle wave period can be measured by the video technique using 'timestacks', time series of images with a sampling interval much shorter than the wave period. Unfortunately however no 'time stacks' were available for the data sets considered in this study. We have therefore used the wave periods predicted by the SWAN model.

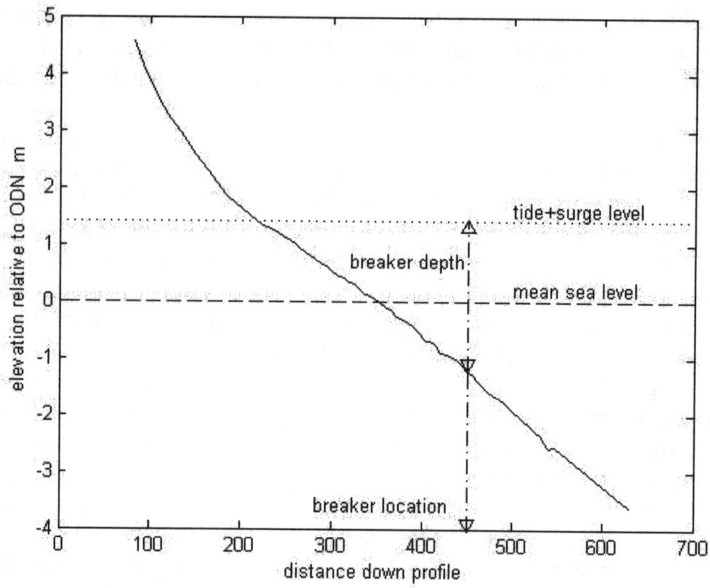

Figure 4: A schematic of the method used to determine breaker depth from the measured cross-shore position of the outer edge of the surf zone. The solid line is the depth profile. The breaker depth for an offshore distance of 450m is shown as the profile depth below mean sea level plus the tide and surge water level.

The maximum wave heights were then converted to equivalent significant wave heights, H_s, by assuming a Rayleigh distribution of wave heights [13, 14] using an equation derived by Longuet-Higgins [15]:

$$H_{max} = [\sqrt{\ln N} + 0.2886 / (\sqrt{\ln N}) - 0.247 / (\sqrt{\ln N})^{3/2}]\, H_s / 1.416 \qquad (4)$$

where N is the number of waves over the 10 minute interval, given by 600/T where T is the wave period in seconds.

It is important to note that the definition of significant breaker height, H_s, computed in this way is based on the location where the highest wave over a 10-minute period begins to break, whereas the Met Office model predictions are based on the location where depth-limited breaking causes the significant wave height to begin to decrease towards the shore. In comparing the predicted and video-derived breaker heights we are therefore assuming that these two definitions are equivalent. The validity of this assumption will be discussed in Section 6.

5 Results

Two months have been selected to test the Met Office predictions. August 2005 is characterised by bright images and relatively low wave conditions. It was chosen for initial analysis since the edge of the surf zone was generally within the field of view of camera 1 alone (for tide height above the mean water level). January 2001 provided a greater test of the methods, generally with less illumination, larger waves and inclusion of camera 2.

Figure 5 shows the result of comparing predicted and Argus-observed breaking wave heights for both months. In this case the Goda equation, equation 1, was used to obtain the observed breaker height. The overall agreement is good, with a best-fit slope of 0.93 and a regression coefficient of 0.82. The slope shows that overall the predictions underestimate the observed heights by 7%, and the 95% bounds on the slope (0.92 to 0.98) suggest that this underestimation is marginally statistically significant. Separate regressions of the August and January data (see Table 1) give very similar slopes but lower regression coefficients, as might be expected for the more limited wave height ranges. Table 1 also shows that observed wave heights computed using the alternative equations 2 and 3 give very similar results. Use of a constant breaker index, on the other hand, results in significantly lower regression coefficients. Table 1 also shows that marginal improvement in the overall fit is obtained by averaging measured breaker heights from three profiles separated alongshore by 30m and by comparing daily-averaged rather than hourly wave heights. However the improvement is generally small.

Despite the overall good agreement, there is considerable scatter in the individual points. Proportionately the scatter is larger at lower wave heights, amounting to approximately a factor of 2 for heights less than 1m, but reduces slightly with increasing wave height, to a factor of around 1.5 for heights above 2m. Overall the rms scatter about the best fit line is approximately 0.28m. In the

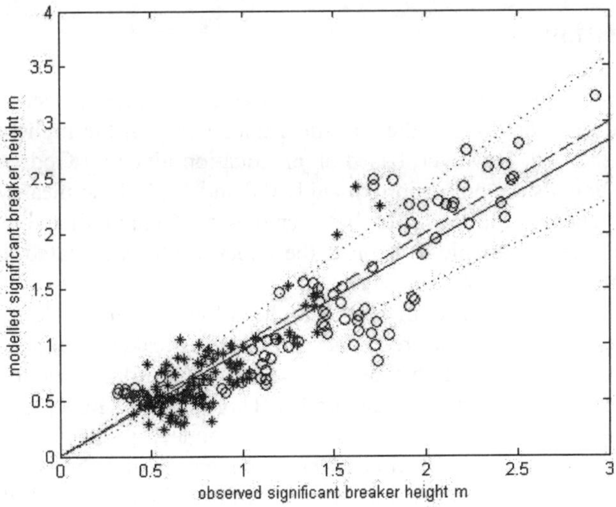

Figure 5: Plot of modelled vs. observed significant breaker heights. The asterisks are for August 2005 and the open circles for January 2001. The solid line is the regression fit through the origin and the dashed line the 1:1 line. The dotted lines show 95% confidence bounds.

Table 1: Linear regression fits between observed and predicted breaker heights.

Conditions	Equation	Regression Coeff R^2	Slope
All data	1	0.82	0.93
August	1	0.67	0.92
January	1	0.79	0.93
All data	2 & 3		
August	2 & 3	0.67	0.94
January	2 & 3	0.77	0.99
August	Constant γ=0.29	0.56	1.2
January	Constant γ=0.29	0.55	1.3
Alongshore averages All data	1	0.83	0.96
August	1	0.59	0.87
January	1	0.79	0.98
Daily averages All data	1	0.82	0.96
August	1	0.74	1.00
January	1	0.81	0.95

following section we investigate possible reasons for this scatter, leading to some suggestions for further work.

6 Discussion

We have investigated several potential causes of the scatter seen in figure 5. Perhaps the most obvious is the statistical uncertainty in the highest wave in a ten minute interval. However, based on an equation given by Goda [8] the 95% confidence bounds are only approximately 0.8 and 1.27 of the mean value so are too small to account for the scatter between observed and predicted wave heights and furthermore are unable to explain the wider scatter associated with smaller wave heights.

A second potential cause of scatter is the difference in the definition of the break point for the measured and predicted breaker heights. We have investigated the impact of this using a variety of available models for wave breaking, including Battjes and Janssen [9], Thornton and Guza [13], Baldock et al. [14] and Lippmann et al. [16]. These models produce very different relationships between the fraction of breaking waves and the mean surf zone wave heights but we find that, for the range of conditions studied, the resulting differences in measured wave height are less than 30% and in most cases considerably smaller, again too small to account for the observed scatter.

Other possible causes include errors in tide and surge levels, errors in the cross-shore beach profile and uncertainty about the appropriate breaker index. However each of these factors is found to have only a minor influence. We conclude therefore that the primary cause of the scatter is error in the predictions rather than the measurements.

This conclusion appears to be generally consistent with previous assessments of the accuracy of the Met Office offshore wave predictions involving comparison between buoy observations and predictions at neighbouring model grid points, and suggests that the nearshore modelling system does not significantly add to the errors inherent in the offshore modelling system. Bidlot and Holt [4] use data from 28 buoys located around the globe but mainly on continental shelves and find good overall agreement but, as with Figure 5, considerable scatter in individual values. Interestingly the slope of their least squares fit is 0.90, again suggesting a small overall underestimation of wave heights; they also find that wave heights above 4-5 m metres tend to be overestimated, a trend also consistent with our data. Wind speed error is suggested as the primary source of the model errors. Bradbury et al. [17] report similar intercomparisons for buoys in much shallow water depths (2.5m to 12.7m) along the southern coast of England.

In conclusion, the new video technique is found to measure breaker heights to an accuracy of at least 30% and this might be considerably improved by appropriate temporal and/or spatial smoothing. This level of uncertainty is likely to be acceptable for most practical applications. Further improvement awaits the development of more accurate representations of the link between breaker height and the fraction of breaking waves, either by refined models or direct observations. The Met Office predictions are good on average but underestimate breaker heights by around 7%, in approximate agreement with comparisons offshore. However individual predictions can be in error by up to a factor of 2 for wave heights around 1m, falling to a factor of around 1.5 at 3m.

References

[1] Ruessink, B. G., Y. Kuriyama, et al. (2007). "Modeling cross-shore sandbar behavior on the timescale of weeks." Journal of Geophysical Research **112**.

[2] Ruessink, B. G., J. R. Miles, et al. (2001). "Modeling the alongshore current on barred beaches." J. Geophys. Res. **106**(C10): 22451-22464.

[3] Golding, B. (1983). "A wave prediction system for real time sea state forecasting." Q. J. R. Meteorol. Soc. **109**: 393-416.

[4] Bidlot, J. R. and M. W. Holt (1999). "Numerical wave modelling at operational weather centres." Coastal Engineering **37**(3-4): 409-429.

[5] Booij, N., R. C. Ris, et al. (1999). "A third-generation wave model for coastal regions - 1. Model description and validation." Journal of Geophysical Research-Oceans **104**(C4): 7649-7666.

[6] Ris, R. C., L. H. Holthuijsen, et al. (1999). "A third-generation wave model for **104**(C4): 7667-7681.

[7] Goda, Y. (1975). "Irregular wave deformation in the surf zone." Coast. Eng. Jpn. **18**: 13-26.

[8] Goda, Y. (1985). Random seas and design of maritime structures. Tokyo, University of Tokyo Press.

[9] Battjes, J. A. and J. P. F. M. Janssen (1978). Energy loss and set-up due to breaking of random waves. Proceedings of the 16th Conference on Coastal Engineering: 569-587.

[10] Battjes, J. A. and M. J. F. Stive (1985). "Calibration and verification of a dissipation model for random breaking waves." Journal of Geophysical Research-Oceans **90**(C5): 9159-9167.

[11] Holman, R. A. and J. Stanley (2007). "The history and technical capabilities of Argus." Coastal Engineering **54**: 477-491.

[12] Davidson, M. A., D. A. Huntley, et al. (1997). The evaluation of large scale (km) intertidal beach morphology on a macrotidal beach using video images. Coastal Dynamics '97, Plymouth, UK, ASCE.

[13] Thornton, E. B. and R. T. Guza (1983). "Transformation of wave height distribution." Journal of Geophysical Research **88**(C10): 5925-5938.

[14] Baldock, T. E., P. Holmes, et al. (1998). "Cross-shore hydrodynamics within an unsaturated surf zone." Coastal Engineering **34**(3-4): 173-196.

[15] Longuet-Higgins, M. S. (1952). "On the statistical distribution of the heights of sea waves." Journal of Marine Research **9**(3): 245-266.

[16] Lippmann, T. C., A. H. Brookins and E. B. Thornton (1996). "Wave energy transformation on natural profiles." Coastal Engineering **27**: 1-20.

[17] Bradbury A.P., Mason T.E. and Holt M.W., (2004). "Comparison of the Performance of the Met Office UK-Waters Wave Model with a Network of Shallow Water Moored Buoy Data." Proc. 8th International Workshop on Wave Hindcasting and Forecasting, Hawaii.

The sea-defence function of micro-tidal temperate coastal wetlands

I. Möller[1], J. Lendzion[2], T. Spencer[1], A. Hayes[1] & S. Zerbe[3]
[1]*Cambridge Coastal Research Unit, Department of Geography, University of Cambridge, UK*
[2]*Institute of Botany and Landscape Ecology, University of Greifswald, Germany*
[3]*Faculty of Science and Technology, Free University of Bozen, Italy*

Abstract

Global environmental change poses a growing challenge for the management of low-lying coastal environments. The challenge is to (a) recognise and quantify the ecological functions of such environments, and (b) develop management approaches that allow those functions to be maintained in the context of global change. Meeting this challenge is particularly important on micro-tidal shorelines, where the ecological sensitivity to sea level rise and changing climatic conditions (e.g. storm frequency and intensity) is likely to be high.

This study addresses the need to quantify the wave-dissipating function of these types of coastal wetland. Previous studies have focused on tidal coasts and salt marsh or mangrove vegetation and have highlighted relationships between coastal wetland vegetation type, water depths, and observed wave energy reduction. Prior to this study, however, no data was available on the sea-defence function of coastal grasslands and reed beds, where irregular inundation by meteorologically driven storm surges dominates over tidal inundation.

Results are presented of wave and vegetation monitoring along three cross-shore transects at the fringes of reed beds and coastal brackish grasslands on the German Baltic shoreline. The data highlight significant differences in the sea-defence function of these two types of micro-tidal coastal habitat, highlighting important differences in the likely response to future climatic (and sea level) changes and raising questions around how these functions might be maintained, enhanced, or restored in the context of environmental change.

Keywords: sea level rise, wave dissipation, baltic coastal wetlands, coastal management.

WIT Transactions on Ecology and the Environment, Vol 126, © 2009 WIT Press
www.witpress.com, ISSN 1743-3541 (on-line)
doi:10.2495/CP090051

1 Introduction

Coastal wetlands have been identified as fulfilling a range of key ecological, morphological, and sea-defence functions within the coastal zone (Allen and Pye [1]). The importance of understanding these functions is likely to increase in the context of global environmental change, in particular relative sea-level rise, wave climate change, and the need for new coastal management approaches to accommodate these environmental shifts.

Particular concern has been raised about the future functioning of intertidal coastal habitats under an acceleration in sea level, as recorded in the last decade and the expected acceleration over the remainder of the twenty first century. A proper assessment of the vulnerability and sustainability of these systems needs to establish the range of both coastal marsh types present and the process environments within which they are found. Thus, for example, there is a need for the more systematic investigation of the notion that wetlands maintaining equilibrium under a macro-tidal range may have greater resilience towards the impacts of sea-level rise than a system operating under a narrow range of tidal fluctuations (e.g. French [2]). Such micro-tidal systems depend to a greater degree on storm-driven sedimentation, raising issues about the interactions between storminess, wave action, sediment supply and system maintenance (e.g. Bartholdy *et al.* [3]). Ultimately, therefore, the existence of coastal wetlands, and the provision of their valuable ecological functions, are crucially dependent upon shallow water hydrodynamics.

One important component in this regard is the wave dissipating function of vegetation present at the coastal fringe. Previous studies of wave attenuation have shown that both micro- and macro-tidal coastal saltmarshes significantly attenuate incident wave energy (e.g. Knutson *et al.* [4] and Möller *et al.* [5]), thus protecting landward habitats from wave impact. Cliffed marsh to mudflat transitions, often the result of differential erosion at the fringes of the vegetated upper intertidal profile, may themselves significantly alter cross-shore wave energy distribution (Möller and Spencer [6]). As well as influencing coastal marsh ecology, any alteration in wave energy levels due to the presence of vegetation, or cliffed marsh edges, has implications for sediment erosion, transport, and deposition and thus for the future functioning of the coastal wetland.

2 Field site

Although most Baltic salt grasslands are small, associated with emergent rocky shores and waters of low salinity, they do occur on sedimentary shorelines outside the influence of glacial rebound. Tidal variations in the Baltic Sea are insignificant and water-level fluctuations are determined by atmospheric forcing and, in some areas, by river discharge. Average water level variations between spring and summer in the central Baltic Sea are 23 cm (Tyler [7]) and average winter sea surface heights (December to February) have been modelled to be up to 90 cm higher than average annual sea surface heights for the period 1961 to

1991 (Meier *et al.* [8]). Gillner [9] records an extreme annual variation in water level of 370 cm at the Polish–German border.

The two locations at which field measurements were made are situated on the southwestern fringes of the Greifswalder Bodden, eastern Germany, to the south of the Island of Rügen, southern shore of the Baltic Sea (Figure 1). The area lies in the region of salinities of 10–5 parts per thousand (Jeschke [10]), considerably less than the transition area from fully marine conditions between Denmark and south-west Sweden but greater than the salinities recorded in the Gulf of Finland and Gulf of Bothnia (Tyler [7]). Along these shores, brackish water reeds (*Phragmites australis*) are common. The salt grasslands not dominated by reeds are mostly of anthropo-zoogenic origin and have developed after semi-continuous grazing (Jeschke [10]). Restoration projects in the past decade have been focussed on the maintenance of specific habitats (Seiberling *et al.* [11]); here grazing is currently applied to maintain coastal biodiversity.

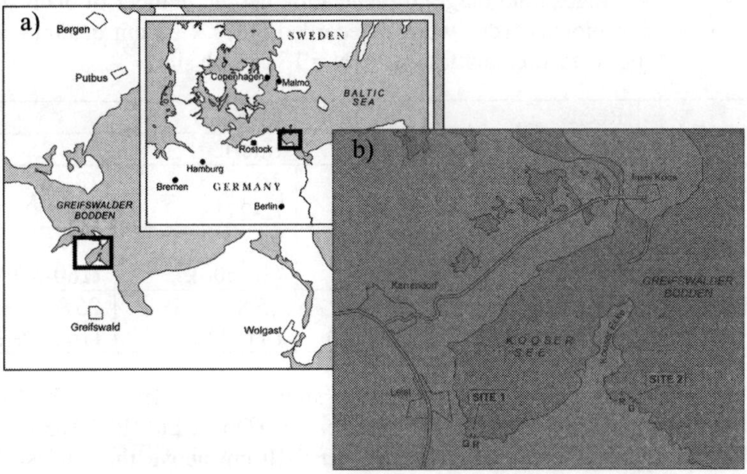

Figure 1: Location of (a) Greifswalder Bodden in the context of the Baltic Sea, and (b) wave monitoring Site 1 and Site 2, with reed (R) and grassland (G) transects, at the southern margin of the Kooser See, Greifswalder Bodden

The two locations differ with respect to their degree of exposure and height of cliff at the transition from unvegetated mudflat to vegetated salt grassland surface. Site 1, located on the southern margin of the Kooser See, was relatively sheltered, with limited fetch (1-3 km from a N to NE direction) and shallow water depths throughout the Kooser See (≤ 1 m). Site 2, located on the western side of the 'Kooser Ecke' (see Figure 1), was much more exposed, with fetch distances of between 20 km (from a N direction) and >> 400 km (from a NNE direction) and with the 2 m depth contour located within one kilometre of the shore.

3 Methodology

Wave conditions and water levels were recorded using pressure sensor technology at five locations along a shore-normal transect within mudflat to i) reed bed and ii) salt grassland transitions at each of the two field monitoring sites (see Figure 1).

Vegetation characteristics recorded in 16 20x20 m quadrats selected randomly along each transect are summarised in Table 1. This shows that the reed vegetation at both sites was more or less similar in terms of its height (210 to 320 cm), density (8 to 28 numbers of stems per quadrat), and dry biomass (33 to 100 g per quadrat). Salt grassland vegetation differed somewhat between the two sites, with Site 2 showing a greater height (11-38 cm compared to 2-27 cm), density (160-610 compared to 30-600 stems per quadrat), and biomass (17-39 compared to 7-32 g per quadrat).

Table 1: Averages (and range) of vegetation height, number of plant stems, and biomass (dry weight) recorded in 20 x 20 cm quadrats at all four measurement transects (N = 16 at each site).

	Reeds		Meadow	
	Site 1	Site 2	Site 1	Site 2
Height (cm)	257 (210-320)	252 (215-310)	10 (2-27)	18 (11-38)
Number of stems	27 (8-30)	19 (11-28)	242 (30-600)	360 (160-610)
Biomass (g)	84.5 (35.8-100.2)	65.9 (32.5-78.0)	15.8 (6.6-32.2)	25.9 (16.5-39.1)

Wave recording followed a methodology similar to that described by Möller [12] and Möller and Spencer [6]. Pressure sensors (Druck Ltd 'PTX1830') were mounted and firmly secured horizontally *ca* 5-10 cm above the mud surface, with data acquisition controlled by Campbell Ltd 'CR10' data loggers.

Figure 2 shows the cross-shore morphology along the reed bed wave recording transects for each of the sites shown in Figure 1. Due to the narrower fringing reed belt at Site 1, transect distances between the individual wave sensors were shortened relative to the transect at Site 2, but both sites were characterised by locally reduced elevations in the central section of the reed fringe (the location of the fourth wave recording sensor).

Water level was measured at the fourth (from seaward) sensor every three hours. If the water level exceeded 5 cm above the sensor, then wave records were obtained at the four most seaward sensors on the transect. In addition, wave records were obtained at all five sensors on the transect at 11:30 each day, whenever the fourth sensor measured water depths of more than 5 cm above the sensor.

Sub-surface water pressure was recorded at a frequency of 4 Hz and, for accurate subsequent Fast Fourier Transform (FFT) analysis, a total of 2^{12} (4096) pressure readings were obtained as part of each wave 'burst' (each ' burst' thus

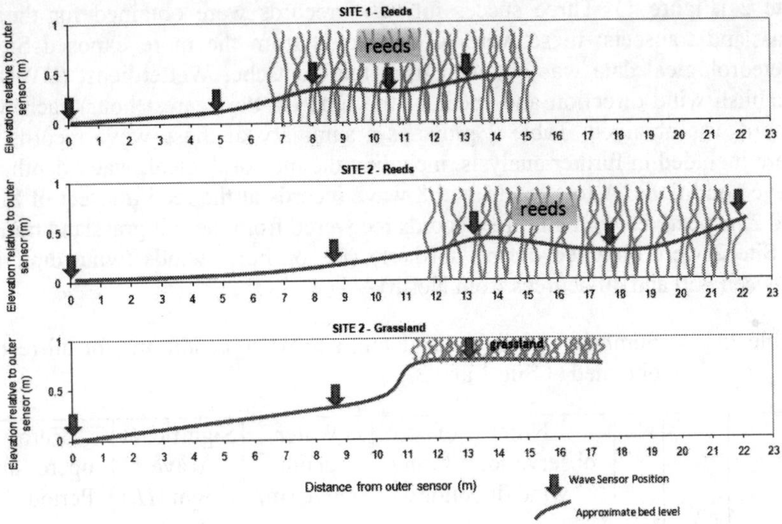

Figure 2: Surface elevation and horizontal distances between wave recording sensors along shore-normal transects at both sites (location of reed vegetation zone is indicated schematically) (for location of sites, see Figure 1).

lasting 17 minutes). A Matlab programme was used to post-process pressure records. The post-processing involved (i) de-trending of the pressure signal to eliminate any low frequency linear water level trend within the time series; (ii) determination of mean pressure / hydrostatic water depth; (iii) application of FFT algorithm for determination of wave frequency spectrum; (iv) frequency and depth dependent correction of the pressure signal to account for signal attenuation with depth; and (v) computation of accurate wave amplitude spectrum. The accuracy of this methodology has been proven by video-calibration in previously reported studies (see Möller *et al* [13]).

The final amplitude spectrum was used to derive the zero-upcrossing wave period (T_z), significant wave height (H_s), root-mean-square wave height (H_{rms}), and total spectral energy (E_{tot}).

4 Results

While the field wave recording equipment was deployed for the period October 2008 to February 2009, a series of factors affected data recovery. These included data logger malfunctioning, animal damage to the pressure sensor cables, and the freezing of the Baltic Sea along the coast from early January to late February 2009, and meant that data recovery during the monitoring period was somewhat limited. Nevertheless, a total of 61 wave records were obtained for the wave recording transect through the reed vegetation at the sheltered Site 1 on the 'Kooser See', and a total of 39 records for the reed transect at the more exposed

Site 2 (Figure 1). Three successful wave records were obtained for the salt grassland transects; these were all recovered from the more exposed Site 2. Meteorological data was obtained from the Deutscher Wetterdienst (DWD) to establish wind direction and speed at the time (to the nearest hour) each wave record was obtained. Table 2 provides a summary of those wave records that were included in further analysis, including the meteorological, water depth, and wave conditions. A total of 15 and 8 wave records at the reed transect of Site 1 and 2 respectively and all three records recovered from the salt grassland transect at Site 2 were obtained during northerly (i.e. onshore) winds (wind directions between 320 and 40 degrees from North).

Table 2: Summary wind, water depth, and wave conditions for all records obtained at Site 1 and 2.

	Site	N	Number of observations during wind direction			Water depth range (m)		Significant wave height (H_s) range (m)		Zero-upcrossing Period (T_z) range (s)	
			N'ly	WNW/ ENE	S'ly	Min	Max	Min	Max	Min	Max
reed	1	61	15	32	14	0.49	1.43	0.01	0.20	0.93	2.99
reed	2	39	8	21	10	0.84	1.55	0.01	0.27	1.12	3.10
grass-land	2	3	3	0	0	1.08	1.30	0.26	0.31	1.87	2.09

Water depths at the outermost pressure sensor during wave records ranged from 0.49 m (the minimum within the data set at the reed transect of Site 1) to 1.55 m (the maximum within the data set at the reed transect of Site 2). Significant wave heights (H_s) at the outermost sensor ranged from 0.01 m (the minimum recorded at both reed transects) to 0.31 (the maximum recorded on the salt grassland transect at Site 2) and zero-upcrossing periods (T_z) from 0.93 s (minimum at the reed transect at Site 1) to 3.10 s (maximum at the reed transect at Site 2).

4.1 Incident wave conditions

A comparison of recorded water depth and incident wave heights at the outer sensors (seaward of the vegetation) shows a marked difference between the more sheltered Site 1 and the more exposed Site 2 (Figure 3). Not only was the observed water depth range and maximum incident wave height larger at the more exposed Site 2 than Site 1, but incident wave heights were also more evenly distributed across the wave height range at the more exposed Site 2 (Figure 4a). As a result, the wave height (H_{rms}) to water depth ratio was more consistent (at around 0.2) at Site 2, compared to Site 1, with a significant relationship between water depth and wave height at Site 2 ($r^2 = 0.58$; $p < 0.05$) but not at Site 1.

Given the longer wave periods and greater water depths at Site 2 compared to Site 1 (see Table 2) and the shallow-water relationship between wave length and wave period of

$$L = T\sqrt{gh} \tag{1}$$

where T is the wave period, g the gravitational constant, and h water depth, it can be inferred that recorded incident wave lengths were larger (within the range 3.2 to 9.2 m) at Site 2 than at Site 1 (within the range 2.0 to 7.5 m). In this case, T_z was used due to it being a more stable parameter than the spectrally derived peak wave period, T_p; the difference between these two parameters was < 0.3 s and < 1.6 s at Site 1 and Site 2 respectively. Given the much lower incident wave heights at Site 1, a much greater frequency of low steepness waves ($H_s/L < 0.01$) was thus observed at this location (Figure 4b).

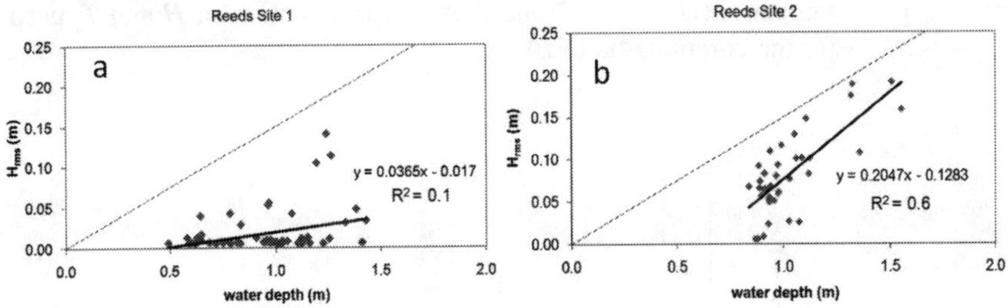

Figure 3: Water depth (h) and root-mean-square wave height (H_{rms}) at the most seaward sensor of the reed transect at (a) Site 1 and (b) Site 2; stippled lines indicate the limiting H_{rms}/h = 0.15 condition observed on the Brouage mudflat by LeHir *et al.* (2000).

For those wave records (N = 8) that coincided with northerly (onshore) wind directions (between 320° to 40° N), wind speeds were well-correlated with wave heights (H_s) (r^2 = 0.6) and wave periods (T_z) (r^2 = 0.70) at Site 2, but less so at Site 1 (r^2 = 0.50 and 0.20 for correlation with wave height and wave period respectively).

Three wave records (on 30th October 2008 at 12:00, 15:00 and 18:00 CET) were successfully recorded along the salt grassland transect, although only the central three sensors were operational along this transect at the time. Water depths during the second and third record were identical at 1.3 m, while the first record was obtained during depths of 1.1 m. Wind directions varied from 10°N to 350°N, and 340°N during the three records shown in Figure 5 respectively, with speeds increasing from 140 ms^{-1} during the first, to 178 ms^{-1} on the second, and 185 ms^{-1} during the third record (reading records from left to right). Incident wave heights (H_s) resulting from those conditions were 0.29, 0.31, and 0.26 m respectively, with wave periods of 1.9, 2.1, and 2.0 s.

4.2 Attenuation over salt grassland

Figure 5 summarises the results of those three records, showing average energy dissipation per metre distance between the outermost and cliff-front sensor

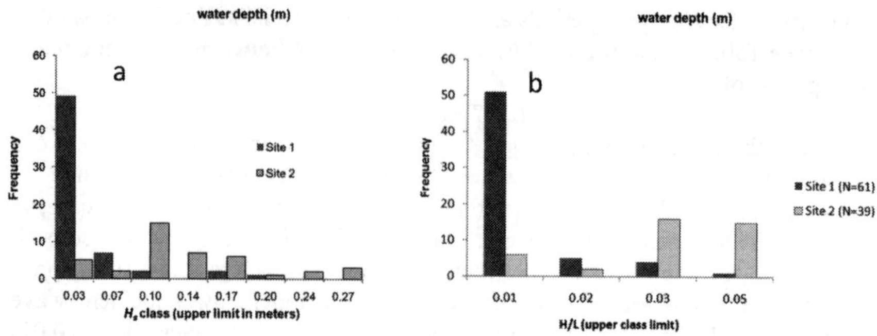

Figure 4: Distribution of (a) significant wave heights (H_s) and (b) wave steepness (H/L, with H_s used as a representation of H and T_z used for the computation of L).

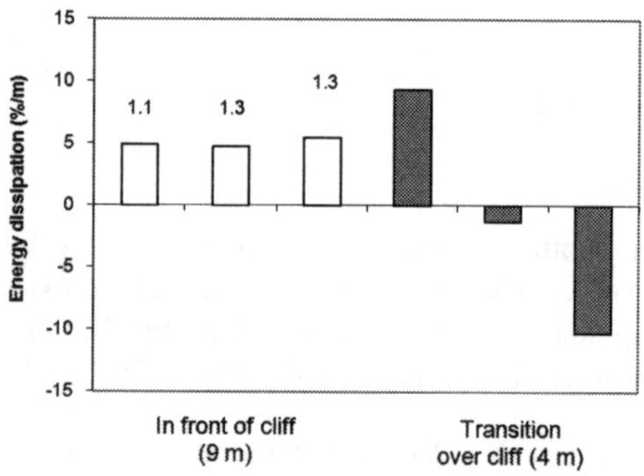

Figure 5: Energy dissipation (%/m averaged over transect section distance (as indicated along x-axis)) along the salt grassland transect (numbers above cliff front bars indicate water depths (in meters)).

(9 metre distance) as well as energy transformations across the cliff and onto the salt grassland (4 metre distance). Wave attenuation in front of the cliff and vegetated section of the salt grassland transect was remarkably consistent at around 5% m^{-1} during all three records (see Figure 5), while energy dissipation across the cliff onto the salt grassland was observed only on the first occasion (9.3% m^{-1}), with energy increases observed across the cliff on the two subsequent occasions (1.3 and 10.3% m^{-1} wave energy increase on the second and third occasion respectively).

4.3 Attenuation in reeds

Wave energy dissipation along each of the reed transects varied considerably between Site 1 and Site 2 and for the varying meteorological conditions encountered. The dataset was divided into wave records obtained during

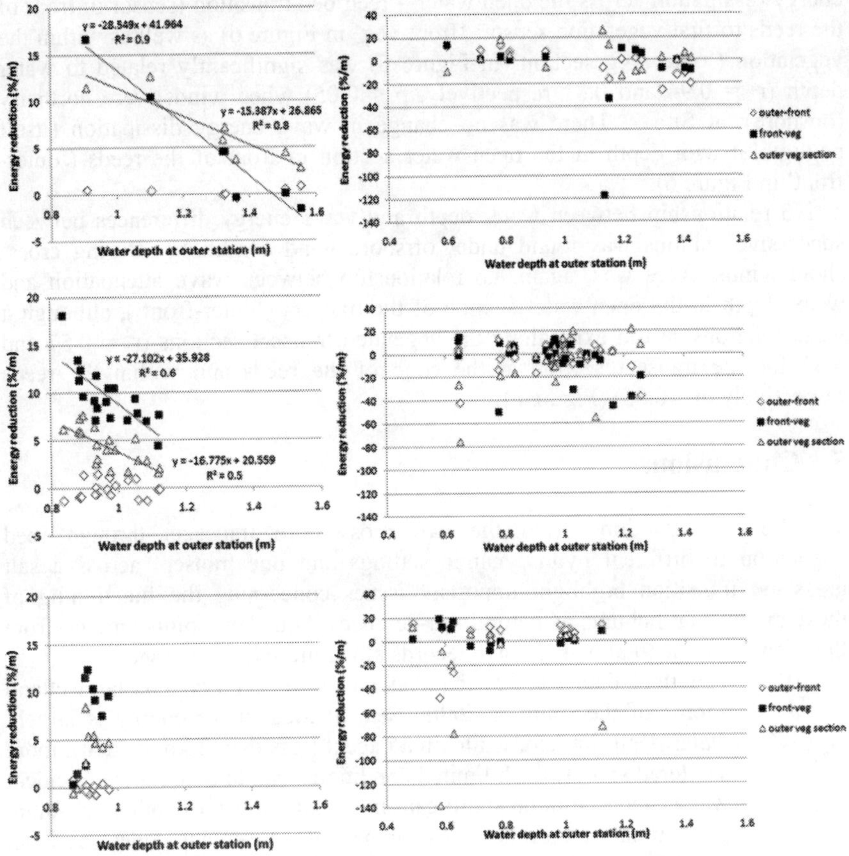

Figure 6: Relationship between water depth at outer sensor and energy reduction for Site 2 (left) and Site 1 (right), separated by wind direction category (top: northerly (onshore) winds, middle: cross-shore winds, bottom: southerly (offshore) winds).

northerly (onshore, 320° to 40° N), southerly (offshore, 100° to 260° N), and cross-shore (50° to 90° N and 270° to 320° N) wind conditions. Energy transformation between individual sensors was calculated as a spatial average (% m^{-1}) over each transect section, to allow comparison between individual transect sections. Previous studies have indicated there to be a degree of control of relative wave height (H_s/h) on wave energy reduction over saltmarshes (e.g. Möller [12]). No such relationship was apparent at either of the reed transects of this study, with the exception of Site 2, where a positive relationship between Hs/h and wave energy dissipation was present when winds were from a southerly (offshore) direction ($r^2 = 0.82$ for the transition from mudflat to reeds; $r^2 = 0.80$ for the most seaward vegetated section). When only water depth is considered as a control on wave energy dissipation, however (Figure 6), wave energy dissipation across the open water – reed bed transition (sensor in front of the reeds to first vegetation sensor; 'front-veg' in Figure 6) as well as within the vegetation ('outer veg section' in Figure 6) was significantly related to water depth ($r^2 = 0.94$ and 0.81 respectively, $p < 0.05$) when winds were northerly (onshore), at Site 2. There was no change in wave energy dissipation (itself negligible) with depth in the open water section in front of the reeds ('outer-front' in Figure 6).

No relationship between water depth and wave energy differences between successive stations was found under offshore wind conditions. During cross-shore winds, there was, again, no relationship between wave attenuation and water depth in the open water section of the transect ('outer-front'), although a weak relationship did exist along the vegetated transect sections ($r^2 = 0.57$ and 0.53 for the transect section at the edge of the reeds and within the reeds respectively, $p < 0.05$) (Figure 6).

5 Conclusions

The field observations from the two cross-shore transects through reed vegetation in different hydrodynamic settings and one transect across a salt grassland transition highlight important issues concerning the functioning of these micro-tidal habitats as natural sea-defences. Four key points emerge from the analysis of the total of 103 wave records along these three transects:

i) The field data suggest that, on the more exposed coastal wetland shores of the southern Baltic Sea, incident wave energy is largely water depth limited, while on adjacent (less than 2 km distant), more sheltered shores, fetch limited conditions lead to much lower incident wave energy with no significant influence of water depth. Any future predictions of morphological or habitat adjustment under sea level rise scenarios (i.e. an increase in water depth; Schäfer *et al.* [14]) must take such depth-dependent process relationships into account.

ii) As has been shown to be the case in macro-tidal saltmarsh settings, micro-tidal grassland cliffs also result in complex wave energy transformation processes across cliffed transitions, with the existence of water depth thresholds that control the transition between energy reflection from, *versus* energy transmission across, the cliff face. Again, the implications are that, under rapid sea level rise, periods of cliff erosion (under high energy wave impact but relatively low water level) are likely to be followed by periods of higher energy landward of the cliff face, with associated re-suspension of sediment and/or adjustment of vegetation composition.

iii) Wave energy dissipation through reed vegetation is significant (up to 26 $\%m^{-1}$ within the vegetation). Its temporal variability, during onshore wind conditions and for wave height to water depth ratios > 0.1, is controlled to a large extent by water depth, rather than incident wave height, relative wave height, or wave steepness.

iv) The depth control on wave energy dissipation is likely to result from vertical variations in the degree of physical obstruction (e.g. biomass) to the progression of waves through the reed vegetation. Any factors that affect this vertical variation in plant matter (such as plant – animal interactions) are thus as critical in determining the potential morphological and ecological impact of waves as considerations of the rate of sea level rise.

Further field studies are planned for the autumn of 2009 to expand the dataset of wave transformation over salt grassland and to more closely investigate the link between water depth, vertical variation in plant biomass, and wave transformation. The intention is also to use the wave measurements presented here in conjunction with broader ecological measurements, currently being undertaken by Greifswald University, to provide an assessment of the likely future ecological and physical functioning and management of coastal micro-tidal brackish grasslands.

References

[1] Allen, J.R.L. & Pye, K., (eds.), *Saltmarshes. Morphodynamics, Conservation and Engineering Significance*. Cambridge University Press: Cambridge, 1992.

[2] French, J.R. Tidal marsh sedimentation and resilience to environmental change: Exploratory modelling of tidal, sea-level and sediment supply forcing in predominantly allochthonous systems. *Marine Geology,* **235**, pp. 119-136, 2006.

[3] Bartholdy J., Christiansen, C., & Kunzendorf, H. Long-term variations in backbarrier saltmarsh deposition on the Skallingen peninsula – the Danish Wadden Sea. *Marine Geology,* **203**, pp. 1-21, 2004.

[4] Knutson, P.L, Brochu, R.A., Seelig, W.N. & Inskeep, M. Wave damping in *Spartina* alterniflora marshes. *Wetlands*, **2**, pp. 87-104, 1982.

[5] Möller, I., Spencer, T., French, J.R., Leggett, D.J. & Dixon, M. Wave transformation over salt marshes: A field and numerical modelling study from North Norfolk, England. *Estuarine, Coastal and Shelf Science,* **49**, pp. 411-426, 1999.

[6] Möller, I. & Spencer, T. Wave dissipation over macro-tidal saltmarshes: Effects of marsh edge typology and vegetation change. *Journal of Coastal Research,* **SI36**, pp. 506-521, 2002.

[7] Tyler, G. Regional aspects of Baltic shore-meadows. *Vegetatio,* **19**, pp. 60-86, 1969.

[8] Meier, H.E.M., Broman, B. & Kjellström, E. Simulated sea level in past and future climates of the Baltic Sea. *Climate Research,* **27**, pp. 59-75, 2004.

[9] Gillner, V. Salt marsh vegetation in southern Sweden. *Acta Phytogeograhica Suecica,* **50**, pp. 97-104, 1965.

[10] Jeschke, L. Vegetationsdynamik des Salzgraslandes im Bereich der Ostseeküste der DDR unter dem Einfluss des Menschen. *Hercynia, N.F.*, **24**, pp. 321-328, 1987.

[11] Seiberling, S., Stock, M. & Thapa, P.P. Renaturierung von Salzgrasländern bzw. Salzwiesen der Küsten. In: Zerbe, S. & Wiegleb, G. (eds.), *Renaturierung von Ökosystemen in Mitteleuropa*. Springer: Heidelberg: pp. 183-208, 2009.

[12] Möller, I. Quantifying saltmarsh vegetation and its effect on wave height dissipation: results from a UK East coast saltmarsh. *Journal of Estuarine, Coastal and Shelf Sciences,* **69(3-4)**, pp. 337-351, 2006.

[13] Moeller, I., Spencer, T. & French, J.R. Wind wave attenuation over saltmarsh surfaces: Preliminary results from Norfolk, England. *Journal of Coastal Research,* **12(4)**, pp. 1009-1016, 1996.

[14] Schäfer, C., Zerbe, S. & Seiberling, S. Mögliche Auswirkungen des Klimawandels auf die Salzgrasländer an der Ostseeküste. Eine Analyse am Beispiel der Küste Mecklenburg-Vorpommerns. *Naturschutz u. Landschaftsplanung,* **40**, pp. 361-366, 2008.

Numerical investigation of sandy beach evolution using an incompressible smoothed particle hydrodynamics method

N. Amanifard[1], S. M. Mahnama[1], S. A. L. Neshaei[2]
& M. A. Mehrdad[2]
[1]*Mechanical Engineering Department, University of Guilan, Iran*
[2]*Civil Engineering Department, University of Guilan, Iran*

Abstract

The current work presents an incompressible smoothed particle hydrodynamics (SPH) model to simulate sandy beach evolution. The Navier-Stokes equations are solved in a Lagrangian framework using a three-step fractional method. In the first step, a temporary velocity field is provided according to the relevant body forces. This velocity field is renewed in the second step to include the viscosity effects. A Poisson equation is employed in the third step as an alternative for the equation of state in order to evaluate pressure. The present method is validated by solving a free surface problem and comparing the computational results with the experimental results, as well as numerical data that is evaluated from the standard SPH method. Then, based on an experimental model, the simulation of sandy beach evolution has been investigated by this method. Comparison of the computed results with previous studies that are reported in coastal engineering references implies the capability of the method for the simulation of such complex flows.

Keywords: smoothed particle hydrodynamics (SPH), Lagrangian method, free surface, sandy beach evolution.

1 Introduction

The best understanding of coastal processes requires a blend of analytical study on the nature of variety of shorelines and making many experimental researches. However, due to the complexity of coastal processes, a number of reliable

www.witpress.com, ISSN 1743-3541 (on-line)
doi:10.2495/CP090061

laboratory and field data are limited. The aim of the current research is to present a numerical method for simulating sandy beach evolution in the coastal zone. For this purpose, in this paper, the smoothed particle hydrodynamics (SPH) method is applied, which uses a purely Lagrangian approach and has been successfully employed in a wide range of applications. The meshless characteristic of SPH makes it unnecessary to use data connectivity for this method, which is not the case for the finite volume and finite element methods. This gives the method a very useful feature when dealing with complex flows that exhibit large deformations and/or free-surfaces.

This method was originally developed by Lucy [1] to solve compressible astrophysical problems. The method was later extended to incompressible flows by Monaghan [2]. Several other researchers have contributed to the method and solved various engineering problems including water wave problems, fluid-structure interaction and interfacial flows [3–5].

In this paper, based on the incompressible SPH method, a three step incompressible SPH algorithm is applied to simulate sandy beach evolution in shorelines. The proposed algorithm is similar to the three step explicit SPH algorithm proposed by Hosseini *et al.* [6] for simulation of incompressible fluid flows. In the first step of this algorithm, the momentum equation is solved in the presence of the body force neglecting all other forces. The calculated temporary velocities are renewed in the second step to include the viscosity effect. A Poisson equation is employed in the third step as an alternative of the equation of state in order to evaluate pressure by projecting the provisional velocity. This Poisson equation considers a trade-off between density and pressure, which is utilized in the third step to impose the incompressibility effect. In order to validate this algorithm, first a wave propagation problem is modeled by this method and the computational results are compared with the experimental data and standard SPH data. Then it is utilized for simulation of an arbitrary pattern of surf zone, which is very similar to the experimental model [7]. The numerical surf zone is much smaller than the experimental one, because of reduced run time and prevention of the numerical divergence.

2 Governing equations

The governing equations for simulating free surface flow in 2-D dimensions are the mass and momentum conservation equations. With regard to fluid particles, they are written in Lagrangian form as:

$$\frac{1}{\rho}\frac{D\rho}{Dt}+\nabla.V=0 \tag{1}$$

$$\frac{DV}{Dt}=-\frac{1}{\rho}\nabla P+g+\frac{1}{\rho}\nabla.\vec{\vec{\tau}} \tag{2}$$

where ρ is the fluid density, V is velocity and p represents the pressure of fluid. Eqn. (1) is in the form of a compressible flow. The purpose is that the deviation of fluid densities at the particle can then be used to enforce incompressibility in the correction step of time integration.

3 SPH formulation

The SPH formulations as developed by Monaghan [8] are obtained by interpolating from a set of points that may be disordered. The interpolation is based on the theory of integral interpolants using kernels that approximate a delta function. The interpolants are analytic functions that can be differentiated without the use of grids. If the points are fixed in position, the equations reduce to finite difference equations, with different forms depending on the interpolation kernel. The SPH equations describe the motion of the interpolating points, which can be thought of as particles. Each particle carries a mass *m*, a velocity V, and other properties, depending on the problem.

Using the above concepts, any quantity of particle *i*, whether scalar or vector, can be approximated by the direct summation of the relevant quantities of its neighboring particles:

$$\phi_i(r_i) = \sum_j m_j \frac{\phi_j(r_j)}{\rho_j(r_j)} W(|r_i - r_j|, h) \tag{3}$$

where *i* and *j* are the reference particle and its neighbor; ϕ_i and ϕ_j are scalar or vector quantities being interpolated; r_i and r_j are the position of particles; W represents the interpolation kernel and h is the smoothing distance. The smoothing kernel used in this literature is a cubic spline, which is most commonly used in hydrodynamic calculations.

4 Three-step incompressible SPH algorithm

In this section, an algorithm is presented to show the sequence of computation of each term in the governing equations. In this paper, a fully explicit three-step algorithm is used. In the first step of this algorithm, the momentum equation is solved in the presence of the body forces neglecting all other forces. As a result, an intermediate velocity is computed as:

$$V^* = V_t + g\Delta t \tag{4}$$

where $g = (g_x, g_y)$ represents the gravity acceleration and $V^* = (u^*, v^*)$ is the first intermediate velocity. Our experience has shown that it is important to impose the body forces in the first step of the solution algorithm, especially in highly viscous fluids. In the second step, the calculated intermediate velocities are employed to compute the divergence of the stress tensor.

In this work, the divergence of the stress tensor in the momentum equation is obtained as:

$$\left(\frac{1}{\rho}\nabla.\overset{\Rightarrow}{\tau}\right)_i=\sum_j m_j\left(\frac{\overset{\Rightarrow}{\tau}_i}{\rho_i^2}+\frac{\overset{\Rightarrow}{\tau}_j}{\rho_j^2}\right).\nabla_i W\left(r_{ij},h\right) \tag{5}$$

At the end of the second step, the velocity components of each particle are updated according to:

$$V^{**}=V^{*}+\left(\frac{1}{\rho}\nabla.\overset{\Rightarrow}{\tau}\right)\Delta t \tag{6}$$

At this stage, each particle is moved according to its second intermediate velocity $V^{**}=\left(u^{**},v^{**}\right)$.

Finally, the velocity of each particle at the end of time-step will be obtained as:

$$\widehat{V}_i=-\Delta t\sum_j m_j\left(\frac{P_i}{\rho_i^{*2}}+\frac{P_j}{\rho_j^2}\right)\nabla_i W_{ij} \tag{7}$$

$$V_{t+\Delta t}=V^{**}+\widehat{V} \tag{8}$$

and the final position of particles is calculated using a central difference scheme in time:

$$r_{t+\Delta t}=r_t+\frac{\Delta t}{2}\left(V_{t+\Delta t}+V_t\right) \tag{9}$$

This completes the computations required for one time-step. The procedure should be repeated for every other time-step until a desired time is reached.

5 Simulation of sandy beach evolution

In this section, deformation of a sandy beach is simulated using the present numerical model. The pattern of this problem is an experimental work that is performed by Mehrdad and Neshaei [7]. The scale of the numerical beach is smaller than the experimental one and so no quantitative comparison is done between the numerical results and the experimental data. In fact, quantitative investigation of this simulation is not our purpose; we are only going to show that this numerical work can be applicable to model these complex problems.

In this paper, two shapes of beaches are considered: an open beach and a beach that is in the vicinity of a seawall.

A major problem in simulating a sandy beach profile is detecting a suitable model for sand. It is here assumed that sand particles behave as a non-Newtonian fluid. There are many rheological models to simulate these fluids. The Bingham model is the simplest and the most known and it is expressed as:

$$\mu_{eff} = \mu_B + \frac{\tau_B}{\dot{\gamma}} \tag{10}$$

where μ_B, τ_B and $\dot{\gamma}$ are viscosity, Bingham yield stress and shear rate, respectively, to calculate the effective viscosity.

In this model, the fluid behaves like a rigid body at shear rates below the yield stress, while it behaves like a Newtonian fluid at shear rates greater than the yield stress. The general Cross model is another rheological model that effectively simulates non-Newtonian fluids:

$$\frac{\mu_0 - \mu_{eff}}{\mu_{eff} - \mu_\infty} = (K\dot{\gamma})^m \tag{11}$$

where μ_0, μ_∞ are viscosity at very low and very high shear rates, respectively; K and m are constant parameters.

By combining eqns. (10) and (11) and taking m as unity, the effective viscosity in the Cross model is defined as:

$$\mu_{eff} = \frac{1000\left(\mu_B + \frac{\mu_B^2}{\tau_B}\dot{\gamma}\right)}{1 + \frac{1000\mu_B}{\tau_B}\dot{\gamma}} \tag{12}$$

In order to avoid numerical instability, μ_0 is frozen at a fixed high value of $1000\,\mu_\infty$. It should be noted that under this condition, the Cross model, unlike the Bingham model, is a continuous variable.

5.1 Simulation of sandy open beach evolution

The numerical open beach is shown in Fig. 1. The overall length of the surf zone is 18m and the maximum depth of water is 0.6m. An inclined beach with a uniform slope of 6%, which is covered by sand particles, is placed at the end of the numerical surf zone. Sinusoidal waves are generated by a wave-maker, which is at the right-hand side of the surf zone. The velocity of the wave-maker particles is computed as:

$$u(t) = \sin(2\pi t) \tag{13}$$

During the simulation the total number of particles is 6471, corresponding to a particle spacing of 0.04m. In the computation a constant time step of 0.00002s is employed. The particle configuration during the evolution of the beach by the effect of waves on sand particles is illustrated in Fig. 2. In this simulation, the following approach of the yield stress and viscosity of sand is adopted by setting $\tau_B = 250pa$ and $\mu_B = 0.1pa.s$.

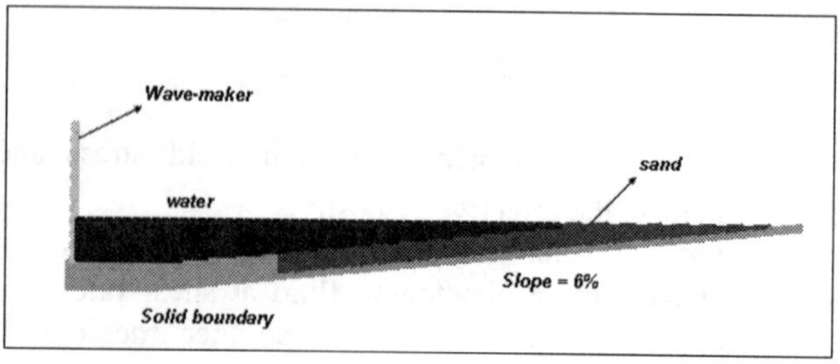

Figure 1: Numerical open beach.

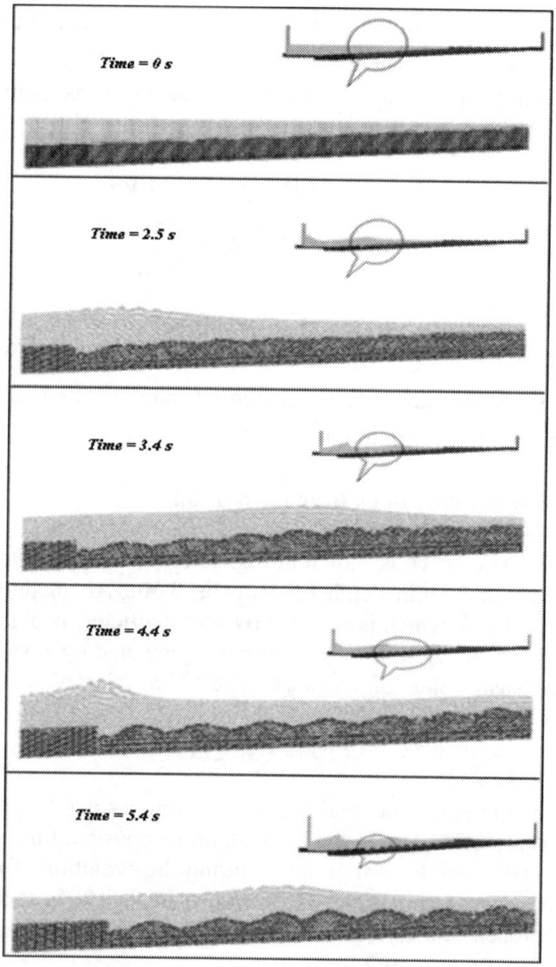

Figure 2: Simulation of sandy open beach at different times.

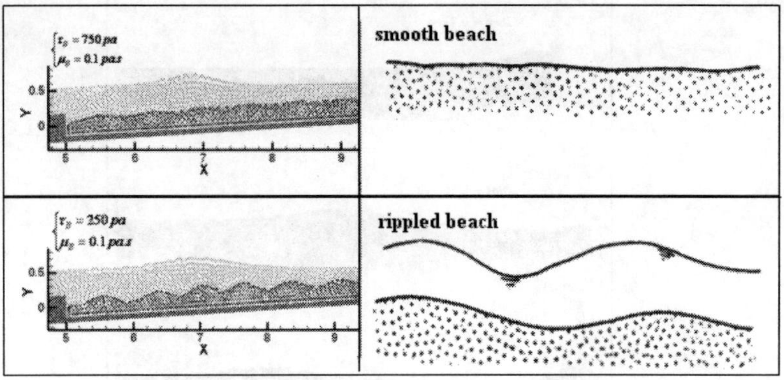

Figure 3: Comparison of beach shapes for different yield stresses of sand.

This shape of the beach is similar to the *rippled beach* that is reported in coastal engineering references. In order to obtain the other shapes of beach, we change τ_B as a changeable value. For this purpose, by setting $\tau_B = 750pa$, the expected resistance of sand particles against the deformation increases, thus a *smooth beach* is formed. Comparison of these two shapes of beach is shown in Fig. 3 (where X is the distance from the shoreline and Y is the water depth).

5.2 Simulation of sandy beach evolution in the vicinity of a seawall

In order to show the effect of a wall and a reflective wave on beach shape, a vertical wall is placed on the right-hand side of the numerical surf zone, which is explained in the previous section. The length of the surf zone is shorter than the previous model because of the formation of a reflective wave. The velocity of the wave-maker is adopted as eqn. (13). The results of this simulation are illustrated in Fig. 4. It can be seen that the reflective waves cause the water to penetrate the lower part of the wall and thus sediments (sands) are pushed back. This problem is very important in coastal engineering in order to design safe structures, which are responsible for resisting violent waves and protecting coastal zones.

6 Conclusion

In this paper, an incompressible SPH method is employed for numerical simulation of some examples of free surface flows. The Navier-Stokes equations are solved in a Lagrangian framework using a three-step fractional method. The main advantages of the proposed algorithm is simulation of free surface flows with large deformation naturally (without imposing any special condition) and more accurately than the previous Standard SPH method. Moreover, the computational time spent to reach the final results is not very long.

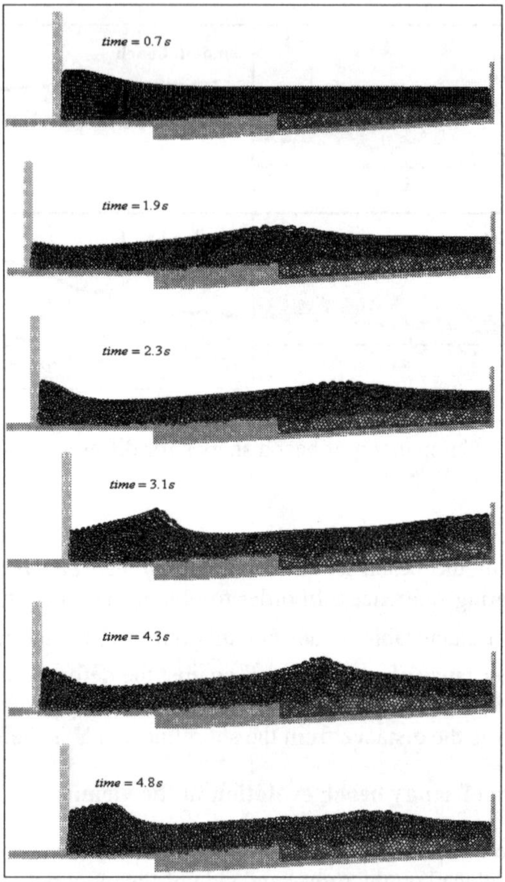

Figure 4: Simulation of a sandy beach in the vicinity of a seawall.

According to existing information, the SPH method has not already been applied to the simulation of sandy beach evolution, so the current SPH method is employed for the simulation of sandy beach deformation in various conditions. Following the previous studies about beach shapes that are reported in the coastal engineering references, the shape of computational beaches is very similar to the real ones qualitatively. This result demonstrates the capability of the method for the simulation of such flows.

In summary, in this paper, following an experimental model, the simulation of sandy beach evolution by an incompressible SPH method has been investigated. Comparison of the computed results with previous studies that are reported in coastal engineering references implies the capability of the method for simulation of such conditions. It is believed that the present work can be regarded as a basis for future researches into this problem and other similar applications.

References

[1] Lucy, L.B., 1977. A numerical approach to the testing of the fission hypothesis. Astron. J. 82, 1013–1024.

[2] Monaghan, J.J., 1994. Simulating free surface flows with SPH, Computational Physics 110, 399–406.

[3] Dalrymple, R.A., Rogers, B.D., 2006. Numerical modeling of water waves with the SPH method, Coastal engineering 53, 141-147.

[4] Farahani, M.H., Amanifard, N., Pouryoussefi, Gh., 2008(b). Numerical simulation of a pulsatory flow moving through flexible walls using smoothed particle hydrodynamics. Proceeding of 2008 International Conference of Mechanical Engineering (ICME 2008), 2008 World Congress of Engineering, London, UK, pp 1337-1341, July, 2008.

[5] Hosseini, S.M., Amanifard, N., 2007. Presenting a modified SPH algorithm for numerical studies of fluid-structure interaction problems, IJE Trans B: Applications 20, 167-178.

[6] Hosseini, S.M., Manzari, M.T., Hannani, S.K., 2007. A fully explicit three step SPH algorithm for simulation of non-Newtonian fluid flow. Numerical Methods for Heat and Fluid Flow 17, 715– 735.

[7] Mehrdad, M.A., Neshaei, M.A.L., 2004. Hydrodynamics of the surf zone in the vicinity of a partially reflective seawall. Civil Engineering 2, No. 3.

[8] Monaghan, J.J., 1992. Smoothed particle hydrodynamics. Annu. Rev. Astron. Astrophys 30, 543–574.

Section 3
Extreme events and sea level rise

A model to predict the coastal sea level variations and surge

M. M. F. de Oliveira & N. F. F. Ebecken
COPPE/Federal University of Rio de Janeiro, Brazil

Abstract

This paper presents a methodology to predict the coastal sea level variations and surge using a neural network model. Although drastic storm surge typically does not occur along the coastal waters of Brazil, these events can cause some damage to coastal regions. A strong storm surge occurred along the southeast coast of Brazil in March 1998 caused severe flooding in these coastal areas, destroying some coastline constructs. Operational forecasting of high sea levels (storm surges) might be important in the southeast coastal of Brazil, where there are registered sea level variations above the astronomical tide predictions that can consistently impact coastal zones in this area. The aim of this study is to develop an empirical prediction of storm surge by determining the relationship of the wind and pressure fields to storm surge. This proposed model can be used to as complement of the standard constant harmonic model to improve the prediction of the sea level variations.
Keywords: storm surge, sea level, neural networks.

1 Introduction

In the South Atlantic Ocean, along the Brazilian coastline, there are few tide gauge records with long series to analyze and predict surge events. Characteristics of the meteorological tide variations along the Southeast coast of Brazil have been studied by Marone and Camargo [18]. Castro and Lee [3] presented a study about the sea level fluctuations due to the wind-driven forces in the southeast continental shelf. Ribeiro [25] investigated a surge caused by the passage of a cyclone along the Rio de Janeiro coastline that raised the sea level 0.60 m above the mean sea level datum, causing damage to coastal communities along the Guanabara Bay. Netto and Lana [20] studied the superficial sediment

WIT Transactions on Ecology and the Environment, Vol 126, © 2009 WIT Press
www.witpress.com, ISSN 1743-3541 (on-line)
doi:10.2495/CP090071

characteristics of tidal flats in Paranaguá Bay. de Mesquita [7, 8] verified a similar behavior of the mean sea level oscillations along this area of the Brazilian coastline. Mantovanelli *et al.* [17] verified the tidal velocity and duration as a determinant of water transport and residual flow in Paranaguá Bay estuary. Dalazoana *et al.* [6] studied the mean sea level variations using longer tide gauge temporal series from Cananéia and Fiscal Island (State of Rio de Janeiro) tide gauge and satellite altimetry to establish analysis methods applicable to Brazilian vertical datum region.

The classical method of harmonic analysis is used to predict the astronomical tides. Tidal curves appear as periodic oscillations and can be described in terms of amplitude, period or frequency. The harmonic analysis is based in tidal variations represented by N harmonic constituents of the tide (Doodson and Warburg [9]). Normally, 365 days of hourly data at a point are needed to extract the constituents with adequate separation of closely spaced constituents using the least squares method. These constituents can then be used to provide reliable predictions for future tides at the respective point (Franco [12]).

Predictions for reference stations are prepared from the astronomical arguments, using local constituents determined by previous analysis and do not take into account meteorological influences. Thus, the observed and predicted values of the sea level are normally different. Numerical model developed to predict surges are still considered insufficient due to the complexity between the non-linear processes involved. These models require a large amount of tidal and meteorological data, collecting many factors as central pressure, speed of the cyclone, rainfall and coastal topography (Lee [16]).

Nowadays, the neural network model (NNM) has been widely applied to modeling non-linear dynamic systems, using time series that translate the physical relations between the input variables (predictors) and the phenomenon that will be modeled. Eisner and Tsonis [10] developed some methodologies for making short-term predictions of nonlinear time-series data, using a neural network model. These authors discuss the implication of these methodologies in the studies of weather and climate.

The NNM has some important characteristics such as generalization, parallelism, non-linearity, adaptability, robustness and others. These models have been used in some fields of science and engineering. Sztobryn [27] applied NNM in hydrological forecasting where the variation of water level is only wind generated. The results were successfully compared with observed sea level and others routine methods. Lee [16] applied a NNM for forecasting storm surge in Taiwan related to the passage of three typhoons over the region. The results indicate that NNM is efficiently capable of learning and predicting from these events. Tseng *et al.* [28] used a typhoon-surge forecasting model developed with a backpropagation neural network in the coastal of north-eastern Taiwan. To compare the better forecasting model, four models were applied and tested under different compositions of the input variables. For coastal and harbor engineering applications, Chang and Lin [4] simulated tides at multi-points considering tide-generating forces. The NNM proposed is applicable for multi-points tidal prediction in which the tidal type is similar to that of the original point.

The southeast coast of Brazil is sufficiently affected by cold fronts over 3–5 day periods. An important event that sometimes occurs due to combination of tides and surges is the rising of the sea level with waves that reached the coastline. There are few NNM applications to predict the variability of sea level along the Brazilian coastline focused on the surge events. The relationships describing the response of the coastal sea level due to the influence of cold fronts was analyzed using cross-correlations and spectral density between the tide gauge series and meteorological variables. Maxima values and time lags of both analyses were proposed as inputs of the sea level forecast model.

This paper presents a methodology to predict the coastal sea level variations and surge using a NNM.

2 Study area

The study area lies within the Cananéia estuary (24°50'–25°05'S/47°45'–48°00'W); southeast coastal region of Brazil in São Paulo State. This region is located on the continental shelf, which is wider than the shelf of the northern coast. The average width and declivity, near Cananéia city in São Paulo State is about 218 km and 46 cm/km, respectively (Filippo [11]). The isobaths are oriented from southwest to northeast, parallel to the coastline with 45° northern direction. It has wide coastal plains, long beach barriers, and large estuaries (Angulo and Lessa [1]). The Cananéia Estuary is an important biological reserve and contains federal and state Environmental Protected Areas (SMA, 1990/1996). This estuarine system covers an area of 135 km^2 and is surrounded by a large mangrove area with high concentrations of nutrients (Besnard [2]).

The South American continent is affected by both tropical and extra tropical regions weather systems. The most severe weather systems in South America are cold fronts, intense extra tropical cyclones near the east coast causing intense winds, upper level cyclonic vortices (ULCV), in some cases responsible for cyclogenesis and frontogenesis, South Atlantic Convergence Zone (SACZ), squall lines, mesoscale convective complexes and the Low Level Jet (LLJ). This region is influenced by persistent high-pressure over the South Atlantic Ocean which enhances northeast flow across the area. This circulation is disturbed, periodically, by the passage of frontal systems caused by migrating anticyclones that move from the southwest across the northeast in the southeast coast of Brazil. In this region is verified the presence of strong cyclogenesis activity (Gan and Rao [14]; Seluchi [26]) associated with ULCV that reach through the South America west coast causing instability in the east and northeast sector. Gan & Rao [14] has verified two regions of persistent cyclogenesis over South America; one over the San Matias Golf in Argentina (42.5°S, 62.5°W) and another over Uruguay (31.5°S, 55°W). The climate is subtropical humid and during the El Niño–South Oscillation (ENSO) phenomenon great climatic disturbances occur in this region, leading to abundant rain.

3 Statistical analysis

The atmospheric pressure, wind and tide gauge time series were analyzed statistically by estimating the center of the distribution (mean and median), variances, standard deviation, asymmetry and kurtosis. From the percentiles analyses could be identified few outliers in the sea level record with the box plot graphic. They were substituted with the average values between the previous and the following hourly data. Before fitting, both series were used for the period from January 1997 to December 1998 to study the coastal sea level response related to the meteorological conditions as well as the behavior of the coastal sea level in this Brazilian region.

3.1 Filtering data using a low-pass filter

The present study was focused on the oscillations in sea level caused by frequency lower than astronomical driven forces related to the passage of frontal systems which have periods around 3 to 5 days. Tides and inertial motions usually cause a high-frequency noise in sea level records used to analyze low-frequency motion in the ocean. To eliminate diurnal and shorter-periods tides oscillations from input data set, the Thompson low-pass filter, a symmetric digital filter, was used.

Hourly observed sea level records were then filtered to remove the oscillations or noises related to tidal frequencies. For the reanalysis data set was used the same filter, considering the 6-hourly intervals. After filtering, the hourly sea level series was replaced at 6-hourly interval as the reanalysis data and so, both data sets could be compared for the same time interval and frequencies.

3.2 Series analysis in the time and frequency domain

Storm surge is usually considered to be driven by two processes: the extreme wind stress and atmospheric pressure. Therefore, cross-correlations between the filtered sea level, atmospheric pressure, zonal and meridional wind stresses were calculated.

4 Neural Network Modelling (NNM)

In this paper, different training methods were applied to find the best performance: Radial Basis Functions (RBF), a network particularly adapted to approximation function. The hidden layer is defined by radial basis functions and the learning fits a non-linear surface accordingly some stochastic criteria; Generalized Regression Neural Network (GRNN), a method for estimating the joint probability density function (pdf) of x and y as in standard regression technique, given only a training set (Cigizoglu and Alp [5]); and NN feed forward - MLP. It was used the supervised learning and back propagation algorithm.

All samples were used with intervals of 6-hourly (LT) between the observations and this data set was selected in 50% for training, 25% for testing and 25% for validation. The input variables for the NN training were atmospheric variables, filtered sea level series of previous hours, and observed wind for the actual time.

5 Statistical analysis results

The filtered records generated the time series of the sea level response in low frequency related with the meteorological systems which were used in the NN (Fig. 1).

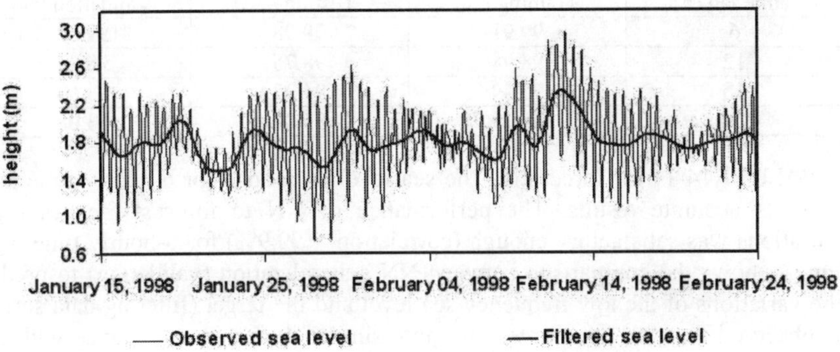

Figure 1: Oscillations of observed and filtered sea level time series.

In order to verify the inter-correlations of the sea level response and meteorological variables in the same frequency, values of cross-spectral densities and coherence between the series were analyzed. It was found that peaks of energy and high coherence for periods from 5 to 3 days were related to passages of cold fronts over the region.

6 NNM performance

The maximum values of cross-correlations described previously were used as input model. Then, a time lag was considered with respect to the sea level response and the meteorological variables. Autocorrelations of the low frequency sea level and wind speed for the current time was also used. Therefore, pressure, zws, mws, 18, 12, and 6-hourly filtered sea level and wind speed predictors) as input vectors. The filtered sea level relating to 6-hourly after was used as output variable.

Table 1 shows the best performances of the NNM with the correlation coefficients (r). The MLP with 7-14-1 layers produced the best results.

The back propagation algorithm was used for the NNM training. The activation function used in the hidden and output units was the hyperbolic tangent function. Table 2 shows the correlation coefficients to the selected pairs for training, testing and validation for 6, 12, 18 and 24-hourly simulations. In both the stages a high correlation was observed.

Table 1: Correlation Coefficients (r).

NNM	Training (%)	Validation (%)
RBF	95.36	93.63
GRNN	98.18	97.65
MLP(7-9-1)	95.73	93.83
MLP(7-11-1)	95.81	94.17
MLP(7-14-1)	99.90	99.87

Table 2: Correlation coefficients (r) for 6, 12, 18 and 24 hourly simulations.

Time lag (h)	Training (%)	Testing (%)	Validation (%)
6	99.91	99.98	99.87
12	98.98	98.75	98.44
18	95.49	94.32	93.64
24	88.19	85.06	83.08

MLP (7-14-1) for forecasting the sea level variations for 6-hourly time lags presents accurate results. The performance of NN to forecast the sea level variations was satisfactory enough (correlation = 99.9%) for 6-hourly time lags. Fig. 2 shows the comparison between NN generalization (validation) to predict the variations of the low frequency sea level and the target (filtering data set). It is observed that the two curves are quite similar, being in accordance with the statistic results shown in Table 2.

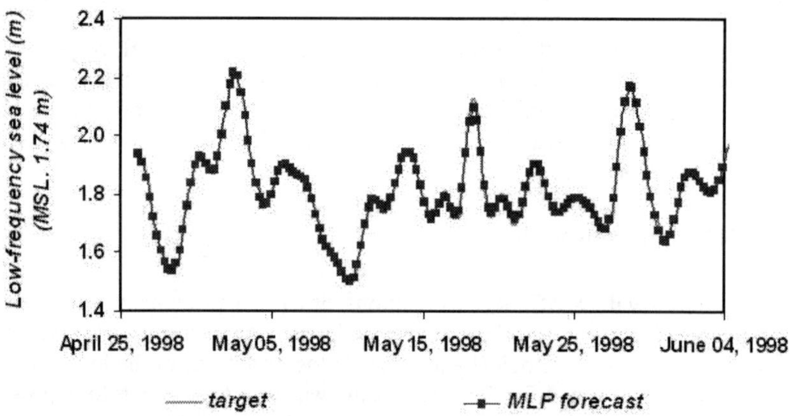

Figure 2: Oscillations of the low-frequency sea level and simulated MLP curve.

Learning rate and momentum parameters affect the speed of the convergence of the back propagation algorithm. The stopping criterion was based on the error to be minimized to improve the performance of the network. The model attained the best performance for 700 epochs in which the training error is 0,008276 and validation error is 0,008531 with learning rate of 0.01 and momentum of 0.9.

The scatter plots shown in Fig. 3 have small disparity illustrating that NN has a small error in learning stage than in the validation stage according to the correlation coefficients present in Table 2. This is a common result in establishing a NN. Normally the simulation performance of predictors is evaluated by the root mean square (RMS) or the square of correlation coefficient (R2) that is called coefficient of determination. Small RMS and large R^2 values indicate that the simulation performance is good (Chang 2006).

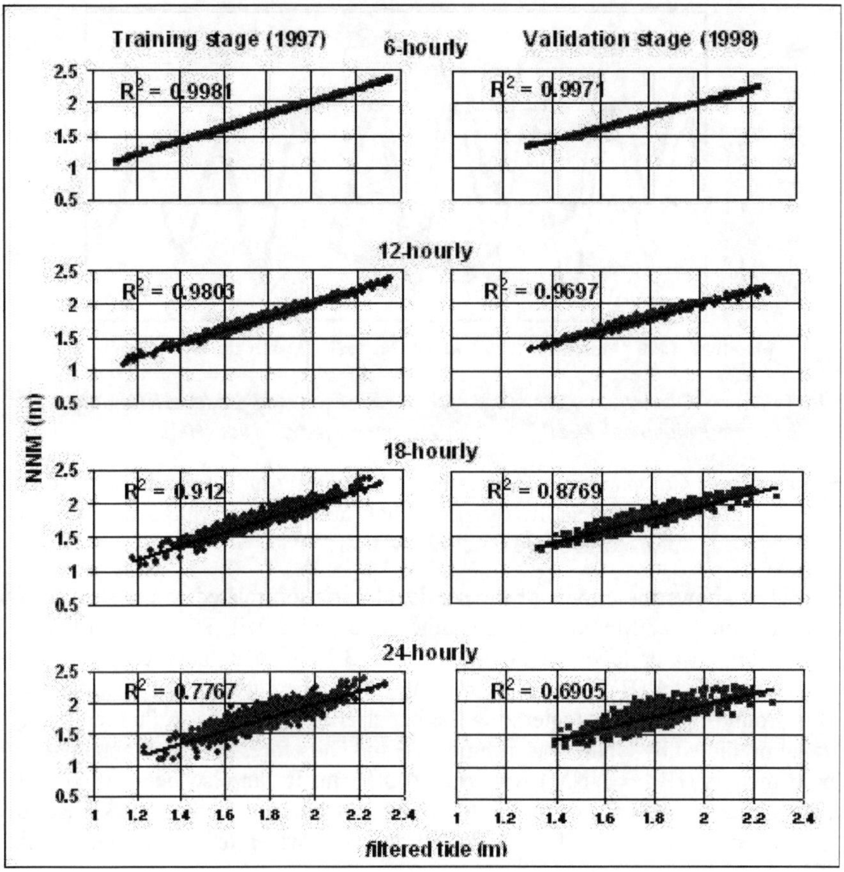

Figure 3: Scatter plots of simulated and filtered sea level data in training stage for 1997 and validation stage for 1998.

The left column of this figure shows the target and desired output simulated by NN in training stage for 1997. This column indicates that there is little disparity between filtered and simulated values for 6 and 12-hourly training in which R^2 is 0.9981 and 0.9803, respectively. The R^2 values for 18 and 24-hourly are around 0.912 and 0.7767, respectively, showing that the NNM preserves the influences of physical process such as pressure and wind in the sea level variations. The right column shows the scatter plots for 1998 in the validation stage. Small differences between the two stages are verified. The R^2 values for 6, 12 and 18-hourly present similar results with the learning stage. For 24-hourly forecasting, the R^2 presented values lower than for testing stage. It can be related to the correlation between the predictors and output.

Strong (~15 m.s-[1]) south-westerly (190-260 degrees) winds blowing during 3 to 5 days, over the ocean parallel the coastline, is the most conducive wind vector for producing storm surge along the southeast Brazilian coast.

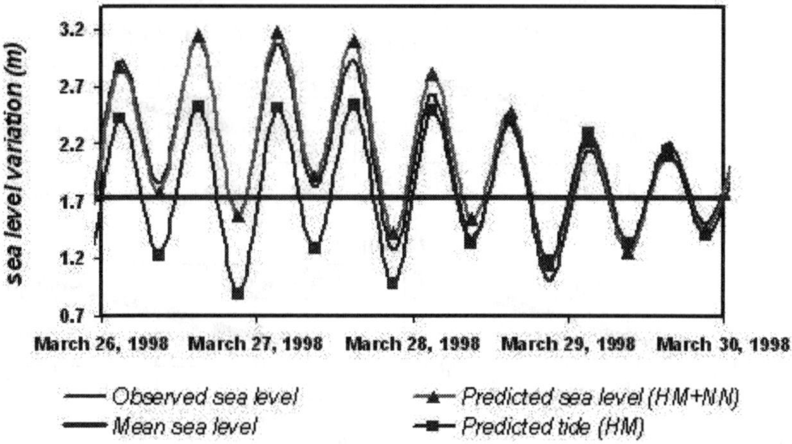

Figure 4: Comparison between observed coastal sea level and predicted with both models. It is verified that the HM undervalues the prediction for a storm surge on 26-28 March 1998.

Fig. 4 shows the curves of the sea level variation related to the storm surge occurred on 26-28 March 1998 in southeast coastal Brazil. The value of the peak of the high water level on 26 March was 3.13 m and the predict tide with Harmonic Model (HM) was 2.53 m. The difference between the maxima peaks was around 0.60 m, characterizing the occurrence of a surge in this region. The value predicted by NNM was around 0.63 m. Therefore, the value obtained with both models (HM + NNM) was around 3.16 m. It can also be verified in the Figure that some peaks of the high water predicted with both models are above the observed sea level. The values of the low water level are quite similar (Fig. 4).

7 Conclusion

Conventional numerical model developed to predict surges are still considered insufficient due to the complexity between the non-linear processes involved. In this paper, an alternative methodology based on the structure of neural network model to predict coastal sea level variations related to meteorological events was proposed.

Pre-processing of the data series in the time and frequency domain allowed defining the input of the neural network model. Maxima correlations in the physical process could determine the time lag between the meteorological variables and the sea level response.

The results indicate that the MLP architecture of the network developed in this work could generalize satisfactorily the non-linear behavior of the sea level fluctuations due to the interactions ocean-atmosphere at Cananéia tide gauge station. This model presented the best performance with correlation coefficient around 99% for 6-hourly time lag simulation and it can be efficient to forecast storm surge. The results obtained for 24-hourly time lag simulations around 83% of correlation coefficient (r) suggest that this model could be still used for forecasting the low-frequency sea level to this time lag with good performance. Forecasting for periods larger than 24-hourly could be improved, considering hydrodynamic variables such as river discharges. The results indicate that the NNM can also be useful as complement for the standard harmonic model (HM) and thus to improve the sea level forecast.

The proposed NNM for predicting the surge level can be further applied to other locations along the Brazilian coast or in others sites in the world. In addition, this NNM could be developed in conjunction a numerical ocean model (e.g. Princeton Ocean Model - POM) to improve forecasting water levels at the key locations.

Acknowledgements

The authors acknowledge the Brazilian Research Agencies CNPq and FAPERJ for their financial support of this work.

References

[1] Angulo, R. J. & Lessa, G. C., The Brazilian sea-level curves: A critical review emphasis on curves from Paranagua and Cananeia regions. *Mar. Geol.,* 140, 141-166, 1997.

[2] Besnard, W., General aspects about the Cananeia-Iguape region-1. *Institute. Paulista Oceanogr.,* 1, 9-26, 1950.

[3] Castro, B. M., & Lee T. N., Wind-forced sea level variability on the southeast Brazilian shelf. *J. Geophsys.* Res., 100, 16 045-16 056, 1995.

[4] Chang, H.-K., & Lin L.-C., Multi-point tidal prediction using artificial neural network with tide-generating forces. *Coastal Eng.,* 53, 857-864, 2006.

[5] Cigizoglu, H. R., & Alp M., Generalized regression neural network in modelling river sediment yield. *Adv. Eng. Software,* 37, 63-68, 2006.

[6] Dalazoana, R., Luz, R. T. & de Freitas S. R. C., Mean sea level studies from tide gauge and satellite altimetry time series looking for the integration of Brazilian Network to SIRGAS. *Rev. Bras. Cartogr.,* 57, 140-153, 2005.

[7] de Mesquita, A. R., Tides: Circulation and sea level in the southeastern coast of Brazil). http:// www.mares.io.usp.br/sudeste/sudeste.html, 2008.

[8] de Mesquita, Sea level variations along the Brazilian coast: A short review. *Brazilian Symp. on Sandy Beache.* http://www.mares.io.usp.br/praias/praias.html.

[9] Doodson, A. T., and H. D. Warburg, *Manual of Tides of the Admiralty).* 2nd ed. Brazilian Navy, 1944.

[10] Eisner, J. B., & Tsonis A. A., Nonlinear prediction, chaos and noise. *Bull. Atner. Meteor. Soc.,* 73, 49-60, 1992.

[11] Filippo, A. M., Variability of sea level as a function of meteorological events of low frequency. Ph.D. thesis, Fluminense Federal University, Rio de Janeiro State, Brazil, 2003.

[12] Franco, A. S., *Tides: Fundamentals Analysis and Prediction.* IPT, 1981.

[13] Fu, L.-M., *Neural Networks in Computer intelligence.* McGraw-Hill, 1994.

[14] Gan, M. A., & Rao V. B., Surface cyclogenesis over South America, *Mon. Wea. Rev.,* 119, 1293-1302, 1991.

[15] Kistler, R. *et al.*, The NCEP-NCAR 50-Year Reanalysis: Monthly means CD-ROM and documentation. Bull. *Amer. Meteor. Soc.,* 82, 247-267, 2001.

[16] Lee, T. L., Neural network prediction of a storm surge. *Ocean Eng.,* 33, 483-494, 2006.

[17] Mantovanelli, A. *et al.*, Combined tidal velocity and duration asymmetries as a determinant of water transport and residual flow in Paranagua Bay estuary. *Estuarine* Coastal *Shelf Sci.,* 59, 523-537, 2004.

[18] Marone, E., & Camargo R., Meteorological tides in the coast of the state of Parana: The event of 18 August 1993. *Nerntica,* 8, 1-2,1994.

[19] McPhaden, M. El Nino: The child prodigy of 1997-98. Na*ture,* 398, 559-562, 1999.

[20] Netto. S. A., & Lana P. C., Influence of *Spartina Alterni*flora on superficial sediment characteristics of tidal flats in Paranagua Bay (South-eastern. Brazil). *Estuarine Coastal Shelf Sci.,* 44, 641-648, 1997.

[21] Paiva. A. M., Study of sea level variations in Arraial do Cabo, Rio de Janeiro. his Research Rep. COPPE. Federal University of Rio de Janeiro, 1993.

[22] Pore, N. A., The relation of wind and pressure to extratropical storm surge at Atlantic City. *J. Appl. Meteor.,* 3, 155-163, 1964.

[23] Pugh, D. T., *Tides, Surges and Mean Sea Level.* John Wiley and Sons,1987.

[24] Rao, V. B., & Rada K., Characteristics of rainfall over Brazil: Annual variations and connections with the southern oscillation. *Theor. Appl. Clitnatol.,* 42, 81-91, 1990.

[25] Ribeiro, C. E. P., A new adaptive technique for directional analysis of waves. *Proceedings of 11 Meeting on Waves and Seas*, IEAPM, 182-193, 1997.

[26] Seluchi, M. E., Diagnostics and prognostics of synoptic situations conducive to cyclogenesis on the east of South America *Geofis. Int.,* 34, 171 186, 1995.

[27] Sztobryn, M., Forecast of storm surge by means of artificial neural network. *J. Sea Res.,* 49, 317-322, 2003.

[28] Tseng, C. M., Jan C. D., Wang J. S. & Wang C. M., Application of artificial neural network in typhoon surge forecasting. *Ocean Eng.,* 34, 1757-1768, 2007.

On a joint distribution of two successive surf parameters

D. Myrhaug[1] & H. Rue[2]
[1]*Department of Marine Technology,*
Norwegian University of Science and Technology, Trondheim, Norway
[2]*Department of Mathematical Sciences,*
Norwegian University of Science and Technology, Trondheim, Norway

Abstract

A joint distribution of two successive surf parameters is provided, and it is represented by a bivariate lognormal distribution. Consequently the joint distribution of two successive breaker indices is represented by a bivariate lognormal distribution. The application of the surf parameter distribution is exemplified by estimating the probability of two successive breakers on slopes by using wave parameters corresponding to typical field conditions.
Keywords: bivariate lognormal distribution, surf zone, surf parameter, breaker index, breaking waves.

1 Introduction

The surf parameter, also often referred to as the surf similarity parameter or the Iribarren number, is used to characterize surf zone processes. It is given by the ratio between the slope of a beach or a structure and the square root of the wave steepness in deep water as introduced by Iribarren and Nogales [1] and used later by Battjes [2]. Shallow water regions where waves break are referred to as the surf zone, and the different breakers on slopes are defined and classified in terms of the surf parameter. It also appears that the surf parameter enters in many empirical and theoretical models for wave-induced phenomena in the surf zone. The breaking of waves is associated with large loss of energy. Within the surf zone along beaches the wave energy flux from offshore is dissipated into turbulence and heat, and consequently the wave height decreases towards the

WIT Transactions on Ecology and the Environment, Vol 126, © 2009 WIT Press
www.witpress.com, ISSN 1743-3541 (on-line)
doi:10.2495/CP090081

shoreline. Wave-breaking also results in strong currents along the shoreline and thereby affects the nearshore circulation. The high intensity of turbulence caused by wave-breaking is also responsible for the intense sediment transport in the surf zone. Wave run-up on beaches and coastal structures such as, e.g., breakwaters, seawalls and artificial reefs are characterized by using the surf parameter. Examples of the relevance and importance of the surf parameter are found in e.g. Herbich [3] and Silvester and Hsu [4].

Tayfun [5] presented a study on the development of approximate theoretical forms of the distributions of wave steepness and surf parameter. The approach is based on assuming the random wave process to be long crested and narrow banded. The results are compared with data from measurements at sea representing two severe storms. Both the wave steepness and the surf parameter are lognormal distributed. The resulting statistics for the surf parameter are applied to breakers at normal incidence on sloping beaches. As stated by Tayfun [5], the joint statistics of wave steepness with wave heights or crest heights, or the wave steepness with wave heights or crest heights above a specified threshold may be appropriate in formulating risks of e.g. capsizing of vessels, overtopping, and slamming forces on seawalls, etc. However, the wave steepness is of interest by itself, particularly in relation with many of the surf zone processes.

In a subsequent discussion Myrhaug and Fouques [6] pointed out that other data sets may result in other distributions. This is exemplified by using the data referred to by Tayfun [5]; the data used by Myrhaug and Kjeldsen [7, 8], Myrhaug and Rue [9], and Myrhaug and Kvålsvold [10]. These papers discuss various aspects of wave steepness statistics using data from a large population of waves obtained by waverider buoys at three different deep water locations on the Norwegian continental shelf. Myrhaug and Fouques [6] found that the wave steepness is Weibull distributed in the right tail and otherwise lognormal distributed, and that the surf parameter is lognormal distributed in the right tail and otherwise Fréchet distributed.

Myrhaug and Rue [9] used the Weibull model to study the statistics of two successive wave steepness parameters with the focus on steep waves, while Myrhaug and Kjeldsen [7], and Myrhaug [11] discussed the joint distribution of wave height and wave steepness. To our knowledge, no studies on the joint distribution of two successive surf parameters, i.e. the values of the surf parameter for two successive waves, are available in the open literature. This is the subject of the present paper, which should represent a useful tool for the assessment of various wave-induced phenomena in the surf zone.

Here the marginal distribution of the surf parameter is taken as the lognormal distribution, as found by Tayfun [5]. The joint distribution of two successive surf parameters is then represented by a two-dimensional lognormal distribution. The application of the results is illustrated by an example; the probability of two successive breaking waves on slopes are given.

2 Background

The surf parameter is defined as $\xi = m/\sqrt{s}$ where $m = \tan\theta$ is the slope with an angle θ with the horizontal, $s = H/(g/2\pi)T^2$ is the wave steepness in deep water, H is the wave height, T is the wave period, and g is the acceleration of gravity. In the forthcoming the surf parameter is normalized, i.e. $y = \xi/\xi_{rms}$, by defining $\xi_{rms} = m/\sqrt{s_{rms}}$ where s_{rms} is the root-mean-square (*rms*) value of s, which will be discussed further in Section 4. The basis for the present approach is that the marginal distribution of the normalized surf parameter follows the lognormal distribution with the probability density function (*pdf*)

$$p(y) = \frac{1}{\sqrt{2\pi} y \sigma_z} \exp\left[-\frac{(\ln y - \mu_z)^2}{2\sigma_z^2}\right] \tag{1}$$

where μ_z and σ_z^2 is the expected value and the variance, respectively, of $z = \ln y$. This is in accordance with the results in Tayfun [5]; further details are given in Section 4.

3 Statistics of the joint behaviour of two successive surf parameters

Now the joint distribution of the surf parameter for two successive waves is considered. There are several numbers of possible forms of two-dimensional distributions where marginal distributions are given by the lognormal distribution in Eq. (1). Let $y_1 = \xi_1/\xi_{rms}$ and $y_2 = \xi_2/\xi_{rms}$ denote the variables which are normalized with the same parameter ξ_{rms}. Here y_1 and y_2 are associated with the first wave and the next succeeding wave, respectively. By introducing $z_1 = \ln y_1$ and $z_2 = \ln y_2$ in the two-dimensional Gaussian distribution in Eq. (A1) (see the Appendix), it can be transformed to a two-dimensional distribution with the marginal distributions given by Eq. (1). This change of variables gives the following joint *pdf* of normalized variables (see e.g. Johnson and Kotz [12])

$$p(y_1, y_2) = \frac{1}{2\pi y_1 y_2 \sigma_z^2 \sqrt{1-\rho_{z_1 z_2}^2}} \cdot \exp\left[-\frac{(\ln y_1 - \mu_z)^2 + (\ln y_2 - \mu_z)^2 - 2\rho_{z_1 z_2}(\ln y_1 - \mu_z)(\ln y_2 - \mu_z)}{2\sigma_z^2 (1-\rho_{z_1 z_2}^2)}\right] \tag{2}$$

where

$$\mu_z = E[\ln y_1] = E[\ln y_2] \tag{3}$$

$$\sigma_z^2 = Var[\ln y_1] = Var[\ln y_2] \tag{4}$$

The correlation coefficients $\rho_{z_1 z_2}$ and $\rho_{y_1 y_2}$ are related by

$$\rho_{y_1 y_2} = \frac{E[y_1 y_2] - \mu_y^2}{\sigma_y^2} = \frac{e^{\rho_{z_1 z_2}\sigma_z^2} - 1}{e^{\sigma_z^2} - 1} \tag{5}$$

by utilizing that

$$\begin{aligned} E[y_1\, y_2] &= \int_0^\infty \int_0^\infty y_1 y_2 p(y_1, y_2)\, dy_1\, dy_2 \\ &= \exp\left[2\mu_z + \sigma_z^2 \left(1 + \rho_{z_1 z_2}\right)\right] \end{aligned} \tag{6}$$

$$\mu_y = e^{\mu_z + \frac{1}{2}\sigma_z^2} \tag{7}$$

$$\sigma_y^2 = e^{2\mu_z + \sigma_z^2}\left(e^{\sigma_z^2} - 1\right) \tag{8}$$

Or, an alternative to Eq. (5) is

$$\rho_{z_1 z_2} = \frac{1}{\sigma_z^2} \ln\left[1 + \rho_{y_1 y_2}\left(e^{\sigma_z^2} - 1\right)\right] \tag{9}$$

The conditional *pdf* of y_2 given y_1 is also lognormal distributed, given by (Johnson and Kotz [12])

$$p(y_2 \mid y_1) = \frac{p(y_1, y_2)}{p(y_1)} = \frac{1}{\sqrt{2\pi}\, y_2 \hat{\sigma}_z} \exp\left[-\frac{\left(\ln y_2 - \hat{\mu}_z\right)^2}{2\hat{\sigma}_z^{\,2}}\right] \tag{10}$$

where

$$\hat{\mu}_z = \mu_z + \rho_{z_1 z_2}\left(\ln y_1 - \mu_z\right) \tag{11}$$

$$\hat{\sigma}_z^{\,2} = \sigma_z^2\left(1 - \rho_{z_1 z_2}^2\right) \tag{12}$$

The mean (expected) value of y_2 given y_1 is given by (Johnson and Kotz [12])

$$\begin{aligned} E[y_2 \mid y_1] &= \int_0^\infty y_2 p(y_2 \mid y_1)\, dy_2 \\ &= y_1^{\rho_{z_1 z_2}} \exp\left[\frac{1}{2}\sigma_z^2\left(1 - \rho_{z_1 z_2}^2\right) + \mu_z\left(1 - \rho_{z_1 z_2}\right)\right] \end{aligned} \tag{13}$$

A quantity of interest is the probability of the surf parameter of a wave to be in an interval y_l to y_h when the surf parameter of the previous wave has been in the same interval. This probability is given as

$$P = \frac{\int_{y_l}^{y_h}\int_{y_l}^{y_h} p(y_1, y_2)\, dy_1 dy_2}{\int_{y_l}^{y_h} p(y_1)\, dy_1} = \frac{\int_{y_l}^{y_h}\left[\int_{y_l}^{y_h} p(y_2 \mid y_1)\, dy_2\right] p(y_1)\, dy_1}{\Phi(u_h) - \Phi(u_l)} \tag{14}$$

by using the relationship $p(y_1, y_2) = p(y_2 \mid y_1)\, p(y_1)$, and where Φ denotes the standard Gaussian cumulative distribution function (*cdf*), i.e.

$$\Phi(v)=\frac{1}{\sqrt{2\pi}}\int_{-\infty}^{v} e^{-t^2/2}dt \tag{15}$$

and

$$u_l=\frac{\ln y_l-\mu_z}{\sigma_z}\ ,\ u_h=\frac{\ln y_h-\mu_z}{\sigma_z} \tag{16}$$

The evaluation of the inner integral in the nominator of Eq. (14) follows by using Eqs. (10) to (12), giving

$$P=\frac{\frac{1}{\sqrt{2\pi}}\int_{u_l}^{u_h}\left[\Phi(\hat{u}_h)-\Phi(\hat{u}_l)\right]\exp\left(-\frac{1}{2}u_1^2\right)du_1}{\Phi(u_h)-\Phi(u_l)} \tag{17}$$

where

$$\hat{u}_l=\frac{u_l-\rho_{z_1z_2}u_1}{\sqrt{1-\rho_{z_1z_2}^2}}\ ,\ \hat{u}_h=\frac{u_h-\rho_{z_1z_2}u_1}{\sqrt{1-\rho_{z_1z_2}^2}} \tag{18}$$

$$u_1=\frac{\ln y_1-\mu_z}{\sigma_z} \tag{19}$$

4 Example of application

In this example the lognormal distribution of the surf parameter proposed by Tayfun [5] is adopted, given by the *pdf*

$$p(\xi)=\frac{1}{\sqrt{2\pi}\xi\sigma_{\ln\xi}}\exp\left[-\frac{\left(\ln\xi-\mu_{\ln\xi}\right)^2}{2\sigma_{\ln\xi}^2}\right] \tag{20}$$

where

$$\mu_{\ln\xi}=\ln\left(\frac{2m}{\sqrt{\alpha}}\right)\ ,\ \sigma_{\ln\xi}^2=\frac{1}{4}\ln\left(\frac{4}{\pi}\right) \tag{21}$$

and α is a parameter related to the wave steepness of the sea state s_{rms}. From Tayfun [5] it appears that his theoretical value of the sea state steepness parameter, $s_{rms}=0.318\alpha$, is very close to $s_{rms}=17.6H_s/4gT_z^2$ used by Myrhaug and Rue [9]; thus giving $\alpha=1.41H_s/T_z^2$. Here H_s is the significant wave height and T_z is the mean zero-crossing wave period.

Now a change of variables from ξ to $y=\xi/\xi_{rms}$ gives the lognormal *pdf* in Eq. (1) with $\mu_z=\mu_{\ln\xi}-\ln\xi_{rms}=\ln\left(2\sqrt{0.318}\right)$ and $\sigma_z^2=\sigma_{\ln\xi}^2$, giving

$$\mu_z=0.120\ \ ,\ \sigma_z=0.246 \tag{22}$$

Figure 1(a) shows the mean (expected) value of $y_2=\xi_2/\xi_{rms}$ given $y_1=\xi_1/\xi_{rms}$ versus y_1 according to Eq. (13) for a range of $\rho_{z_1z_2}$ values from 0 to 0.9. From Fig. 1(a) it appears that $E\left[y_2\,|\,y_1\right]$ approaches y_1 as $\rho_{z_1z_2}$

increases. It should be noted that $E\left[y_2 \mid y_1\right]=y_1$ for $\rho_{z_1 z_2}=1$; see Eq. (13). For zero correlation the mean value of y_2 given y_1 is always constant, i.e. Eq. (13) reduces to Eq. (7).

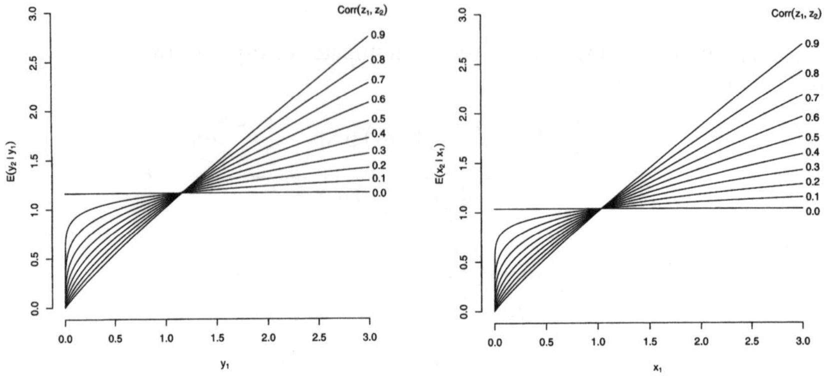

Figure 1: (a) (left) Conditional expected value of $y_2=\xi_2 / \xi_{rms}$ given $y_1=\xi_1 / \xi_{rms}$ versus y_1 for different values of $\rho_{z_1 z_2}=\operatorname{Corr}\left(z_1, z_2\right)$. (b) (right) Conditional expected value of $x_2=h_{2b} / h_{brms}$ given $x_1=h_{1b} / h_{brms}$ versus x_1 for different values of $\rho_{z_1 z_2}=\operatorname{Corr}\left(z_1, z_2\right)$.

Furthermore, a sea state specified by $H_s=7$ metres and $T_z=7$ seconds is chosen, representing a "steep" sea state which has been measured at a deep water location on the Norwegian continental shelf (Krogstad [13]). Thus, $s_{rms}=0.064$ and $\alpha=0.20$, giving $\xi_{rms}=3.95\,\mathrm{m}$.

Moreover, in this example breaking waves on slopes will be considered. Types of breaking waves are defined in terms of the surf parameter, classified as (see e.g. Tayfun [5])

$$
\begin{array}{ll}
\text{spilling if} & \xi \leq 0.5 \\
\text{plunging for} & 0.5<\xi \leq 3 \\
\text{collapsing for} & 3<\xi \leq 3.5 \\
\text{surging if} & 3.5<\xi
\end{array} \tag{23}
$$

Thus, by taking $y_1=\xi_1 / \xi_{rms}$, $y_2=\xi_2 / \xi_{rms}$ and μ_z, σ_z from Eq. (22), Eqs. (16) to (19) can be used to calculate the probability of two successive breaking waves on slopes.

Figure 2 shows the probability P of two successive spilling breakers versus the correlation coefficient $\rho_{z_1 z_2}$ for the slopes m = 1/10, 1/5, 1/4, 1/3, 1/2.

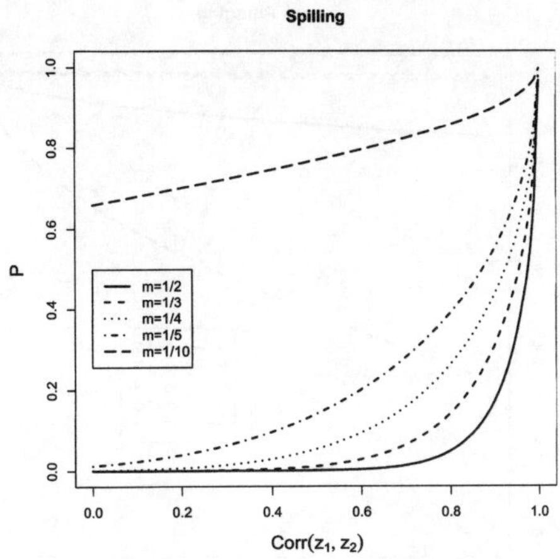

Figure 2: The probability (P) of two successive spilling breakers versus $\rho_{z_1z_2} = \text{Corr}(z_1, z_2)$ for different slopes m.

Similar results are shown in Figures 3, 4 and 5 for plunging, collapsing and surging breakers, respectively.

From Figs. 2 to 5 it appears that P increases as $\rho_{z_1z_2}$ increases for a given slope, which is physically sound. Moreover, from Fig. 2 it appears that P decreases as the slope increases for a given value of $\rho_{z_1z_2}$, which a priori is not quite obvious. However, it can be demonstrated by considering the results for $\rho_{z_1z_2} = 0$, i.e. when y_1 and y_2 are statistically independent. Then the marginal *pdf* of y_1 (and y_2) is given by Eq. (1) and μ_z, σ_z from Eq. (22), and the *pdf* of $\xi = \xi_1 \left(\text{or}\, \xi_2\right)$ for $\xi_{rms} = 3.95\,\text{m}$ is shown in Fig. 6 for the same slopes m as in Figs. 2 to 5. From Fig. 6 it appears that the probability of a spilling breaker for a given slope (i.e. given by the area under the *pdf* for the slope considered corresponding to $0 < \xi \leq 0.5$), is largest for $\text{m} = 1/10$ and decreases as the slope increases. This will also be the case for other values of $\rho_{z_1z_2}$, and consequently the results are as shown in Fig. 2.

From Fig. 3 it appears that the probability of two successive plunging breakers for a given value of $\rho_{z_1z_2}$ increases for the slope in the order m=1/10, 1/2, 1/5, 1/3, 1/4; for the three latter values the differences are very small. The understanding of this is supported by the results for $\rho_{z_1z_2} = 0$ in Fig. 6; for $0.5 < \xi \leq 3.0$ it is observed that the area under the *pdf* is smallest for m=1/10 and that it increases for the slope in the order referred to in the discussion of Fig. 3.

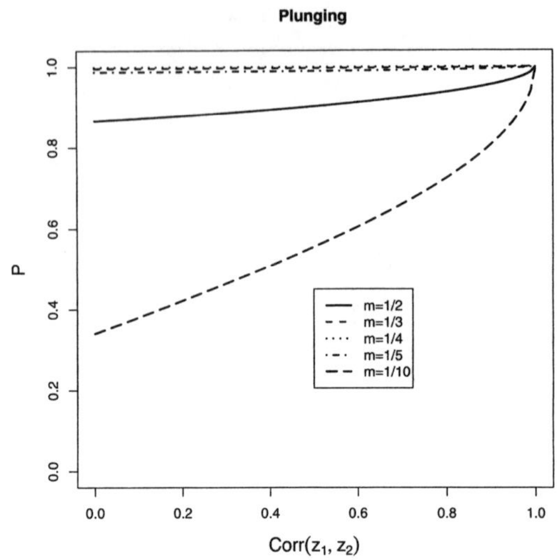

Figure 3: The probability (P) of two successive plunging breakers versus $\rho_{z_1z_2} = \mathrm{Corr}(z_1, z_2)$ for different slopes m.

The results for the collapsing and surging breakers in Figs. 4 and 5, respectively, are similar; it appears that the probability of two successive collapsing and surging breakers increases as the bed slope increases for a given value of $\rho_{z_1z_2}$, which is supported by the results in Fig. 6 for $\rho_{z_1z_2} = 0$.

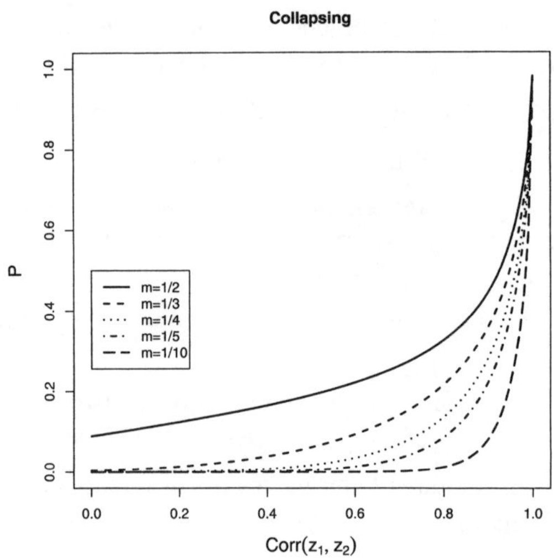

Figure 4: The probability (P) of two successive collapsing breakers versus $\rho_{z_1z_2} = \mathrm{Corr}(z_1, z_2)$ for different slopes m.

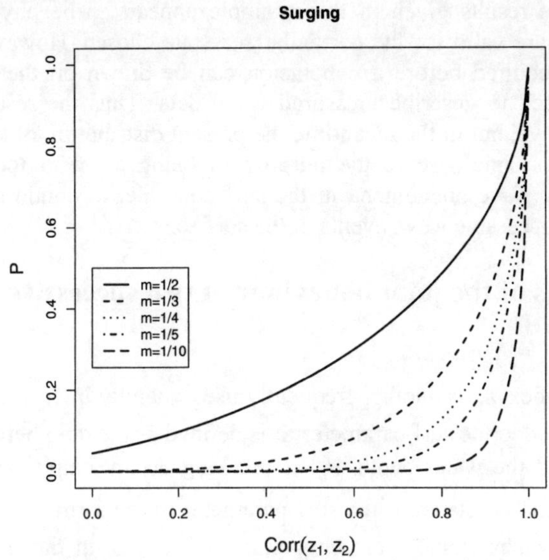

Figure 5: The probability (P) of two successive surging breakers versus $\rho_{z_1 z_2} = \mathrm{Corr}(z_1, z_2)$ for different slopes m.

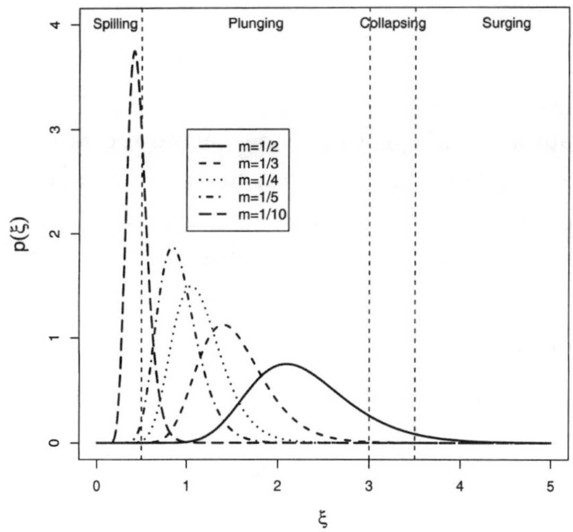

Figure 6: The *pdf* of the surf parameter $\xi = \xi_1$ (or ξ_2) for $\xi_{rms} = 3.95$ m and different slopes m.

Overall, the results given in this example appears to be physically sound, although they are valid for the particular sea state chosen. However, validation with data is required before a conclusion can be drawn on the ability of the present approach to describe measured wave data. Thus the results should be taken as tentative, but in the meantime the present distribution of two successive surf parameters should serve the purpose of being a useful tool for making assessments of wave phenomena in the surf zone, i.e. to obtain an estimate of two extreme successive wave events in the surf zone.

5 Statistics of the joint behaviour of two successive breaker indices

The breaker index h_b is another frequently used quantity in coastal work, which is closely related to the surf parameter. It is defined as the ratio between the wave height H_b and the water depth d_b at breaking. Many empirical relationships exist for h_b; one is related to the surf parameter in the form $h_b \equiv H_b / d_b = a\xi^c$ (Tayfun [5]), where a and c are empirical coefficients. In the forthcoming the breaker index is normalized, i.e. $x = h_b / h_{brms}$, by defining $h_{brms} = a\xi_{rms}^c$ where ξ_{rms} is defined in Section 2, giving $x = (\xi / \xi_{rms})^c = y^c$. The *pdf* of *x* is obtained from Eq. (1) by a change of variable from *y* to *x*, taking the form

$$p(x) = \frac{1}{\sqrt{2\pi}x\sigma} \exp\left[-\frac{(\ln x - \mu)^2}{2\sigma^2}\right] \tag{24}$$

where

$$\mu = c\mu_z \ , \ \sigma^2 = (c\sigma_z)^2 \tag{25}$$

are the mean value and the variance, respectively, of $\ln x$.

Thus, the statistics of the joint behaviour of two successive breaker indices $x_1 = h_{b1} / h_{brms}$ and $x_2 = h_{b2} / h_{brms}$ normalized with the same parameter h_{brms} follow by utulizing the results in Section 3. More specifically it follows that: $p(x_1, x_2)$ is given in Eq. (2); $p(x_2 \mid x_1)$ in Eqs. (10) to (12); $E[x_2 \mid x_1]$ in Eq. (13), by replacing $y_1, y_2, \mu_z, \sigma_z$ with $x_1, x_2, c\mu_z, c\sigma_z$. Moreover, $\rho_{x_1 z_2}$ is given in Eq. (5) (or alternatively $\rho_{z_1 z_2}$ in Eq. (9)) by replacing y_1, y_2, σ_z with $x_1, x_2, c\sigma_z$; this is obtained by utilizing the results in Eqs. (6) to (8) by in addition replacing y, μ_z with $x, c\mu_z$. The results in Eqs. (14) to (19) can be re-arranged accordingly to be valid for two successive breaker indices.

Figure 1(b) shows the mean value of $x_2 = h_{b2} / h_{brms}$ given $x_1 = h_{b1} / h_{brms}$ versus x_1 for a range of $\rho_{z_1 z_1}$ values from 0 to 0.9 according to Eq. (13) by replacing $y_1, y_2, \mu_z, \sigma_z$ with $x_1, x_2, c\mu_z, c\sigma_z$, and $(a,c) = (1.20, 0.27)$ from Kaminsky and Kraus [14]. The results are similar to those given in Fig. 1(a) for the surf parameter except for the shift of the values. Results for the breaker index will not be elaborated further here since they will be similar to those presented for the surf parameter.

6 Summary

A joint distribution of two successive surf parameters is provided, and it is represented by a bivariate lognormal distribution. Consequently the joint distribution of two successive breaker indices is represented by a bivariate lognormal distribution. The application of the surf parameter distribution is exemplified to estimate the probability of two successive breakers on slopes; spilling, plunging, collapsing and surging breakers, by using wave parameters corresponding to typical field conditions. Overall, these results appear to be physically sound, although they are valid for the particular sea state chosen. The results should be taken as tentative, because validation with data is required before a conclusion can be drawn on the ability of the present approach to describe measured wave data. However, in the meantime the bivariate lognormal distribution of two successive surf parameters should serve the purpose as a useful tool for making assessments of wave phenomena in the surf zone, i.e. to obtain an estimate of two extreme successive wave events in the surf zone.

Appendix

The joint *pdf* of two Gaussian random variables z_1 and z_2 with the same mean value μ_z and variance σ_z^2, is given by (Bury [15])

$$p(z_1,z_2)=\frac{1}{2\pi\sigma_z^2\sqrt{1-\rho_{z_1z_2}^2}}\cdot \exp\left[-\frac{(z_1-\mu_z)^2+(z_2-\mu_z)^2-2\rho_{z_1z_2}(z_1-\mu_z)(z_2-\mu_z)}{2\sigma_z^2\left(1-\rho_{z_1z_2}^2\right)}\right] \tag{A1}$$

where the correlation coefficient $\rho_{z_1z_2}$ is given as

$$\rho_{z_1z_2}\equiv\frac{Cov[z_1,z_2]}{\sigma_z^2}=\frac{E[z_1z_2]-\mu_z^2}{\sigma_z^2} \tag{A2}$$

References

[1] Iribarren, C.R. & Nogales, C., Protection des ports, Sect. 2. Comm. 4, 17th Int. Nav. Cong. Lisbon, pp. 31-80, 1949.

[2] Battjes, J.A., Surf similarity. *Proceedings 14th Int. Conf. on Coastal Engineering, ASCE,* New York, Vol. 1, pp. 466-479, 1974.

[3] Herbich, J.B., *Handbook of Coastal and Ocean Engineering.* Volume 1. Wave Phenomena and Coastal Structures. Gulf Publishing Co., Houston, Texas, 1990.

[4] Silvester, R. & Hsu, J.R.C., *Coastal Stabilization*, World Scientific, Singapore, 1997.

[5] Tayfun, M.A., Distributions of wave steepness and surf parameter. *J. Waterway, Port, Coastal, Ocean Eng.*, 132(1), pp. 1-9, 2006.

[6] Myrhaug, D. & Fouques, S., Discussion of "Distributions of wave steepness and surf parameter" by M. Aziz Tayfun. *J. Waterway, Port, Coastal, Ocean Eng.*, 133 (3), pp. 242-243, 2007.

[7] Myrhaug, D. & Kjeldsen, S.P., Parametric modelling of joint probability density distributions for steepness and asymmetry in deep water waves. *Appl. Ocean Res.* 6(4), pp. 207-220, 1984.

[8] Myrhaug, D. & Kjeldsen, S.P., Predictions of occurrences of steep and high waves in deep water. *J. Waterway, Port, Coastal, Ocean Eng.*, 113(2), pp. 122-138, 1987.

[9] Myrhaug, D. & Rue, H., Joint distribution of successive wave steepness parameters. *J. Offshore Mech. Arct. Eng.* 115(3), pp. 191-195, 1993.

[10] Myrhaug, D. & Kvålsvold, J., Comparative study of joint distributions of primary wave characteristics. *J. Offshore Mech. Arct. Eng.*, 117(2), pp. 91-98 , 1995.

[11] Myrhaug, D., Statistics of steep waves in deep water. *J. Marine Environ. Eng.* 1(2), pp. 161-173, 1994.

[12] Johnson, N.L. & Kotz, S., *Distributions in Statistics: Continuous Multivariate Distributions*, John Wiley & Sons, New York, 1972.

[13] Krogstad, H.E., Height and period distributions of extreme waves. *Appl. Ocean Res.*, 7(3), pp. 158-165, 1985.

[14] Kaminsky, G.M. & Kraus, N.C., Evaluation of depth-limited wave breaking criteria. *Proceedings 2nd Int. Symp. on Wave Measurements and Analysis, ASCE*, New York, pp. 180-193, 1994.

[15] Bury, K.V., *Statistical Models in Applied Science*. John Wiley & Sons, New York, 1975.

Simulation of storm surge and overland flows using geographical information system applications

S. Aliabadi, M. Akbar & R. Patel
Northrop Grumman Center for High Performance Computing of Ship Systems Engineering, Jackson State University, USA

Abstract

This study focuses on a few aspects of an on-going project at Jackson State University regarding homeland security in the state of Mississippi. The project proposes an integrated tool for multi-scale storm surge and overland flow (flood) forecast due to hurricane, as well as evaluation of the flood damage on coastal infrastructure including transportation systems in the Mississippi coast. Three models are executed in sequence to get all the necessary results. Two out of these three codes are extensively parallel to ensure real time forecast to deal with the emergency evacuation days before the hurricane strikes the coast. The results from the models are fed into Geographical Information Systems (GIS) for visualization, analysis and decision-making.

Keywords: multi-scale hurricane simulation, meteorological data, overland flow, parallel computation.

1 Introduction

In this study, we present an integrated modelling scheme of a hurricane from its approach to landfall and associated water surge and flooding in the coastal regions. Using the most updated meteorological data days before a hurricane strikes, the ground wind speed, pressure, rain, etc can be predicted using the open source parallel code Weather Research and Forecasting (WRF) [1]. We obtain wind speed and pressure data from WRF, which are used as input to another open source parallel code ADvanced CIRCulation (ADCIRC). ADCIRC is a model for oceanic, coastal and estuarine waters [2]. We use ADCIRC results

WIT Transactions on Ecology and the Environment, Vol 126, © 2009 WIT Press
www.witpress.com, ISSN 1743-3541 (on-line)
doi:10.2495/CP090091

to model the coastal area flooding phenomena using our finite element method based CaMEL Overland flow solver [3]. The water surge values simulated from ADCIRC along the shoreline is used as the Dirichlet boundary condition input to CaMEL Overland. The rain data predicted from WRF is used as the source term in this solver. The graphical presentation in fig. 1 shows the integration of the whole simulation process.

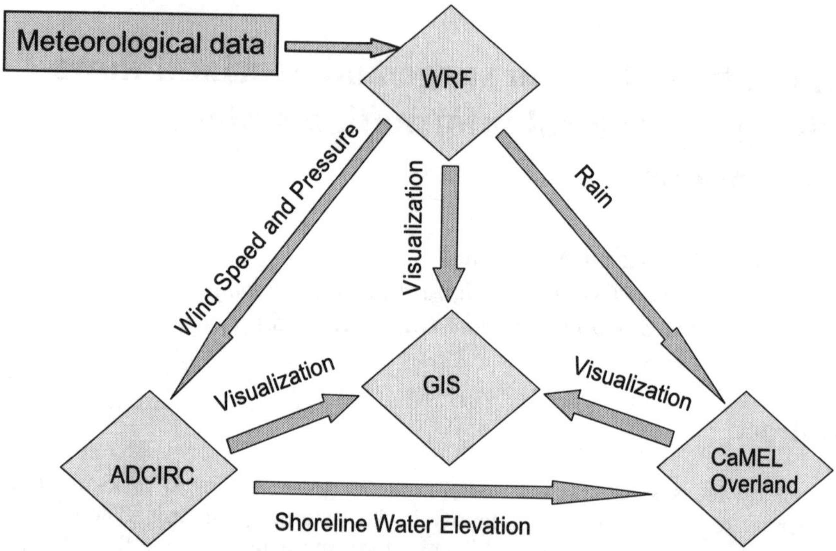

Figure 1: Graphical representation of our integrated modelling scheme.

In an actual hurricane case WRF, ADCIRC, and CaMEL Overland codes must be executed in sequence two to three days before its landfall, most likely every 6 to 12 hrs. Repeated simulations of the codes are needed because the more recent meteorological data we use the better accuracy we obtain from WRF. The accuracy of WRF results propagate into ADCIRC and CaMEL Overland simulations through the wind and rain input. Therefore, parallel implementation of the codes is absolutely necessary to ensure real time hurricane and flood forecast.

As a case study in the present research, we have chosen hurricane Katrina (2005) and its flooding impact on the Mississippi coastal region.

2 Model implementation

WRF is a parallel model, which is designed to serve both operational forecasting and atmospheric research needs. It is suitable for a broad spectrum of applications across scales ranging from meters to thousands of kilometers. It allows researchers the ability to conduct simulations reflecting either real data or idealized configurations. We have studied the parallel implementation of WRF extensively in our parallel cluster, which has Intel Xeon processors and total of 80 cores. The speed up of WRF in our cluster is displayed in fig. 2 (a). WRF uses structured mesh with the option of multiple nested domains. We used a single domain with 300 grid points in east-west, and 220 grid points in south-north. Each segment was 8 km.

Using the WRF wind speed and pressure data as input, ocean water surge is simulated using two-dimensional depth integrated (2DDI) model of ADCIRC. ADCIRC is a highly developed computer program for solving the equations of motion for a moving fluid on a rotating earth. These equations have been formulated using the traditional hydrostatic pressure and Boussinesq approximations and have been discretized in space using the finite element method and in time using the finite difference method. The water elevation is obtained from the solution of the depth-integrated continuity equation in Generalized Wave-Continuity Equation (GWCE) form. The speed up of ADCIRC in our parallel cluster is displayed in fig. 2 (b). The ADCIRC grid used in our simulation is the same as Mukai et al. [4], which consists of 254,565 nodes and 492,179 elements. ADCIRC Tidal Database [3], Version ec2001_v2d, is used to extract tide data during the Katrina period. Zero-flux boundary conditions are used on the land boundary, and tidal conditions are used in the ocean boundary.

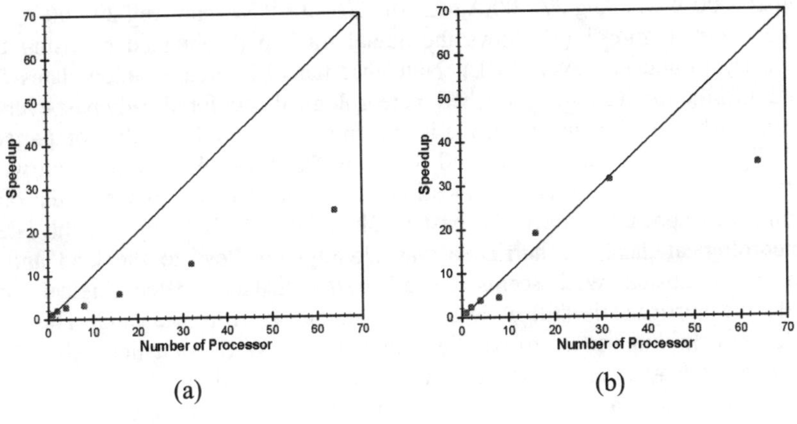

Figure 2: Code speedup with respect to the number of processors in our cluster. (a) WRF, (b) ADCIRC.

After the ADCIRC simulation, we model the coastal area water surge phenomena using our CaMEL Overland flow code. We have solved diffusive wave or Richard's equation, as shown in (1), by the Galerkin finite element method [3]. The time dependent water surge values simulated from ADCIRC along the shoreline is used as the Dirichlet boundary input in the model. The rain data predicted from WRF is used as the source term in the model. Since the execution of CaMEL Overland code is very fast, we have not made any attempt to make it parallel yet. We, however, will make it parallel soon.

$$\frac{\partial h}{\partial t} - \nabla \cdot \left(K \nabla h \right) = \dot{q} \tag{1}$$

where, h, K, $\dot{q}$ are water elevation, diffusion coefficient, and source terms, respectively.

3 Results and discussion

WRF simulation results heavily depend on the meteorological data. Hurricane may take unexpected turns, which only the latest meteorological data may reflect. Hurricane landfall location has a huge impact on ocean water surge. Experience suggests that water rises rapidly if the hurricane hits Louisiana coast, most likely due to the converging funnel effect of complicated land structure. On the contrary, hurricane hitting the Alabama coast is most likely to cause much lesser water surge. The computer modelling done by other researchers suggests that the funnel effect in Louisiana area may increase the surge by 20 to 40 percent [5]. This funnel-effect fact is particularly very much applicable for Katrina (Aug 23-31, 2005) type hurricanes with twisted track paths.

Figure 3 shows the comparison of Katrina simulation and actual track path. Figure 3(a), (b), (c), and (d) show the WRF simulated track paths starting from Aug 26 - 00 A.M., Aug 27 - 00 A.M., Aug 27 – 12 P.M., and Aug 29 - 00 A.M., respectively. Figure 3 (e) shows the actual track path obtained by using the Planetary Boundary Layer (PBL). Note that the PBL method interpolates the wind information from the published meteorological data for already past events. The published track path of Katrina is shown in fig. 3 (f). From the comparison with fig. 3 (e) and (f), fig. 3(a) and (b) show that these WRF simulation were started too early. These landfall locations are somewhat east of the actual one. Figure 3(c) appears to have the best result. Although fig. 3(d) had the latest meteorological data, the hurricane was already too close to the land and it appears to subside. WRF seems to work best with latest meteorological data, while the hurricane is still at least 24 hr far away from the land. This fact underlines the importance of parallel simulation of WRF for quick delivery to facilitate r safe and quick evacuation during an actual hurricane.

Katrina ocean water elevation plots from ADCIRC with different wind speed and pressure input from WRF or Planetary Boundary Layer (PBL) are displayed in fig. 4. Figure 4 (a)-(c) use WRF wind input with different starting date and time, while fig. 4(d) uses actual Katrina wind data provided by PBL. From the comparison of fig. 4(a) – (c) with fig. 4(d), it is evident that the starting date of WRF simulation has huge impact on the results. It is because of the fact that the latest meteorological data in WRF generates more accurate wind speed, pressure, and landfall location of hurricane. The impact subsequently is carried to ADCIRC and CaMEL Overland codes. In addition to that, ADCIRC simulation has to be done for several days around the hurricane period, typically for 5-7 days, to get reasonably good results. Longer simulation period captures both short and long ocean waves. All the facts mentioned above emphasize the

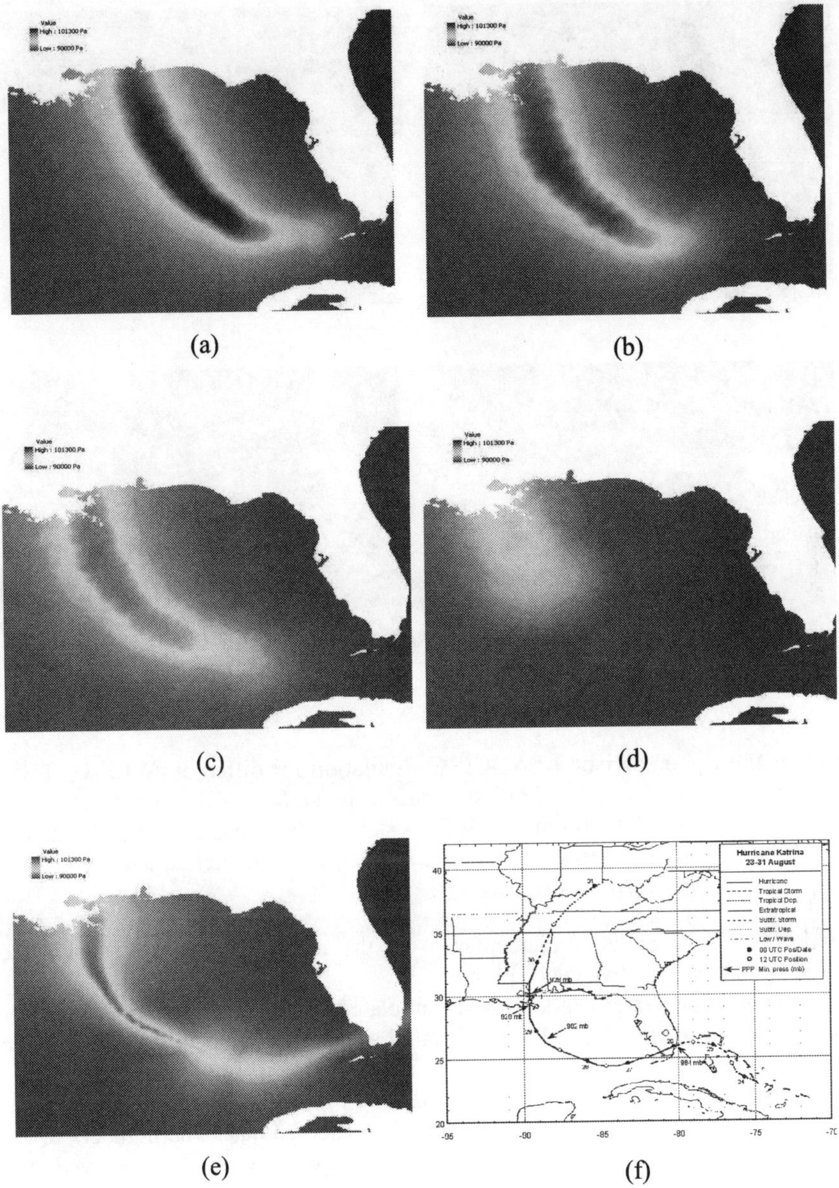

Figure 3: Katrina wind pressure plots (a) WRF simulation starting from Aug 26, 00 A.M., (b) WRF simulation starting from Aug 27, 00 A.M., (c) WRF simulation starting from Aug 27, 12 P.M., (d) WRF simulation starting from Aug 29, 00 A.M., (e) Actual, using Planetary Boundary Layer (PBL) (Aug 23 to Aug 31), (f) Published track path.

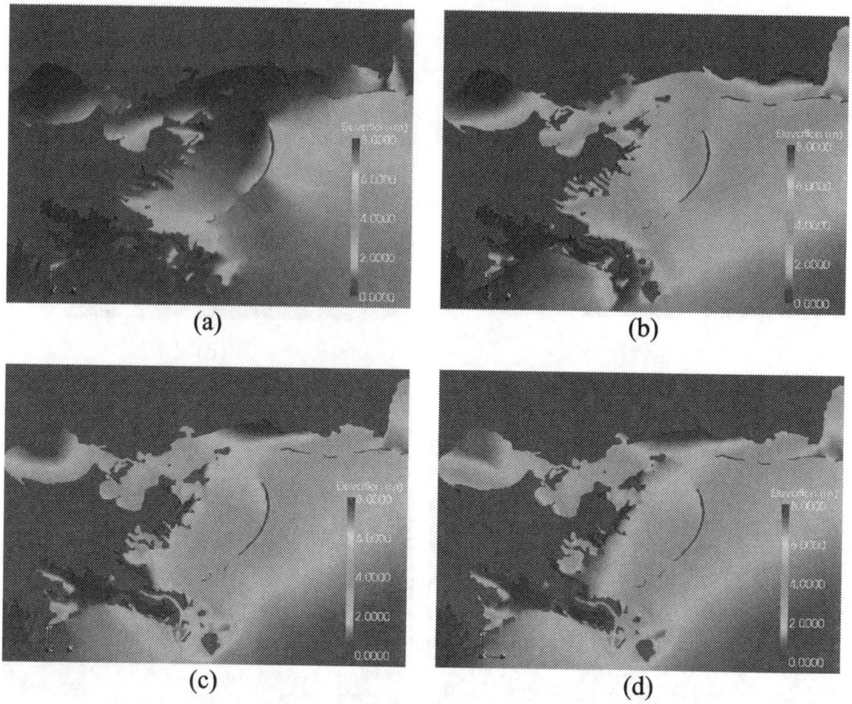

Figure 4: Comparison of ADCIRC simulation for different (WRF vs. PBL) wind speed and pressure data as input. (a) WRF - starting date Aug 26, 2005, 00 hr, (b) WRF - starting date Aug 27, 2005, 00 hr, (c) WRF - starting date Aug 27, 2005, 12 hr (d) Actual wind data from PBL (Aug 24 to Aug 31).

importance of parallel implementation of the codes for real time hurricane forecast to help effective evacuation.

Katrina High Water Mark (HWM) simulated from CaMEL Overland code is displayed in fig. 5(a). This plot is generated using WRF wind speed and pressure data without rain source terms. HWM values were recorded at many observation locations after the Katrina. The simulated HWM values are interpolated at the 32 observation locations and compared with the observed data. The numbers in the legend 'inter values' mean the difference range between the observed and simulated HWM in meters. For example, if the 'inter values' reads -2 in an observation location, it means the simulation under-predicts the HWM and the difference is within 2 meters. This result can be used to predict whether any structure in the domain will be flooded, damaged, or unaffected because of the water surge. This is one of the most important information that will help setting up the evacuation plan.

Figure 5(b) shows the comparison of simulated Katrina HWM with the observed ones for all 32 observation stations. The discrepancies in the

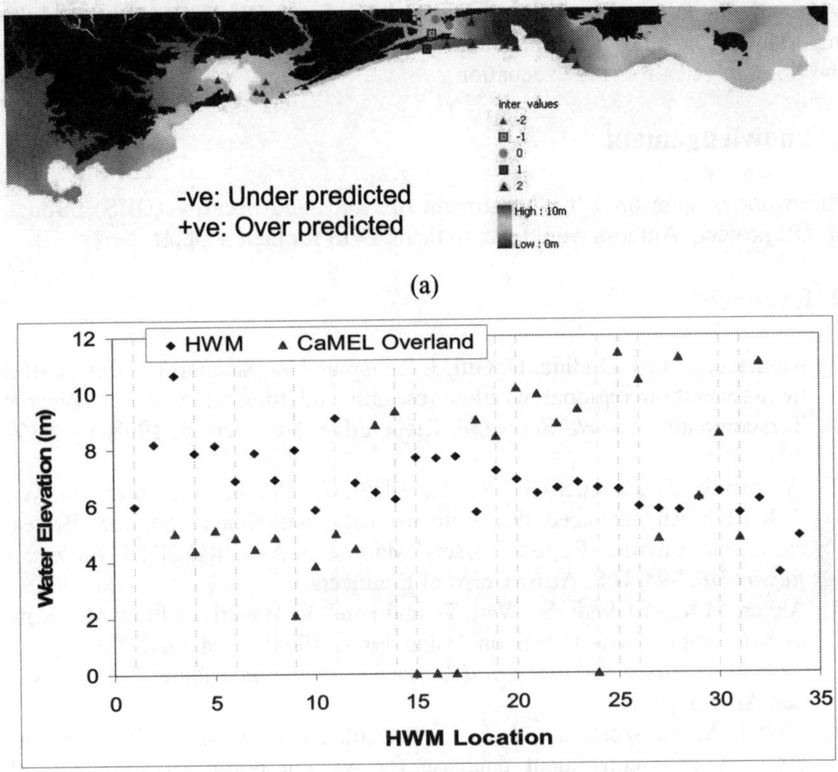

Figure 5: CaMEL Overland model Katrina results. (a) HWM in the coastal region, (b) Comparison of observed and simulated HWM.

comparison arise primarily because the simulation scheme is integrated. That means, error in WRF propagates to ADCIRC and overland models; and error in ADCIRC propagates to overland model. From our experience, getting an exact prediction is hard to achieve with the current technology available. However, we may be able to provide with the possible worst and best case scenarios during an actual hurricane event.

4 Conclusions

We have presented an integrated modelling scheme of a hurricane from its approach to landfall and associated water surge and flooding in the coastal regions using WRF, ADCIRC, and CaMEL Overland codes. We have demonstrated that repeated simulations of the codes are needed because the more recent meteorological data we use, in general, the better accuracy we obtain from WRF. The accuracy of WRF results propagate into ADCIRC and CaMEL

Overland simulations through the wind and rain input. We have emphasized the importance of parallel implementation of the codes to ensure real time hurricane and flood forecast for safe evacuation.

Acknowledgement

This work is sponsored by Department of Homeland Security (DHS) through SERRI project. Authors would like to thank DHS for their support.

References

[1] Michalakes, J., J. Dudhia, D. Gill, J. Klemp and W. Skamarock: Design of a next-generation regional weather research and forecast model : Towards Teracomputing, *World Scientific*, River Edge, New Jersey, 1998, pp. 117-124.

[2] Westerink, J. J., Luettich, R. A., Blain, C. A., & Scheffner, N. W., ADCIRC: An advanced three-dimensional circulation model for shelves, coasts and estuaries. Report 2: Users' Manual for ADCIRC-2DDI. *Technical Report DRP-94*, U.S. Army Corps of Engineers.

[3] Akbar, M.K., Aliabadi, S., Wan, T., and Patel, R. "Overland Flow Modeling of Mississippi Coastal Region Using Finite Element Method", Accepted, *19th AIAA Computational Fluid Dynamics Conference*, June 22-25, 2009, San Antonio, TX.

[4] Mukai, A. Y., Westerink, J. J., Luettic, Jr., R. A., & Mark, D., Eastcoast 2001, A tidal constituent database for western north Atlantic, Gulf of Mexico, and Caribbean Sea. *U.S. Army Corps of Engineers, ERDC/CHL TR-02-24.*

[5] Day, J., Ford, M, Kemp, P., Lopez, J. Mister Go Must Go - A Guide for the Army Corps Congressionally-Directed Closure of the Mississippi River Gulf Outlet. December 4, 2006. (http://www.edf.org/documents/5665_Report%20-%20Mister%20Must%20Go.pdf)

Decadal changes in wave climate and sea level regime: the main causes of the recent intensification of coastal geomorphic processes along the coasts of Western Estonia?

Ü. Suursaar & T. Kullas
Estonian Marine Institute, University of Tartu, Estonia

Abstract

Strong erosion of accumulative shores and redistribution of matter has been observed along the practically tideless west coast of Estonia during the last half century. This indicates manifestations of climate change, such as increased storminess and high sea level. The objective of the paper is to presents an analysis of sea level data obtained from the Estonian tide gauges over the period 1842–2008 and to discuss the results of a hindcast simulation of wave conditions in the fetch-limited nearshore location of West Estonia for the period 1966–2006. After adjusting the historical sea level time series to take into account land uplift, the series of mean sea level display an upward trend of between 1.5 and 2.7 mm/yr, and 4–11 mm/yr for annual maximum sea levels. The results for wave climate show some quasi-periodic cycles, including a rise in mean wave height in the 1970s and 1980s, and decrease since 1990. Although both the average wind speed and the wave height show a decrease since 1990s, the annual sea level maxima, as well as the wave maxima continue to increase. Major coastal geomorphic changes seem to occur precisely as a combined result of such relatively infrequent but extreme wintertime wavestorms and storm surges.

Keywords: sea level, wave hindcast, wave climate, trends, storm surges, climate change, Baltic Sea.

1 Introduction

Estonia lies in the eastern section of the semi-enclosed Baltic Sea (Fig. 1). With amplitudes of M2 and K1 waves measuring less than 5 cm, the main

WIT Transactions on Ecology and the Environment, Vol 126,
www.witpress.com, ISSN 1743-3541 (on-line)
doi:10.2495/CP090101

hydrodynamic agents acting on seashores are waves and meteorologically forced currents and sea level variations [1]. The frequent occurrence of cyclones passing from west to east, and their corresponding changes in wind direction and speed [2], create considerable sea level fluctuations in some suitably exposed bays [3]. While more than 95% of the sea level data falls within the range of –50 and +60 cm in relation to the Kronstadt zero benchmark (which is nearly equal to the long-term mean sea level for the Estonian coast), there are a few exceptional (up to 275 cm) storm surges on record [4]. Also, the coastal sea near the islands of Saaremaa and Hiiumaa has the roughest wave regime in Estonian coastal waters, where wave heights can reach 9–10 m [5].

Some geomorphically interesting sections of the coast lie there, where vitalization of shore processes and redistribution of sediment has been observed during the last decades [6,7]. Illustrious changes in shoreline positions have occurred in the accumulative gravel spits, such as at Küdema (Fig. 1b) [8] and Cape Kelba (Fig. 1c). The coastal formations of the Kelba Spit have been under investigation since the 1960s. The spit is mainly comprised of granite shingles and pebbles, and the beach ridges of different age reach 3.8 m above sea level [7]. The recent documentation of shoreline positions show that elongations of such spits occur by gradual formation of new accumulative beach ridges, and most probably, during strong storms [1,4].

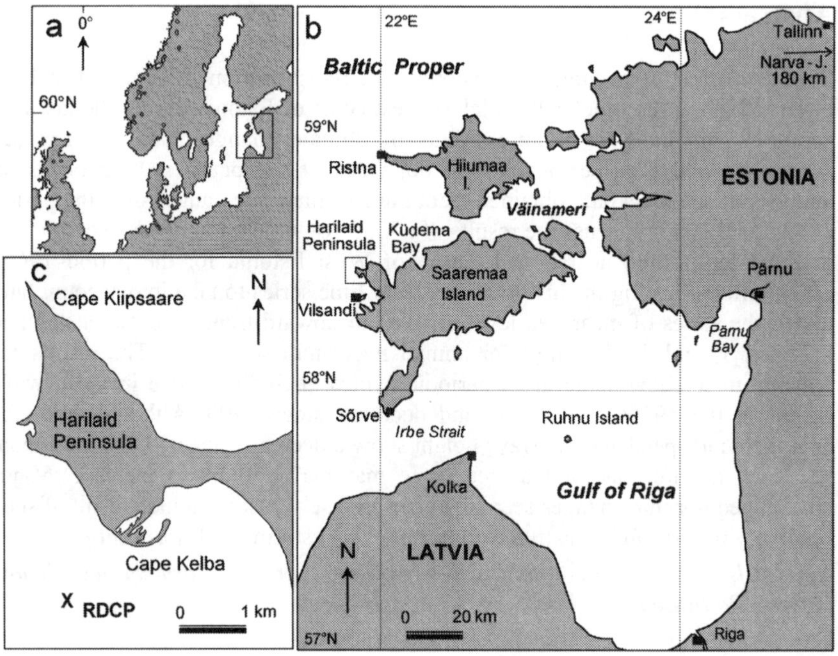

Figure 1: Map of the study area. The location of wave measurements and hindcast is marked with RDCP on (c).

As the coastal section under study is exposed to the Baltic marine winds, it can be hypothesized that changes in the wind [9], wave [10], and sea level regime are probably reflected in the historical changes in shoreline position and contour. A question arises: is there something particular to the recent developments in sea level and wave regimes that can explain the intensified coastal geomorphic processes in the study area?

The objectives of the paper are (1) to investigate the long-term changes both in mean water level and in extreme sea level events in the Estonian coastal sea; (2) to discuss, based on wave hindcast for 1966–2006, recent changes in mean and extreme wave properties along the western coast of Estonia; and (3) to analyse the relationships between the wind and wave climate, sea level regime and coastal processes.

2 Material and methods

2.1 Data sets and statistical methods for sea level and winds

The Estonian Meteorological and Hydrological Institute (EMHI) currently operates 21 meteorological stations and 12 tide gauges. Most of the tide gauges are equipped with tide poles and have a sampling frequency of 2 or 3 times a day. Automatic tide gauges of the EMHI, which provide hourly data, are located at Pärnu, Narva-Jõesuu and Ristna (Fig. 1). In Tallinn, the capital city of Estonia, regular sea level measurements started in 1809. The near continuous data sets are available from 1842, but the measurements were discontinued in 1996 due to construction work at the Tallinn Port. We used the data on monthly mean and extreme sea levels up to 2008 from these four stations. They represent relative sea level values in regard to the Kronstadt datum and are based, since 1951, on hourly measurements. The monthly values for earlier periods were obtained on the basis of daily data.

Data about land uplift for the studied stations was taken from the map compiled on the basis of the precise levellings in 1933–1943, 1956–1970 and 1977–1985 [11]. The radial crustal movements in Estonia are mainly influenced by regional Fennoscandian postglacial rebound and the uplift rates vary between 0.5 and 2.8 mm/yr along the Estonian coast. Tendencies in time series were analysed using linear regression analysis. A trend slope indicates a mean change per year, the change by trend line is calculated by multiplying the slope by the number of years.

For supplying the wave model with wind speed and direction, we acquired data from the Vilsandi meteorological station. This is the closest station to the calibration site for our wave study, just 7 km south of the Harilaid Peninsula (Fig. 1b). The station is on the western coast of the island. It has the most open location of all the Estonian weather stations, and satisfactorily represents both the scalar and the directional properties of the marine wind regimes in the northern Baltic Proper [12]. The wind data consisted primarily of hourly measurements from December 2006 until May 2007, which were used for the wave model calibration in relation to special wave measurements [1]. Then, all the historical wind data since 1966, available in digital form with a time interval of 3 hours, were used for the long-term hindcast of waves. The wind has been measured with "weathercocks" (wind vanes of Wild's design) in 1966–1976, automatic anemorhumbometers in 1976–2003, and MILOS-520 automatic weather complexes since September 2003. The older data in the database has been slightly corrected for homogeneity [13].

2.2 Wave measurements and modelling

The long-term calculations of wave parameters with the use of the SMB model were performed for the coastal region of Vilsandi-Harilaid (Fig. 1b). The location of the case study of hydrodynamic measurements and modelling was 1–1.5 km off the coast of the Harilaid Peninsula (Fig. 1c). The SMB-model, also called the significant wave method, is based on the fetch-limited equations of Sverdrup, Munk, and Bretschneider [14–16]. It calculates the significant wave height, wave period and wavelength for the chosen location under the assumption that the wind properties are constant over the entire fetch area. As strong winds are mostly homogeneous in the Baltic Proper [12] and both the reaction and memory time of a large part of the wave fields in this basin are relatively short [5], such simple models are valuable tools for rapid estimates of the wave statistics and the first approximation of the wave time series.

For comparison and calibration of the SMB-model, an oceanographic measuring complex RDCP-600 from AADI Aanderaa Instruments was deployed at a seabed depth of 14 m between 20 December 2006 and 23 May 2007 [1]. During this 5-month comparison period, the SMB model reproduced the wave parameters at the single point with acceptable accuracy. The model output required very moderate calibration to yield as high correlation coefficient as 0.88, low RMSE (0.233) and nearly equal average and maximum values of calculated and measured wave properties [13]. The calibrated model was further used in multiyear (1966–2006) wave hindcast. An additional comparison of hindcasts for the year 1996 between the SMB model and 3rd generation WAM model showed that considering the limitations of such SMB-type models, it performed surprisingly well and can be used for local long-term hindcasts [17].

3 Results and discussion

3.1 Decadal variations in mean and extreme sea level

With regard to the mean relative sea level of a location, the main factors influencing its variations are: global sea level change, the land uplift or subsidence and changes in the water balance of the particular sub-basin. Time series of annual mean sea levels show mostly increasing tendencies in Estonia (Fig. 2, Table 1). After adjusting these rates to account for local land uplift rates, we calculated sea level rise rates of 1.5 mm/yr in Tallinn (1842–1995), 1.7 mm/yr in Narva-Jõesuu (1899–2008), 1.7 mm/yr in Ristna (1950–2008) and 2.7 mm/yr in Pärnu (1924–2008). The trend estimates for annual maximum sea levels are much higher and vary between 4 and 10 mm/yr for different tide gauges and study periods (Fig. 3, Table 1). However, moving averages of annual mean sea level series show 30–40 year cycles (Fig. 2). These cycles, with amplitudes of about 5 cm, can influence linear trend estimates, which slightly depend on the period chosen.

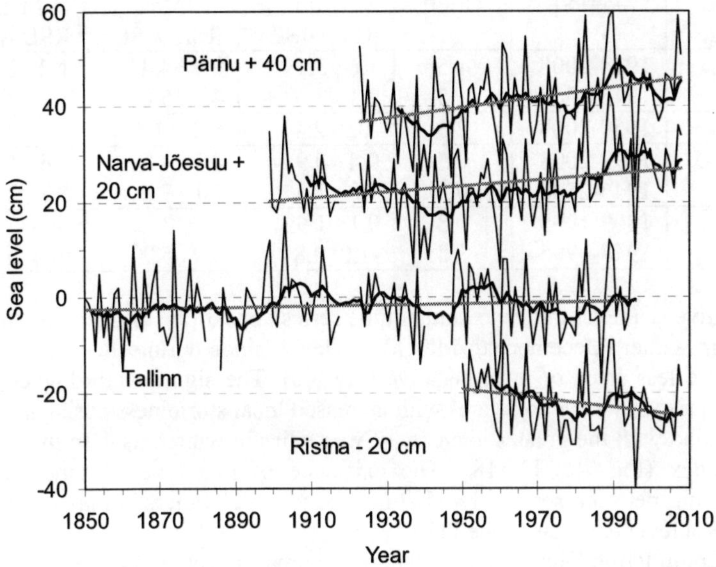

Figure 2: Decadal variations in annual mean relative sea levels (until 2008) together with 11-year moving averages and linear trendlines at four Estonian tide gauges. The series are not corrected with land uplift rates, which are different in different gauges (see Table 1). The different series are shifted in relation to each other for clarity.

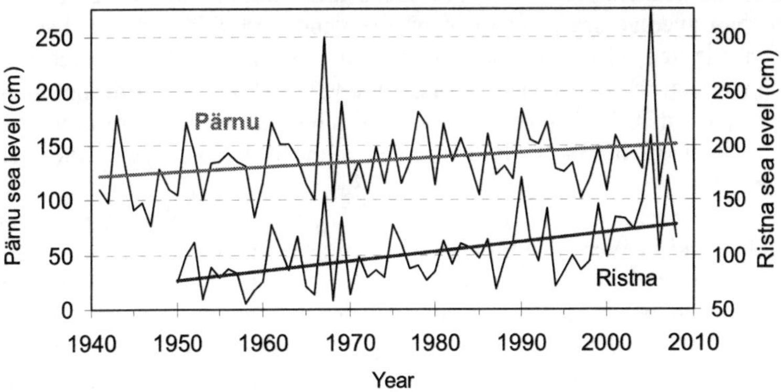

Figure 3: Decadal changes of annual maximum sea levels at Ristna and Pärnu (excerpt).

www.witpress.com, ISSN 1743-3541 (on-line)

Table 1: Trends for annual minimum, maximum and mean sea levels in different tide gauges and study periods. Actual sea level (ASL) rise rate is a sum of rates of relative sea level (RSL) and local land uplift (mm/yr).

Tide gauge	Period	Uplift	Min RSL	Min ASL	Max RSL	Max ASL	Mean RSL	Mean ASL
Pärnu	1924-2008	1.5	1.6	3.1	2.9	4.4	1.2	2.7
	1946-2008	1.5	2.1	3.6	4.2	5.8	1.1	2.6
Ristna	1950-2008	2.6	-0.2	2.4	8.8	11.4	-0.9	1.7
Narva-J.	1899-2008	0.5	0.4	0.9	4.3	4.8	0.6	1.1
	1946-2008	0.5	1.0	1.5	7.1	7.6	1.7	2.2
Tallinn	1899-1995	1.8	0.1	1.9	1.9	3.7	-0.3	1.5
	1947-1995	1.8	-1.0	0.8	3.9	5.7	-0.3	1.5

Positive sea level trends in annual time series appear due to the more positive trends in winter (December to March) sea level, since during the summer such trends are less steep or even negative (Fig. 4b). The significantly higher mean sea level rise in winter correlates with increased local storminess during the same months and with the greater intensity of westerlies in winter, as described by the NAO-index (Fig. 4c) [2, 18]. The existence of this type of time variable relationship between sea level and atmospheric circulation is common for other Baltic Sea level data sets as well [19].

Except in Pärnu, our mean sea level rise estimates (Table 1) are roughly equal to or insignificantly higher than the most recent global sea level rise estimates, which are around 1.7 mm/yr according to [20]. The excessive Pärnu sea level rise rate (Table 1, Figs. 2,3) over the global estimates can be explained by a mechanism, which was explained in a hydrodynamic modelling experiment [3]. These results indicate that in case of an obvious decadal trend in wind conditions the sea level change rates of a semi-enclosed basin may deviate from the global estimates. A positive trend in wind speed and storminess should result in a steeper than average sea level trend on the windward side and one that is less steep on the leeward side. Indeed, a clear increase in the westerly wind component occurred between 1950 and 1990 [21, 22]. Although further increase in westerly winds is anticipated, it is also possible that the increase in westerlies and winter NAO in 1950–90 could be followed by a relative increase in easterlies [23].

Trend analyses have shown that storm surges are becoming higher both in Estonia [18 and in Western Europe [23]. Pärnu sea level records (Figs. 2, 3, 4a) identify 30 individual events higher than the critical value of 150 cm, 24 of which occurred between the months of October–March. The two highest sea level events off the Estonian coast (since 1923) were both registered at Pärnu: 253 cm on 19 October 1967 and 275 cm on 9 January 2005. At the same time, the average wind speed has probably decreased over the last 50 years in Estonia (Fig. 5a). However, it is important that westerlies have increased and easterlies decreased. Also, both frequency and intensity of high wind speed events have increased (Fig. 5a) [13].

3.2 Decadal variations in wave condition

The mean wave heights and wave periods show quasi-periodic cycles with the last high stage in 1980–1995 and a slightly decreasing overall trend –0.001 m per year in 1966–2006 (Fig. 5).

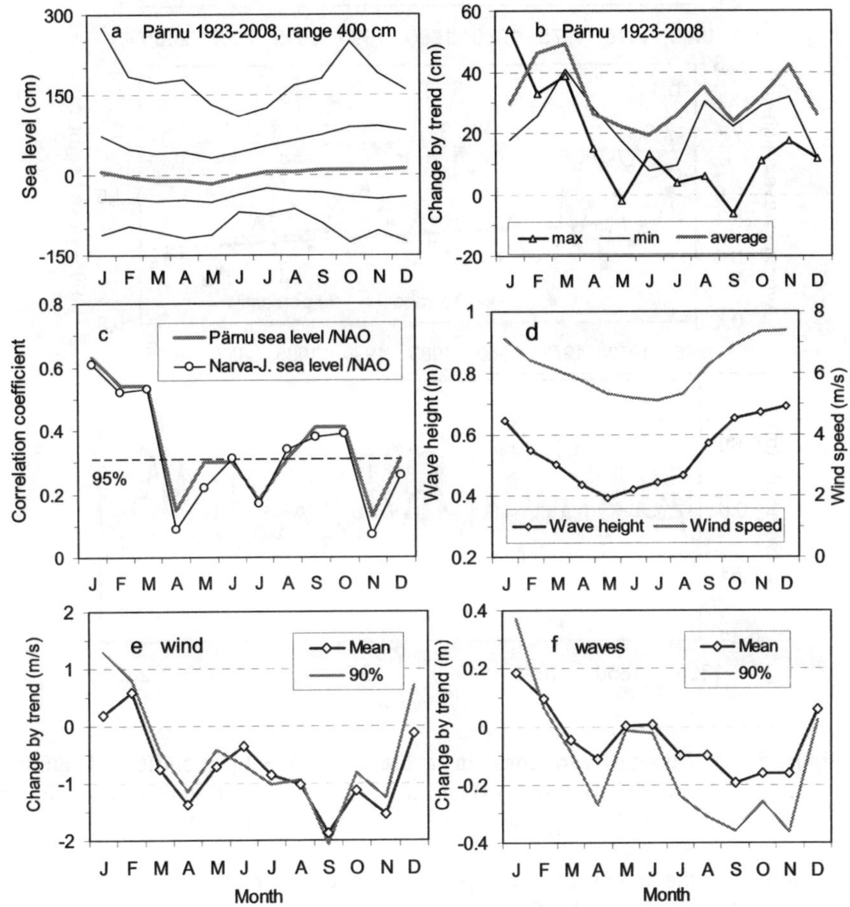

Figure 4: Seasonal variations in annual sea level statistics (a; lines from top: absolute maximum, average maximum, mean, average minimum and absolute minimum), sea level changes by trend over the period (b), correlations with atmospheric circulation data (c), mean wave heights and wind speeds (d), changes by trend in wind speed (e) and wave statistics (f) over the period 1966–2006.

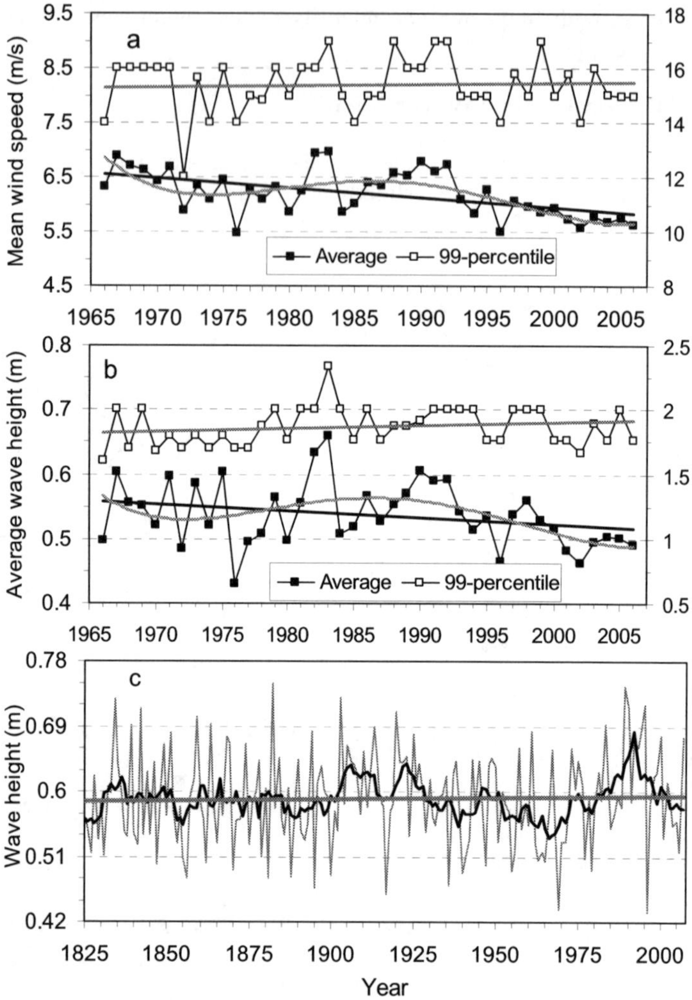

Figure 5: Decadal variations in averages and 99-percentiles of annual samples for winds (a) and waves (b). NAO-based reconstruction of wintertime significant wave heights along West Estonian coast (c).

However, the trends of wave properties are different in different months (Fig. 4f). For example, the December to March subset of mean wave heights shows an increasing trend of 0.0017 m per year. On the basis of annual series of the 90 and 99 percentiles, as well as annual maxima, the trend is clearly increasing (Fig. 5b) [13]. Thus, the trends in the average properties of wave fields and in extreme wave conditions are different along the western coast of Estonia.

Such tendencies in wave heights and variability seem to correspond to the long-term tendencies in mean and extreme wind speeds at the Vilsandi meteorological station (Figs. 4d,5). The time series of waves are strongly correlated with wind speed, the W-E wind component, and the NAO index (Table 2). Generally, the Iceland-Gibraltar version of the NAO index [21] is expected to best describe the variations in the Estonian climate [2]. Since monthly NAO data have been available since the 1820s (and even closely similar historical wind data does not exist), we tried to reconstruct wave conditions at the selected location back to 19th century. The highest correlation (r=0.72) between the NAO-index and the wave data in 1966–2006 was found from December to March. Based on this regression, a reconstruction of wave conditions is shown in Fig. 5c. There, an increase in wintertime wave activity is visible in 1970s, 1980s and 1990s. Quite possibly, the observed "vitalization of shore processes" falls into this period. But we cannot assume that this tendency will last forever (Fig. 5c).

Also we must stress here: though statistically significant, the outcome is strongly dependent on the quality of the NAO reconstructions and the NAO-based reconstruction of past (1825–2006) wave climates should be treated with the utmost caution. Although the wave conditions in fetch-limited areas are supposed to be highly site-dependent, some recent changes in regional wind climate and cyclone trajectories [2, 22] may have had a similar effect on wave conditions in different, though not necessarily adjacent marine areas. The main factor seems to be the similar exposure of a location to a certain direction, the importance of which in wind distribution is changing. For example, high wave events increased during 1958–1992 in the SE section of the North Sea along the windward coasts of Germany, Netherlands and Denmark [24] under conditions of increased westerlies and cyclonic activity [22, 25].

3.3 Relationships between atmospheric circulation, local wind climate, hydrodynamic regime and coastal geomorphic processes

Wind speed, storminess, sea level variations and wave statistics are correlated to each other and to the NAO-index on the windward coast of Estonia (Table 2) [2, 18].

Table 2: Correlation coefficients between some atmospheric forcing factors and selected average and extreme wave and sea level statistics (annual data samples 1966–2006).

	Wave statistics			Ristna sea level	
	Av.	90%	99%	Av.	Max.
NAO, months I–XII	0.48	0.35	0.41	0.56	0.18
NAO, months VIII–II	0.61	0.50	0.43	0.63	0.11
Wind speed modulus	0.87	0.52	0.38	0.63	-0.10
Wind speed 90%-ile	0.76	0.65	0.56	0.53	0.15
Wind speed 99%-ile	0.67	0.56	0.57	0.52	0.18
W-E wind component	0.85	0.79	0.60	0.70	0.23
S-N wind component	0.11	-0.02	0.07	0.09	0.12

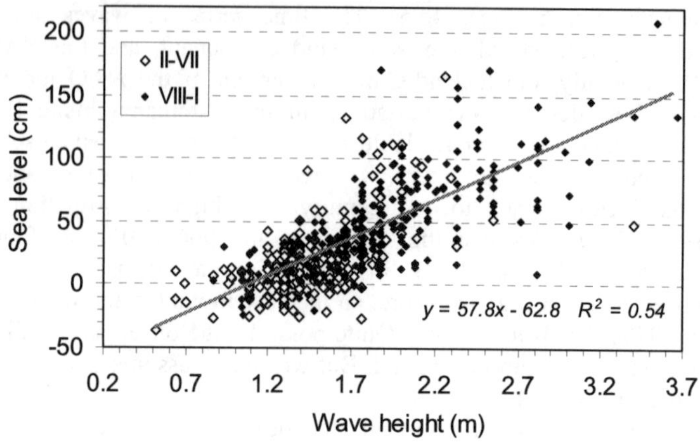

Figure 6: Dependence between monthly maximum sea levels at Ristna and corresponding monthly 99%-ile wave heights near Harilaid from 1966 to January 2007 grouped by two half years (II-VII and VIII-I; the regression is given for the whole data set).

The increase in W and S winds and the decrease in N and E winds imply certain shifts in the air-pressure systems and cyclone trajectories above the Baltic Sea; this is also expressed by recent tendencies in the NAO-index [2, 21].

Interestingly enough, neither the mean wind speed (Fig. 5a), mean wave heights (Fig. 5b), nor mean sea level regime (Fig. 2) had changed in a way that can explain all the documented "vitalization" of shore processes. The explanation obviously lies behind the extreme events (Figs. 3, 5), which have become both more frequent and more prominent. Also, such events are not distributed randomly over the seasonal cycle (Fig. 4). As a rule, powerful wave storms occur in autumn or winter months, when also the sea level is higher (Fig. 6). As wave energy is roughly proportional to the wave height squared, the energy of extremely strong storms and its impact on the coastal zone are many times higher than that of ordinary storms. Compared to the relatively very small energies of typical conditions, that energy is released in a broader zone and 1–3 meters above the average sea level (Fig. 6). On the other hand, small velocities below a certain threshold value are not able to erode, suspend, and transport sediment at all. They are "wasted" in geomorphic point of view.

In Estonia, the coastal processes in the past were largely influenced by isostatic land uplift (1–3 mm/yr) and marine regression. By now it is entirely compensated by sea level rise (Table 1). Despite some visible quasi-periodic cycles in climatic variables, in a longer perspective, the anticipated warmer winters, higher mean and extreme sea levels, more frequent strong storms, decreasing sea ice extent and duration [2–6, 23] are expected to intensify the shore processes in the future.

4 Conclusions

Although the mean properties of wind speed, wave heights and sea level have not changed recently in a way that can explain the observed intensified shore processes, the extreme statistics of these parameters all show significant increases. Powerful wave storms occur in autumn or winter months, when also the sea level is higher. Major coastal geomorphic changes occur precisely as a combined result of such relatively infrequent but extreme wintertime wavestorms and storm surges. As wave energy is roughly proportional to the wave height squared, the energy of extremely strong storms and its impact on the coastal zone can be increasingly higher.

Acknowledgements

The study was supported by the Estonian Science Foundation through grants No. 7609 and 7564, and target financed theme SF0180104s08.

References

[1] Suursaar, Ü., Jaagus, J., Kont, A., Rivis, R. & Tõnisson H., Field observations on hydrodynamic and coastal geomorphic processes off Harilaid Peninsula (Baltic Sea) in winter and spring 2006–2007. *Estuarine Coastal and Shelf Science*, **80**, pp. 31–41, 2008.

[2] Jaagus, J., Post, P. & Tomingas, O., Changes in storminess on the western coast of Estonia in relation to large-scale atmospheric circulation. *Climate Research*, **36**, pp. 29–40, 2008.

[3] Suursaar, Ü. & Kullas, T., Influence of wind climate changes on the mean sea level and current regime in the coastal waters of west Estonia, Baltic Sea. *Oceanologia*, **48**, pp. 361–383, 2006.

[4] Tõnisson, H., Orviku, K., Jaagus, J., Suursaar, Ü., Kont, A. & Rivis, R., Coastal Damages on Saaremaa Island, Estonia, Caused by the Extreme Storm and Flooding on January 9, 2005. *Journal of Coastal Research,* **24**, pp. 602–614, 2008.

[5] Broman, B., Hammarklint, T., Rannat, K., Soomere, T. & Valdmann, A., Trends and extremes of wave fields in the northern part of the Baltic Proper. *Oceanologia,* **48** (s), pp. 165–184, 2006.

[6] Orviku, K., Jaagus, J., Kont, A., Ratas, U. & Rivis, R., Increasing activity of coastal processes associated with climate change in Estonia. *Journal of Coastal Research*, **19**, pp. 364–375, 2003.

[7] Rivis, R., Changes in shoreline positions on the Harilaid Peninsula, West Estonia, during the 20th century. *Proc. Estonian Acad. Sci. Biol. Ecol.*, **53**, pp. 179–193, 2004.

[8] Suursaar, Ü., Tõnisson, H., Kullas, T., Orviku, K., Kont, A., Rivis, R. & Otsmann, M., A study of hydrodynamic and coastal geomorphic processes in Küdema Bay, the Baltic Sea. *Coastal Engineering VII,* eds. C.A. Brebbia & C. Cunha, WIT Press: Southampton, Boston, pp. 187–196, 2005.

[9] Alexandersson, H., Schmidt, T., Iden, K. & Tuomenvirta, H., Long-term variations of the storm climate over NW Europe. *The Global Atmos. Ocean. Syst.*, **6**, pp. 97–120, 1998.

[10] WASA Group, Changing waves and storms in the Northeast Atlantic? *Bull. Am. Meteorol. Soc.*, **79**, pp. 741–760, 1998.

[11] Vallner, L., Sildvee, H. & Torim, A., Recent crustal movements in Estonia. *J. Geodyn.*, **9**, pp. 215–223, 1988.

[12] Soomere, T., Extreme wind speeds and spatially uniform wind events in the Baltic Proper. *Proc. Estonian Acad. Sci. Eng.*, **7**, pp. 195–211, 2001.

[13] Suursaar, Ü. & Kullas, T., Decadal variations in wave heights off Cape Kelba, Saaremaa Island, and their relationships with changes in wind climate. *Oceanologia*, **51**, pp. 39–61, 2009.

[14] Seymour, R.J., Estimating wave generation in restricted fetches. *J. ASME WW2, May 1977*, pp. 251–263, 1977.

[15] *U.S. Army Coastal Engineering Research Center, Shore Protection Manual, Vol.1,* Third Ed., U.S. Govt. Printing Office, Washington D.C., 719 pp, 1984.

[16] Huttula, T., Suspended sediment transport in Lake Säkylän Pyhäjärvi, *Aqua Fennica*, **24**, pp. 171–185, 1994.

[17] Räämet, A., Suursaar, Ü., Kullas, T. & Soomere, T., Reconsidering uncertainties of wave conditions in the coastal areas of the northern Baltic Sea. *Journal of Coastal Research*, SI **56**, 2009. (in press).

[18] Suursaar, Ü. & Sooäär J., Decadal variations in mean and extreme sea level values along the Estonian coast of the Baltic Sea. *Tellus A*, **59**, pp. 249–260, 2007.

[19] Wakelin, S.L., Woodworth, P.L., Flather, R.A. & Williams, J.A., Sea-level dependence on the NAO over the NW European Continental Shelf. *Geophys. Res. Lett.*, **30**, Art. No. 1403, 2003.

[20] Church, J.A. & White, N.J., A 20th century acceleration in global sea-level rise. *Geophys. Res. Lett.*, **33**, L01602, 2006.

[21] Jones, P.D., Jónsson, T. & Wheeler, D., Extension to the North Atlantic Oscillation using early instrumental pressure observations from Gibraltar and South-West Iceland. *Int. J. Climatol.*, **17**, pp. 1433–1450, 1997.

[22] Siegismund, F. & Schrum, C., Decadal changes in the wind forcing over the North Sea. *Climate Research*, **18**, pp. 39–45, 2001.

[23] Lowe, J.A., Gregory, J.M. & Flather, R.A., Changes in the occurrence of storm surges around the United Kingdom under a future climate scenario using dynamic storm surge model driven by the Hadley Centre climate models. *Climate Dynamics,* **18**, pp. 179–188, 2001.

[24] Weisse, R. & Günther, R., Wave climate and long-term changes for the Southern North Sea obtained from a high-resolution hindcast 1958–2002. *Ocean Dynamics*, **57**, pp. 161–172, 2007.

[25] Bauer, E., Interannual changes of the ocean wave variability in the North Atlantic and in the North Sea. *Climate Research*, **18**, pp. 63–69, 2001.

Section 4
Sea defence and energy recovery

Coastal storm damage reduction program in Salerno Province after the winter 2008 storms

G. Benassai[1], P. Celentano[2] & F. Sessa[3]
[1]*Department of Applied Sciences, University of Naples Parthenope, Italy*
[2]*Coastal Risk, University of Naples Parthenope, Italy*
[3]*Province of Salerno, Coastal Risk, University of Naples Parthenope, Italy*

Abstract

In this paper the coastal protection system of the inhabited coastline near the city of Salerno, designed by the Salerno Province, is discussed. This low urbanized coastline, which has already suffered serious erosion over the last 40 years, had a lot of damage due to the storms of the winter of 2008: inundation of the roadways, vertical beach loss, generalized scour around bulkheads, overwash of infrastructures and recreational facilities.

These huge amounts of damage induced the design of an extensive defence project on the whole inhabited coastline of Salerno, with advanced technologies of beach nourishment, in order to assure flood damage reduction and environmental sustainability.

After analysis of the long-term erosion of the coastal site, the procedure for the choice of the coastal protection works is reported here in some detail. The description of the coastal protection system follows, which consists mainly of a long series of parallel submerged breakwaters in order to reduce the storm wave impact on the most exposed infrastructures (roads, buildings, seawater facilities). The proposed intervention takes into account the suggestions of the international EU projects "Eurosion", "Beachmed" and "Messina".

Keywords: beach erosion, coastal protection works.

1 Introduction

Beach erosion of the whole Mediterranean coastline is becoming of increasing importance, involving serious environmental and socio-economic consequences.

www.witpress.com, ISSN 1743-3541 (on-line)
doi:10.2495/CP090111

The accretion of sandy coastlines had an inversion from the second half of the last century, with a gradual retreat that accelerated in the last two decades.

In the Salerno province the phenomenon is critical mainly in the inhabited coastline near the city of Salerno, where the damage due to both long term and seasonal erosion reached the highest values. These low urbanized coastlines, which already suffered serious erosion in the last 40 years, had a lot of damage in the winter 2008 storms. The situation is even more serious due to the absence of beach protection works in many areas and due to buildings that have now reached the coastline or are protected by a vertical wall without any defence work (fig. 1).

2 Long-term erosion in the Salerno inhabited coastal zone

The inhabited coastline near the city of Salerno is one of the most critical littorals of the whole province, because the beach retreat threatens the urbanized coastal zone. The statistical analysis implemented by ISPRA (formerly APAT) on the erosion of the Italian littorals, based on the comparison between the 1954 and 1998 coastline data, shows that the erosive trend is between 20 and 50 m in 50 years, which is a strong value if compared with the average beach depth still remained, which is at most 40m (fig. 2). Moreover, the last severe storms (December 1999, December 2004, December 2008) increased the erosive trend with marine inundation of large littoral zones, including the state road n. 175 and several recreational facilities.

3 Wave climate, beach characteristics and littoral dynamics

The results of the coastal engineering study have demonstrated that the highest and more frequent waves come from the west and south-west. In particular, the

Figure 1: Salerno inhabited coastline protected by a vertical wall without any defence work.

Figure 2: Long-term erosion of the Salerno coastline (the lost beach area is shown in dark grey).

Table 1: Sediment characteristics at the mouth of river Irno, Salerno.

	sample	depth	D_{50}(mm)	St. Dev.	Skewness	kurtosis
Irno river mouth	1	0	0.891	0.187	-0.175	0.583
	2	1.0	0.199	0.580	-0.269	1.243
	3	2.0	0.196	0.636	-0.127	1.200
	4	2.8	0.164	0.609	-0.032	1.270
	5	4.0	0.087	0.519	-0.031	0.804
	6	5.1	0.126	0.680	0.141	0.986
	7	6.0	0.140	0.708	0.118	1.168
	8	7.0	0.136	0.725	0.195	1.019
	9	8.0	0.135	0.717	0.217	0.991

extreme waves with a return period of 50 years coming between 210°N and 260°N have values between 6.0 and 7.0m, while the ones coming between 160°N and 210°N are lower than 4.0 m.

The sediment analysis showed that the emerged beach is formed by gross sand with D_{50} of approx 1 mm and lower, while the submerged beach evidences finer sediments, with D_{50} even lower than 0.1 mm, as illustrated in tab. 1 by the results of the sediment analysis of the coastline offshore the mouth of the river Irno, close to the city of Salerno.

The littoral transport was calculated on the basis of the longitudinal component of the wave energy flux at the stage of breaking, starting from the inshore wave climate (frequency of sea states classified by height, period and direction).

A useful indicator of littoral stability is the Dean parameter N_0, given by:

$$N_0 = H_0/w_f\,T \qquad (1)$$

where:
H_0 = offshore significant wave height;
T = offshore significant wave period;
w_f = fall speed of the sand particle, given by:

$$w_f = [\{(\rho_s/\rho)-1\}g\}0.7\, D_{50} / (6\, \nu^{0.4})\;\; T \tag{2}$$

The value assumed by N_0, called the Dean number, gives the following possibilities:

$N_0 < 2.4$ very probable accretion;
$N_0 < 3.2$ probable accretion;
$N_0 \geq 3.2$ probable erosion;
$N_0 > 4.0$ very probable erosion.

The N_0 value in tab 2 for the Irno coastal zone demonstrates a clear erosive tendency.

Table 2: Value of the Dean number for the Irno river mouth littoral.

site	D_{50} (mm)	w_f (m/s)	H_0 (m)	T_0 (s)	N_0
Irno river mouth	0.150	0.0178	0.95	6.80	8.22

4 Main damage due to the winter 2008 storms

In recent years, the coastline suffered from severe damage due to marine inundation of sea storms, the last of them in December 2008 and January 2009. These events also caused extensive damage along the coast, consisting in vertical beach loss of several meters, scour around bulkheads and damage to many infrastructures and recreational facilities.

This damage was due to lack of adequate building setbacks from the shoreline in combination with inadequate foundation systems of the bulkheads, so the buildings resulted extremely vulnerable to storm surge and wave-induced damage (fig. 3, fig. 4).

In some other cases, the damage was caused by lack of efficiency of the existing protection system, as in the damage of the Salerno city waterfront (fig. 5).

In this case the lack of efficiency of the breakwater located offshore the gap is evident by the comparison with the crest height and the number of the blocks of the adjacent breakwaters. Probably the higher depth of the offshore breakwater caused higher impact of the breaking waves on the armour layer, with extensive damage of the breakwater and loss of efficiency. This caused the damage to the waterfront which has been occurred only to the portion defended by this inefficient structure.

Figure 3: Storm surge and wave-induced damage to buildings during the winter 2008 storms.

Figure 4: Damage to the waterfront of Salerno after the winter 2008 storms.

5 The coastal protection system

The long-term erosion and the inundation damage, which increased greatly during the winter 2008 storms, induced the Province of Salerno to design an extensive defence project on the whole inhabited coastline of Salerno, in order to assure flood damage reduction and environmental sustainability.

Figure 5: Damage to the waterfront of Salerno after the winter 2008 storms.

The last aspect is quite delicate, because it is known that artificial interventions often show their effects a long time after their construction, so that they often do not constitute convenient solutions for coastal protection, causing imbalance in the littoral transport between contiguous areas. Therefore the choice of the optimal intervention must be achieved through a detailed and careful study and a proper execution must consider gradual and flexible planning operations in order to reach the best solutions.

In this context the project of Salerno Province for the "interventions of defence, qualification and exploitation of the coast of the city of Salerno" is a shoreline protection system which divides the inhabited coastal zone in four parts, providing specific interventions for each of them.

The determining factors involved in the selection of the defence structures can be summed as follows:

- The purpose of the protection system in relation to urgency, efficiency and beach fruition.
- The morphology of the coastal area.
- The regime of sediment transport.
- The sea level variations and meteomarine factors.

Table 3, originally produced by Kobayashi (1987), can be useful to give a first indication of the degree of suitability for each type of intervention.

The most significant factors are the following (see Benassai [3]).

- Urgency (descending order from *a* to *c*).
- Littoral transport patterns (ascending order from *d* to *h*).
- Tide (ascending order from *i* to *j*); the Tyrrenian Sea is enclosed in class *'i'*.
- Morphology of the coastal area which is linked to the dimensions of the main beach features, the height of the active beach and the erodibility of the coast (ascending order from *h* to *l*).

Table 3: Main factors for the choice of a beach protection system.

Protection systems	METEOMARINE AND SOCIOECONOMIC FACTORS												
	Urgency			*Long-shore transport*					*Tide*		*Morphologic instability*		
	a	*b*	*c*	*d*	*e*	*f*	*g*	*h*	*i*	*j*	*k*	*l*	*m*
Groynes	*	1	2	0	1	3	2	2	1	2	2	1	*
Det. Breakwaters	*	0	2	2	3	+	2	+	2	1	2	1	0
Beach revetment	2	+	*	1	0	0	*	*	1	2	2	1	*
Nourishments	3	3	2	3	2	2	1	1	2	2	1	2	3

The degree of suitability is indicated in the following:

- 3 *advisable solution*;
- 2 *suitable solution*;
- 1 *acceptable solution*;
- + *suitable solution for some coastal characteristics, inefficient for others*;
- 0 *inefficient solution*;
- * *unadvisable solution*.

From the first column, it can be seen that a mixed protection system composed by detached breakwaters and beach fills is also suitable for a moderate urgency intervention and it can be recommended also in case of a definite littoral transport pattern. Besides, it is an advised solution in case of low or moderate morphologic instability, which is the situation of Salerno coastline.

In synthesis, the methodology illustrated in tab. 3 gives a suitable solution for artificial nourishment also in emergency conditions as an alternative to beach protection works with a greater visible impact. In fact, if the nourishment is coupled with submerged breakwaters in order to reduce the maintenance cost, it is a convenient solution even when the longshore transport is significant.

As a result of the choice, the coastal storm damage reduction program consist of a long series of parallel submerged breakwaters designed to reduce the storm wave impact on the most exposed infrastructures (roads, buildings, recreational facilities). The submerged breakwaters have been localized on depths ranging between 4 and 5 m, at a distance between 100 and 150 m, with beam width of approximately 12-14m, in order to dissipate most of the breaking wave energy.

The intensity of coastal protection, for both emerged and submerged breakwaters, depends on the relative crest height and on the distance from the shoreline. Fig. 6 gives two cases of emerged breakwaters of different length at the same distance from the shoreline. It is showed that a significant relative length (length made non-dimensional with the breaking zone width) or a too close relative distance (distance made non-dimensional in the same manner) can lead to the formation of a tombolo (fig. 6). This is the reason why the designed breakwaters are alternated with gaps and located closer to the coastline in case of need of major protection, while they can be located at a higher distance when there is no particular infrastructure to be protected (fig. 7).

The hard measures have been coupled with beach fills refurnished by offshore marine sand, taken from the offshore submerged bars (fig. 8). These protection works ensure the necessary defence from the meteomarine factors and minimise the environmental impact. Furthermore, as the defence structures are submerged with gaps of suitable size, they ensure the necessary water circulation and periodic replacement of water, in order to avoid stagnation phenomena

The proposed intervention takes into account the suggestions of the international EU projects "Eurosion", "Beachmed" and "Messina". With reference to the first project, the intervention is classified as soft. The EU project "Beachmed" has demonstrated that this type of intervention (beach fill protected with a submerged breakwater) has proven to suffer from minor losses of borrow material, in comparison with a pure beach fill without submerged barriers.

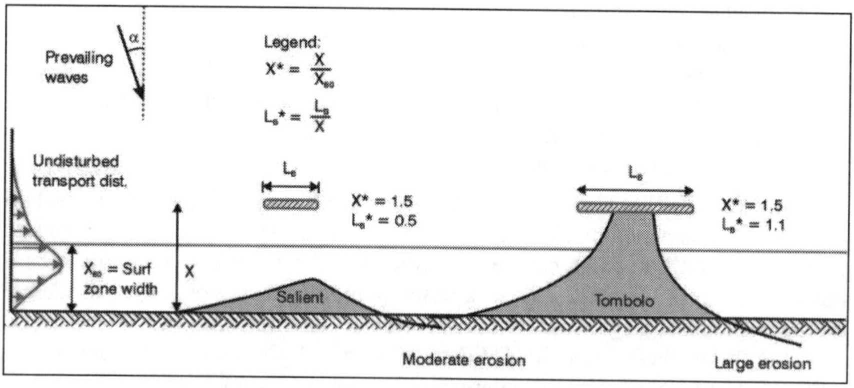

Figure 6: Intensity of protection as a function of the relative length; the same applies for the relative distance.

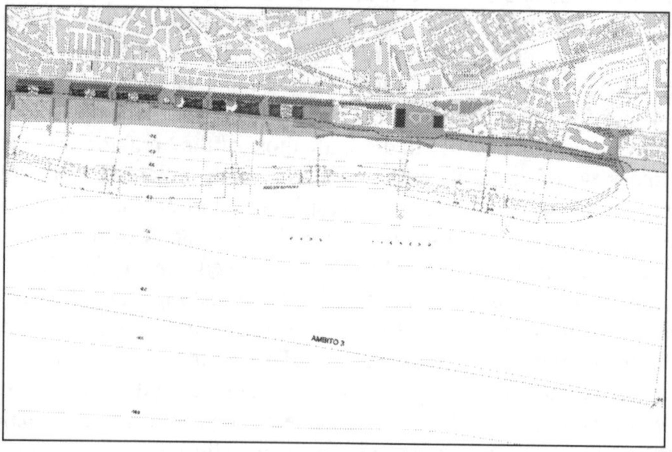

Figure 7: Plan of part of the Salerno shoreline protection works.

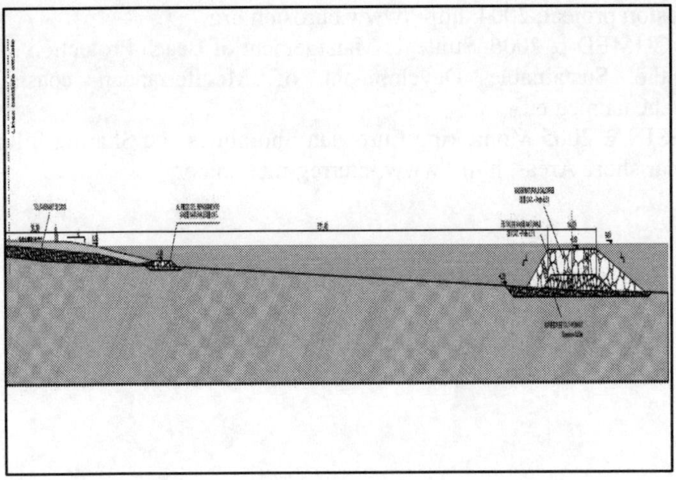

Figure 8: Profile of part of the Salerno shoreline protection works.

6 Conclusions

An extensive defence project on the whole inhabited coastline of Salerno, with advanced technologies of beach nourishment, has been proposed by Salerno Province in order to assure flood damage reduction and environmental sustainability. The system consists mainly of a long series of parallel submerged breakwaters in order to reduce the storm wave impact on the most exposed infrastructures (roads, buildings, seawater facilities), coupled with beach nourishment.

The choice of the coastal protection works has been justified in view of the scientific literature and of the suggestions of the international EU projects "Eurosion", "Beachmed" and "Messina".

After the project realization, planned for the next two years, an intensive monitoring program will be realized in order to verify the project efficiency in terms of beach protection and coastline advance.

References

[1] APAT, Università degli Studi di Roma Tre, 2002. Atlante delle onde nei mari italiani. Istituto Poligrafico e Zecca dello Stato.

[2] APAT, 2007. Atlante delle opere di sistemazione costiera. Manuali e Linee Guida, 44/2007, Roma.

[3] Benassai G., 2006. Introduction to coastal dynamics and shoreline protection. WIT Press, 336 pp.

[4] Provincia di Salerno (2008). Progetto di difesa e riqualificazione della costa di Salerno e di Pontecagnano (In Italian).

[5] USACE, 2002, *Coastal Engineering Manual,* Coastal and Hydraulics Laboratory, Vicksburg, MS. http://chl.erdc.usace.army.mil

[6] EUrosion project, 2004: http://www.eurosion.org
[7] BEACHMED-e, 2006. Strategic Management of Beach Protection Measures for the Sustainable Development of Mediterranean coastal areas www.beachmed.eu
[8] MESSINA, 2005 Monitoring European Shorelines and Sharing Information on near shore Areas. http://www.interreg-messina.org

Wave energy conversion systems: optimal localization procedure

G. Benassai[1], M. Dattero[2] & A. Maffucci[1]
[1]*Department of Applied Sciences, University of Naples Parthenope, Italy*
[2]*Coastal Risk, Department of Applied Sciences, University of Naples Parthenope, Italy*

Abstract

Wave energy is a renewable and pollution-free energy source that has a potential world-wide contribution to the electricity market estimated in the order of 2,000 TWh/year, which represents about 10% of the world's electricity consumption with an investment cost of EUR 820 billion. Sea waves have one of the highest energy densities among the Renewable Energy Sources (RES). Today, the largest problem in harvesting wave energy is obtaining reliability of the technology and bringing the cost down.

The main types of wave energy converters are:

- The oscillating water column, which consists of a partially submerged, hollow structure open to the sea below the water line.
- Overtopping devices that collect the water of incident waves in order to drive one or more low head turbines.
- Point absorbers (floating or mounted on the sea bed), which usually provide a heave motion that is converted by mechanical, magnetic and/or hydraulic systems in linear or rotational motion for driving electrical generators.
- Surging devices that exploit the horizontal particle velocity in a wave to drive a deflector or to generate the pumping effect of a flexible bag facing the wave front.

Obviously the amount of available energy depends on the different distribution of the wave energy on the world coastlines. The most energetic sites are situated on the west coasts of the oceans at middle latitudes. European and Australian west coasts reach amounts of 80 kW/m of available wave energy; in

WIT Transactions on Ecology and the Environment, Vol 126, © 2009 WIT Press
www.witpress.com, ISSN 1743-3541 (on-line)
doi:10.2495/CP090121

fact, the first "wave farms" have been installed in Portugal and Scotland where the energy due to the waves reaches these highest values.

The power in a wave is proportional to the square of the amplitude and to the period of the motion. Therefore, long period (>7–10 s), large amplitude (>2 m) waves have energy fluxes commonly averaging between 40 and 70 kW per m width of oncoming wave. Situated at the end of the long, stormy fetch of the Atlantic, the wave climate along the western coast of Europe is characterized by particularly high energy.

In the Mediterranean basin, the annual power level off the coasts of the European countries varies between 4 and 11 kW/m, the highest values occurring in the area of the south-western Aegean Sea. In the Thyrrenian Sea, this lowest value of 4 kW/m is acceptable as a limit for possible location of an offshore wave farm.

In order to identify the best sites and to optimize the location of new wave farms offshore of the coastlines of the Campania region (South of Italy), the significant wave height data set of ISPRA has been used, on which the Italian Wave Atlas is based. The wave buoys data are useful to calibrate the wave prediction model WWIII implemented by the Department of Applied Sciences of the University Parthenope, which simulates the most significant wave storm events of the last ten years. This procedure has been used to produce in the recent past a risk assessment of the Campanian coastlines to the wave impact, so it can also be used to evaluate the potential wave energy to be converted offshore of the Campanian coasts.

Keywords: wave energy, optimal localization, wave farms.

1 Introduction

Among the sources of clean renewable and inexhaustible energy, the one that probably has in itself the largest margins of development is certainly the exploitation of wave energy. In fact, water is about 800 times denser than air, so the exploitation of wave power could provide an amount of energy much higher than that of wind energy; both wave and current energy conversion are an enormous potential in almost all coastal zones of the world.

The west coasts of Europe are the places where the energy available per meter of coast is among the largest in the world. Recent studies estimated that about 320 GW power are available from the waves on the North-Eastern coastline of the Atlantic Ocean with values of energy per meter of coast ranging from 25 kW/m offshore of the Canary Islands to 70 kW /m offshore the coastlines of Scotland.

In the Mediterranean Sea, the power available is estimated to be about 30 GW, with power values ranging from 4 kW/m to 13 kW/m (southeast Aegean Sea).

The methods being pursued to try to exploit the enormous availability of this kind of energy have been very different; we will make a brief description of the most used systems, classifying them according to the mechanisms of energy conversion.

1.1 Oscillating water column

The OWC device comprises a partly submerged concrete or steel structure, open below the water surface, inside which air is trapped above the water free surface. The oscillating motion of the internal free surface produced by the incident waves makes the air to flow through a turbine that drives an electric generator. The axial-flow Wells turbine, invented in the late 1970's, has the advantage of not requiring rectifying valves. It has been used in almost all prototypes.

Several OWC prototypes have been built in on the shoreline in Norway, China, UK (LIMPET), Portugal (Pico Island); incorporated in a breakwater (in the harbour of Sakata, NW Japan) or placed outside it (Trivandrum, India).

1.2 Overtopping devices

This typology includes the Wave Dragon, one of the systems that achieved in recent years very interesting results. The Wave Dragon is composed of two long arms of 130 mt. each, which are responsible for the waves to reflect and convey to the central unit that is placed at an elevated level from the sea surface, where a series of vertical Kaplan turbines are located.

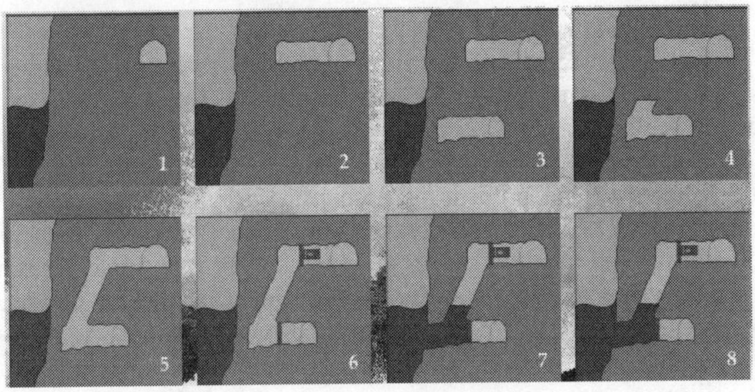

Figure 1: Scheme of an oscillating water column.

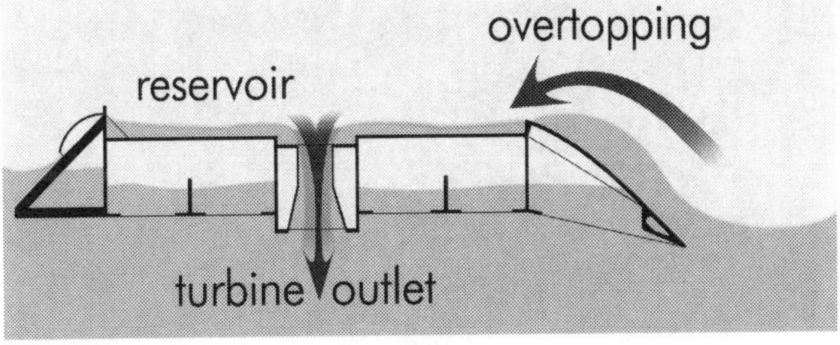

Figure 2: Scheme of the Wave Dragon.

The water that overflows in the basin, driven only by the force of gravity, passes through the Kaplan low head turbines. Today, the maximum power obtainable from any Wave Dragon is of 4 MW, but in future years units of 7 MW will be implemented.

1.3 Archimedes Wave Swing

The AWS consists of an upper part (the floater) of the underwater buoy that moves up and down in the wave while the lower part (the basement or pontoon) stays in position. The periodic changing of pressure in a wave initiates the movement of the upper part. The floater is pushed down under a wave top and moves up under a wave trough. To be able to do this, the interior of the system is pressurised with air and serves as an air spring. The air spring, together with the mass of the moving part, is resonant with the frequency of the wave. The mechanical power required to damp the free oscillation is converted to electrical power by means of a Power Take Off system (PTO). The PTO consists of a linear electrical generator and a gas- (nitrogen-) filled damping cylinder.

2 Suitable devices and locations for installing wave farms in Italy

A preliminary assessment of the most suitable device as a wave energy conversion system passes through its survivability. In fact, wave energy devices are placed in regions of high incident wave power, which is normally related to very rough sea states that have to be considered for the design. The absence of clear and reliable design procedures accounting for loads arising in such an environment makes it difficult to conduct wave energy projects with sufficient planning safety. This implies that, to a greater extent than for other renewable energy technologies, the survivability plays a decisive role for the success of ocean wave energy projects. In this context the AWS has the advantage to be a completely submerged device, which brings along considerable plus for the design stability.

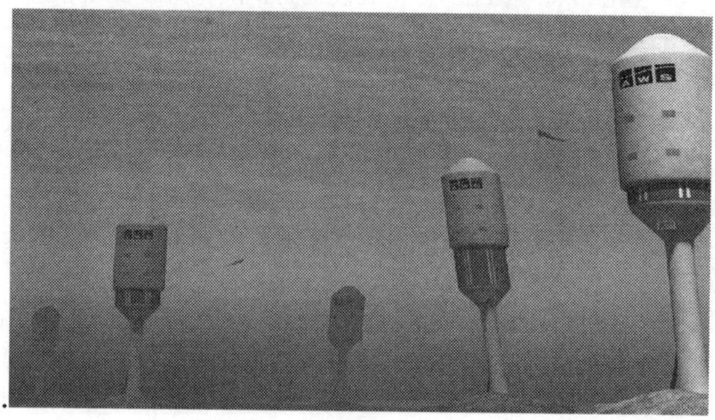

Figure 3: Scheme of a series of AWSs.

A preliminary assessment of the most suitable locations for wave farm installation starts with the spatial analysis of the sea state characteristics in order to establish the potential available energy per meter of coastline. The potential amount of wave power available on the Italian coasts has been established by wave data analysis derived from measurements taken from buoys located along the Italian coastlines belonging to the National Wave Network (RON) which is run by ISPRA.

According to these good quality data, the best energetic coastal sites are found on the Sardinian and Sicilian coasts.

The amount of incident wave power is given by the following equation:

$$J = 0.42 \times Hs^{2} \times T_{P} \quad (1)$$

The 0.42 multiplier in the above equation is exact for any sea state that is well represented by a two-parameter Bretschneider Spectrum, but it could range from 0.3 to 0.5, depending on the shape of the wave spectrum and on the relative amounts of energy in sea and swell components. Tab. 1 shows the energy amount available from waves of given height and period.

Using the data collected in the Italian wave Atlas by ISPRA, we are able to obtain a fairly satisfactory estimate of the available energy in coastal sites where a RON buoy is located. It is well-known that the western coast of Sardinia is the worst location for wave storms, so the best Italian location for exploitation of wave energy. The wave energy calculation implemented for the buoy of Alghero gives an average energy of about 11 kW/m, according to the values expected for the most energetic sites of the Mediterranean.

3 Suitable locations for installing wave farms in Campanian coasts

The buoy of Ponza, located offshore of the coastlines of Lazio, is the closest to the coastlines of Campania region and so its data have been used to analyse the wave energy potential in Campanian coasts.

The value of wave energy calculated for the buoy of Ponza is significantly lower than Alghero, giving an average value of about 4 kW/m. These calculations are valid for the most energetic directions, that is between 225°N and 285°N.

Table 1: Power generation (KW) per classes of wave height and period.

Power Generation (kW) per wave height and period for each unit						
	Wave Period (sec)					
Wave Height	5	6	7	8	9	10
1	23.6	19.7	16.9	14.8	13.1	11.8
2	94.6	78.8	67.5	59.1	52.5	47.3
3	153.9	128.3	109.9	96.2	85.5	77.0
4	189.1	157.6	135.1	118.2	105.1	94.6
5	189.1	157.6	135.1	118.2	105.1	94.6

In order to obtain comparative results of the obtainable wave power for the Campanian coastlines with a sufficient spatial density, the 3rd generation wave model Wave Watch III was run with the simulations of the storms of the last 10 years. It was demonstrated that the data collected by the buoy of Ponza are in good agreement with the simulations derived through the implementation of Wave Watch III model for the most exposed coastal sites of the Campanian coasts. The model was applied to the Gulf of Naples referenced to two classes of wave storms: the first class of sea storms, which is showed more frequently, is exemplified by the storm of December 1999, with waves coming from South-West and West. The second class of storms is exemplified by the storm of December 2004, with waves coming from South.

4 Archimedes Wave Swing possible installations on the Campanian coasts

Either in the most frequent case of waves coming from west - south west, like the sea storm of 1999, or in the less frequent case of waves coming from the south, like the sea storm of 2004, the most suitable sites for the installation of wave farms were found to be the most exposed coastlines, that is the coastlines not sheltered from the Isles of Capri and Ischia. This leads to the offshore coastlines of Ischia and Capri and the offshore coastlines of Sorrento peninsula For example, the sites offshore the Isle of Ischia, illustrated in fig. 5 or the site offshore the Sorrento peninsula, on the Amalfi coast, illustrated in fig. 6, are all suitable because they gain interesting energy values in both types of sea storms.

A comparison between the energy values of Ponza and these coastal sites have shown that, at a first stage of analysis, there are minor differences. So, in absence of available field data on the Campanian coastal sites, it can be assumed that the available data of the buoy of Ponza can be used for the sites offshore the coastal sites of Ischia and Sorrento.

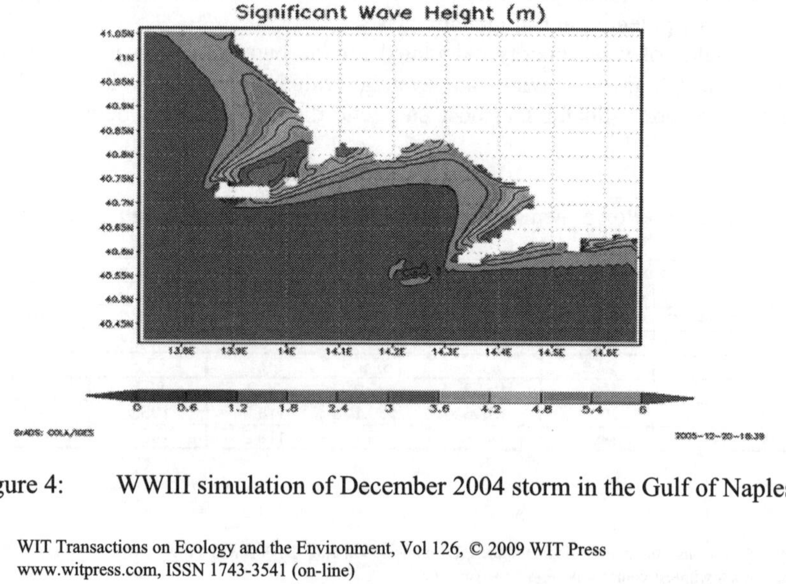

Figure 4: WWIII simulation of December 2004 storm in the Gulf of Naples.

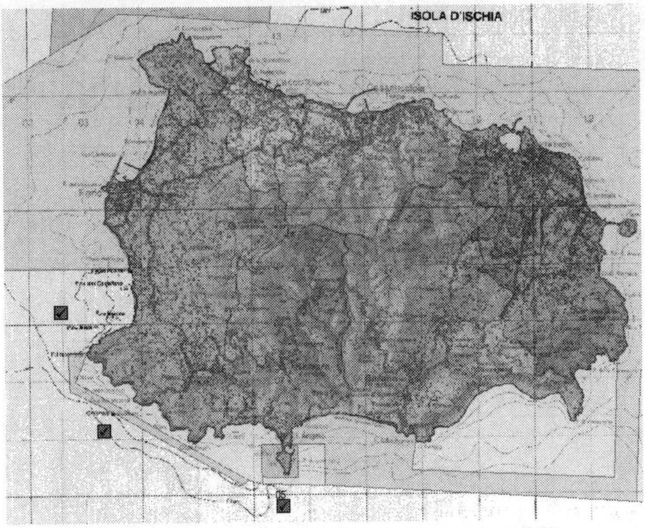

Figure 5: Optimal AWS localization in Campania (offshore of Ischia).

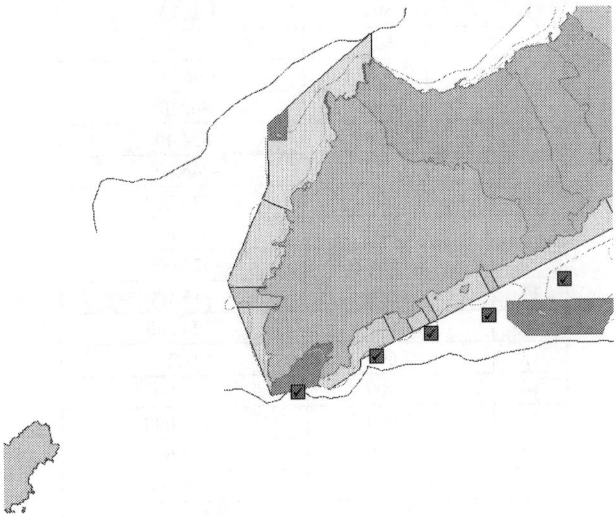

Figure 6: Optimal AWS localization in Campania (offshore of Sorrento).

4.1 AWS economic analysis

As already stated, the use of the AWS on Campanian coasts is particularly suitable for the absence of impact on the landscape because of its location a few meters below the surface; this permits the passage of small and medium vessels above the wave farm.

Table 2: Annual profit for wave energy selling.

ANNUAL PROFIT (€) FOR AWS (200 MW\year)	
1.500.000	AWS investment + installation price
200.000	Kw for each year
0.098	Energy sell price for kw
19.600	Total profit for energy sale
20.000	Profit for the Green Certificates sale
0.34	GSE bonus for kw
68.000	Total profit for the GSE bonus
107.600	Total profit for each year

Table 3: Cash flow for AWS.

CASH FLOW FOR AWS(200 MW\year)		
year	Profit for year (€)	Investment recovered (€)
1	107.600	-1.392.400
2	107.600	-1.284.800
3	107.600	-1.177.200
4	107.600	-1.069.600
5	107.600	-962.000
6	107.600	-854.400
7	107.600	-746.800
8	107.600	-639.200
9	107.600	-531.600
10	107.600	-424.000
11	107.600	-316.400
12	107.600	-208.800
13	107.600	-101.200
14	107.600	+6.400
15	107.600	+114.000
16	107.600	+221.600
17	107.600	+329.200
18	107.600	+436.800
19	107.600	+544.400
20	107.600	+652.200

The average value of available energy of about 4 kW/m was assumed for the following economic calculations. Using the available data for the buoy of Ponza a production of energy of about 160 kW for each AWS can be estimated. The annual energy production can thus be estimated at about 200 MW per year for each device. These results have been used for subsequent calculation.

4.2 AWS cash flow and total profits

The Italian Energetic Agency rates for the energy payment have been considered, on the basis of an annual energy production of 200 MW. Each AWS costs approximately 1.500.000 € including the installation cost. In the next table the rates payout for a device with a production of 200 MW for year have been given.

In table 3 there is the cash flow of the AWS, the investment will be recovered in 14 years and after 20 years (that is the duration of the contract with the energy distributor) the investor will gain a profit of 650.000 €.

5 Conclusions

The conversion of wave energy is now an important resource particularly for the most exposed coastal areas. This is the reason why that the major projects for installation of wave energy farms have been made in Great Britain, Denmark and Portugal which have the most interesting wave energy resources on their coasts. The Archimedes Wave Swing and other devices, such as the Wave Dragon, are already in a competitive position to balance and in some cases overcome in terms of performance and cost the results obtained in recent years by wind power.

Of course even for the most energetic coastal sites, a careful analysis of the costs and the benefits of the investment must be implemented. In this paper we have considered the installation of small wave farms offshore the most exposed coasts of Campania region in order to calculate the pay-back time of the investment.

At present the most suitable device is certainly the Archimedes Wave Swing both for the amount of obtainable energy and for its characteristics of being submerged at a depth higher than 6 meters from the surface, thus limiting the environmental impact. For this device the calculations of pay-back time demonstrated that a period of 14 years is sufficient for cost recovering, and at the end of the period of the contract with the Italian State Energy Agency there is a net gain of 650.000 Euros. These already promising results may improve substantially with a careful cost optimization, which can be achieved with higher efficiency of energy converters and lower cost of energy plant installation.

The present analysis, although not exhaustive, has demonstrated the real potential of wave energy conversion for the considered locations. The proposed solution minimizes the possible interactions between the energy devices and the tourist or trade ship routes, as well as any other constraint regarding marine protected areas, archaeological sites, etc.

References

[1] EPRI - Guidelines for Preliminary Estimation of power production by offshore wave energy conversion devices

[2] L.A. St. Germain, A Case Study of Wave Power Integration into the Ucluelet Area Electrical Grid, Carleton University, 2003

[3] Alain Clément, Pat McCullen, Antònio Falcao, Antonio Fiorentino, Fred Gardner, Karin Hammarlund, George Lemonis, Tony Lewis, Kim Nielsen, Simona Petroncini, M.-Teresa Pontes, Phillippe Schild, Bengt-Olov Sjostrom, Hans Christian Sørensen, Tom Thorpe, *Wave energy in Europe: current status and Perspectives*, Renewable and Sustainable Energy Reviews 6 (2002), 405–431

[4] Atlante delle onde e dei mari italiani – APAT, Università degli Studi di Roma Tre

[5] Thorpe, An Overview of wave energy technologies: Status, Performance and Costs, Wave power: moving towards commercial Viability, 30 November 1999

[6] T. Thorpe, *"Ocean Wave Energy: Energetech"* presented at 5th BASE International Investment Forum, Bonn, Germany, June 2004.

[7] S. H. Salter, *"Wave Power"* Nature, vol. 249, pp. 720-724, 1974.

[8] J. P. Kofoed, P. Frigaard, E. Friis-Madsen, and H. C. Sorensen, *"Prototype Testing of the Wave Energy Converter Wave Dragon"* presented at World Renewable Energy Congress VIII (WREC 2004), 2004.

[9] Y. Washio, H. Osawa, and T. Ogata, "The open sea tests of the offshore floating type wave power device "Mighty Whale" -characteristics of wave energy absorption and power generation," 2001.

[10] A. F. d. O. Falcao, "Control of an Oscillating-Water-Column Wave Power Plant for Maximum Energy Production" Applied Ocean Research, vol. 24, pp. 73-82, 2002. 107

[11] I. Glendenning, *"Wave Power - A Real Alternative?"* Ocean Management, vol. 4, pp. 207-240, 1978.

[12] S. Raghunathan, *"The Wells Air Turbine for Wave Energy Conversion"* Progress in Aerospace Sciences, vol. 31, pp. 335-386, 1995.

[13] T. J. T. Whittaker, *"Learning from the Islay Wave Power Plant"* presented at IEE Colloquium on Wave Power; An Engineering and Commercial Perspective, 13 March 1997.

[14] Wavegen, "Islay Limpet Project Monitoring Final Report" ETSU V/06/00180/00/Rep, 2002.

[15] T. Setoguchi, S. Santhakumar, M. Takao, T. H. Kim, and K. Kaneko, *"A Modified Wells Turbine for Wave Energy Conversion"* Renewable Energy, vol. 28, pp. 79-91, 2003.

[16] V. S. Raju, M. Ravindran, and U. A. Korde, *"Experiments on the Oscillating Water Column Wave Energy System"* OCEANS, vol. 16, pp. 938-943, 1984. 112

Experimental study of multi-functional artificial reef parameters

M. ten Voorde[1,2], J. S. Antunes do Carmo[1], M. G. Neves[2] & A. Mendonça[2]
[1]*IMAR - Institute of Marine Research; University of Coimbra, Portugal*
[2]*LNEC - Laboratório Nacional de Engenharia Civil, Portugal*

Abstract

Portugal is one of many countries in the world to suffer from coastal erosion. Conventional ways of protecting a coastline appear to entail some disadvantages. An innovative and interesting way of protecting a local coastal zone by means of multi-functional artificial reefs avoids some of them.

A multi-functional artificial reef (MFAR) is a submerged breakwater which protects the local coastline and may also enhance the surfing possibilities and the environmental value of the local area. The structure has several positive side-effects: first, it provides an unimpaired visual amenity; second, it offers tourist and economic benefits by improving the surfing.

A 2D physical study is under way at LNEC to investigate the relations between the breaker type and the submergence and the length of a MFAR. Two different geometries were tested for different incident wave characteristics and different water depths.

This paper presents the main characteristics of the physical experiments in progress in a wave flume to analyze the influence of the submergence and the length of a MFAR on the wave breaking type. The main preliminary conclusions are that the length of the reef doesn't have much influence on the breaker type and that the shorter the slope of the reef the more the wave breaks towards the crest.

Keywords: coastal protection, multi-functional artificial reefs, physical experiments, breaker type.

WIT Transactions on Ecology and the Environment, Vol 126, © 2009 WIT Press
www.witpress.com, ISSN 1743-3541 (on-line)
doi:10.2495/CP090131

1 Introduction

The economic importance of coastal zones has been growing in the past few decades, for a variety of reasons which include an increase in the population and related economic activities established near the coastlines. All this development has led to growing numbers of visitors wanting to enjoy a sandy beach on their holidays and practise outdoor sports such as surfing, sailing, fishing, etc. Unfortunately, many coastal zones are now suffering from erosion, and the aspects and characteristics that make the coasts so attractive could be among the causes of their gradual destruction.

In Portugal there are several examples of coastline erosion and degradation. One of these places is the Leirosa agglomeration, located to the south of Figueira da Foz, midway along the Portuguese West Atlantic coast (Antunes do Carmo et al. [1]; Schreck Reis et al. [2]). Heavy protection structures are neither allowed nor planned in the POOC (legal development plan for the coast), and so a multi-functional artificial reef (MFAR) is under study as an alternative measure to protect this very sensitive coastal dune system.

A MFAR (Figure 1, in which S is the submergence of the reef) is a relatively new approach to protecting a coast; it is a submerged breakwater that has several purposes. In addition to protecting the local coastline and improving the surfing possibilities, a MFAR can enhance the environmental value of the area where it is built. The advantages of a MFAR are that the visual impact is low and that with a proper design the down drift erosion can be minimal.

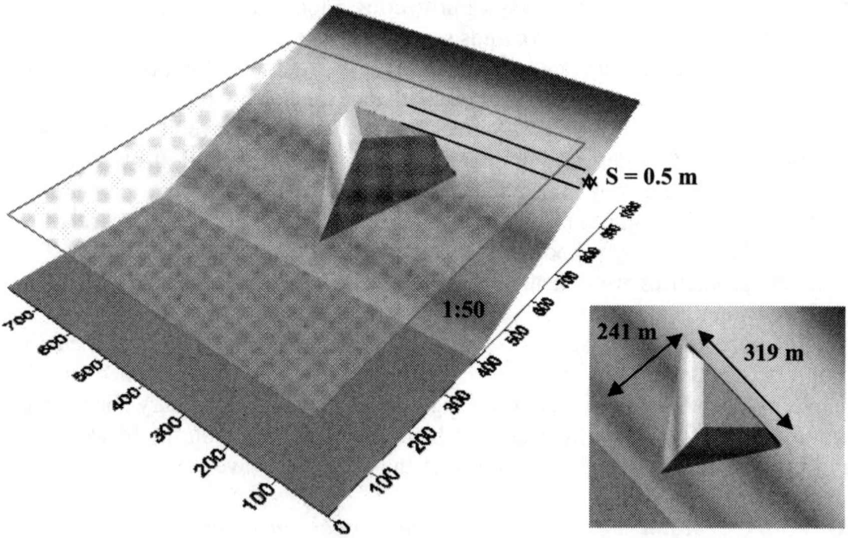

Figure 1: Multi-Functional Artificial Reef (MFAR).

Regarding the functionality of a MFAR, much research has been carried out on surfability, i.e. the possibility to surf a wave (for example Mead and Black [3] and Henriquez [4]). However, no research has yet been done on the influence of the submergence and length of the reef slope on the breaker type, even though this is important for the design of the reef regarding the surfability aspect.

According to Battjes [5] different types of wave breaking are categorized according to the offshore or inshore Iribarren number. The value of the (offshore/inshore) Iribarren number is determined by the bottom slope, the wave height (offshore/inshore) and the offshore wave length. However, Henriquez [4] notes that the submergence of a submerged reef also has an influence on the breaker type.

The length of the slope of the reef could also have an influence on the breaker type. In fact, when the reef has a fixed height and if the same wave conditions are used, deeper submergence automatically means a breaking point nearer the crest of the reef, which means a greater length of the slope experienced by the wave. In order to fill this gap of information, physical experiments have been prepared to investigate the relations between the breaker type and the submergence and the length of one MFAR.

This paper presents the experimental study conducted in a wave flume, to analyse the influences of the submergence and the length of the reef. After this introduction, the theoretical framework is described in which the theory behind the experiments is elucidated. After that, the experimental set-up is described and the main preliminary results are presented and discussed.

2 Theoretical framework

The term 'breaker type' refers to the form of a depth-limited wave at breaking and influences other breaking wave properties. Although there are several classifications of breaker type, it is generally accepted that waves break by spilling, plunging, collapsing, and surging (Galvin [6, 7]).

Battjes [5] used the surf-similarity parameter, ξ_0 (equation (1)), to describe the breaker type on plane slopes and converted transition values determined by Galvin [6, 7] to find:

- surging or collapsing if $\xi_0 > 3.3$;
- plunging if $0.5 < \xi_0 < 3.3$;
- spilling if $\xi_0 < 0.5$.

$$\xi_0 = \frac{s}{\sqrt{\dfrac{H_0}{L_0}}} \tag{1}$$

where ξ_0 is the offshore Iribarren number, s is the bottom slope, H_0 the offshore wave height and L_0 the deep water wave length.

Smith and Kraus [8] performed a laboratory study with waves breaking over bars and over artificial reefs. They constructed the bar of marine plywood and tested six different design seaward angles and four design shoreward angles. They tested several combinations of shoreward and seaward angles of the bar, five regular wave conditions and three random wave conditions. They found for barred profiles the following transition values:

- surging or collapsing if $\xi_0 > 1.2$;
- plunging if $0.4 < \xi_0 < 1.2$;
- spilling if $\xi_0 < 0.4$.

However, their results are for narrow crested artificial reefs, and the influences of the structure submergence and of the length of the slope on the breaker type and on the Iribarren number were not studied. Regarding the breaking behavior on artificial reefs with a smooth slope, as far as the authors know just one study has been conducted in a wave flume by Corbett and Tomlinson [9]. They performed a physical study for Noosa Main Beach in Australia on a permeable reef built of geotextile sand containers to investigate the wave breaking behavior and associated safety issues for an artificial reef.

The analysis of the results of these tests was especially focused on the breaker type, on the breaker wave height and on the breaker location as an indication for the safety of the submerged reef. However, even though several submergences were tested, no analysis of the relation between the breaker type and the corresponding Iribarren number was made. Furthermore, the influences of the length of the slope on the breaker type and the corresponding values of the Iribarren number were not investigated.

A fraction of the local wave length is taken as the minimum length of the reef, namely 1/4 times the local wave length (Ten Voorde et al. [10]).

3 Experimental set-up

In order to get an idea about the influences of the length and of the submergence of the reef on the breaker type and its position, physical model tests were made in one of the LNEC's flume with the following dimensions: 73.0 m long, 3.0 m wide and 2.0 m deep. The model was operating according to Froude's similarity law, with a geometric scale of 1:10.

Two geometries were tested with the structure built of concrete with two different values for the length of the slope of the structure. The seaward slope of the reef was constant and equal to 1:10, a regular value for the side slope of a multi-functional artificial reef. The shoreward slope was 1:3. The slope of the foreshore is 1:50 and its length is 241 m for the lower geometry, and 164 m for the higher geometry, all in prototype scale. These values are at least 1.5 m times the wave length at the wave maker, for each depth tested.

Based on this, on the reef heights and on the reef submergences, the water depth at the wave maker was defined and varied from 7.5 m and 9.1 m, in prototype scale.

Based on the assumption that the minimum length of the reef is 1/4 times the local wave length, in order to get a good surfable wave, the minimum length of the reef in the experiments would be 15 m in prototype or less for reefs starting at a depth of 4 m, for a maximum tested period of 10 s in the experiments. However, as the slope was 1:10, a length of 15 m gives a height of 1.5 m. This values (1.5 m) was assumed to be too low, as larger periods should also break on the reef. A height of 1.9 m was decided to be the minimum value to be tested.

For the higher geometry, the length of the reef was almost twice the value for the lower geometry: 36 m in prototype scale. The width of the reef is 75 m, in prototype scale. This value is at least one time the wave length. This wave length was determined for all periods tested and for the water depth at the start of the highest reef (largest submergence tested). Figure 2 shows the geometry parameters of the reef, and Tables 1 and 2 show the values of these parameters in the model scale. The lower and higher geometry are shown in Figure 3.

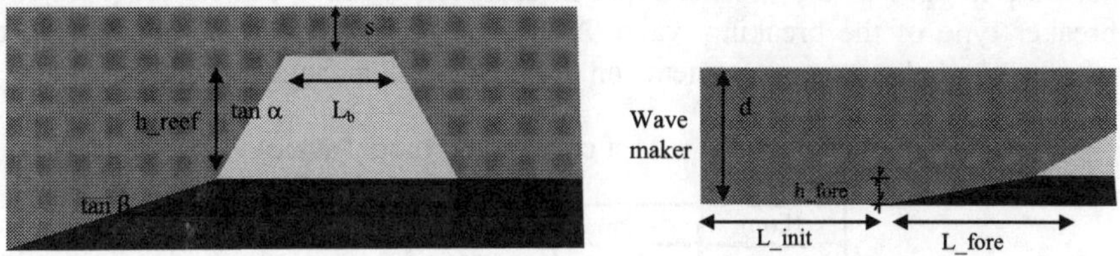

Figure 2: Definition of the geometric reef parameters tested.

Table 1: Values of some structure parameters in the model scale.

Case	Material	h_reef (m)	tan α	tan β	L_s (m)	L_b (m)
1	Concrete	0.19	1:10	1:50	1.9	7.5
2		0.36			3.6	

Table 2: Values of more structure parameters in the model scale.

Case	h_fore (m)	d_{min} (m)	d_{max} (m)	L_init (m)	L_fore (m)
1	0.48	0.75	0.91	18.2	24.2
2	0.33				16.4

For each reef geometry tested, 51 tests were executed, corresponding to 16 combinations of different values of wave height and submergence. Each wave height was tested for three different periods. Each test lasted 320 seconds, which corresponds to, at least, 100 waves.

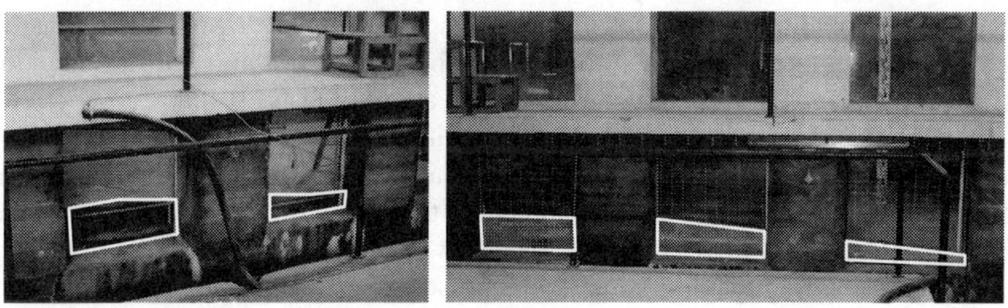

Figure 3: Lower geometry (left) and higher geometry (right).

The choice of the submergence values was based on the following. In order to guarantee breaking conditions, a depth smaller than $0.8*H_b$ was assumed to be necessary (Kaminsky and Kraus [11]), where the wave height at breaking, H_b, was chosen to be at minimum $1.0*H_0$.

For all wave heights, a submergence of $0.8*H_b$ or smaller was tested. However, in order to see if waves with larger submergences than that will also break, some wave heights have been tested with submergences of $(0.8*H_b + 0.04)$ m or $(0.8*H_b + 2*0.04)$ m. Table 3 shows the test conditions.

Since it was the primary purpose of this experimental research to obtain accurate images of the breaker type, in each test a video camera recorded the breaker type of the breaking wave. A grid of 10 by 10 cm was made on the windows to help the measurements on the images taken from the videos.

Table 3: Test conditions (model scale).

H_s(m)	S (m)	h (m)	T_p (s)
0.1	0.08	0.75	2.52 2.84 3.16
	0.12	0.79	
0.15	0.08	0.75	
	0.12	0.79	
	0.16	0.83	
	0.2	0.87	
0.2	0.12	0.79	
	0.16	0.83	
	0.2	0.87	
	0.24	0.91	
0.25	0.16	0.83	
	0.2	0.87	
	0.24	0.91	
0.3	0.2	0.87	
	0.24	0.91	
0.35	0.24	0.91	

Also 8 wave height gauges (WHG) were mounted in order to measure surface elevation at several positions along the wave flume, mostly positioned in the breaker zone.

The data collected would provide mainly wave height records, but it can also be used to obtain mean surface displacement (setup/setdown). Two sensors, gauge 1 and gauge 2, remained throughout the experiment at a seaward location close to the wave paddle, for control and repeatability tests.

Another gauge was located in the beginning of the foreshore slope and another at the beginning of the reef. The remained 4 gauges were mounted in the breaking zone. All tests were repeated with these 4 gauges 0.5 m moved in the direction of the end of the channel. In that way, at eight positions, 0.25 m from each other, in the breaking zone, the time series of the surface elevation were recorded.

4 Preliminary results

From the tests made, some preliminary results have been achieved. In order to analyze the breaking type in a more convenient way, the results were divided into 5 categories, according to the Iribarren number:

- smaller than 0.6,
- from 0.6 to 0.8,
- from 0.8 to 1.0,
- from 1.0 to 1.2 and
- larger than 1.2.

The conclusions are here presented for each category and are illustrated by comparing pictures for two different geometries of the reef with the same submergences that illustrate the results. In these pictures, Hm is the mean wave height, Tm is the mean wave period in the wave gauge at a length of 5.1 m from the wave maker (gauge 1), S is the submergence and Ir is the Iribarren number. For Iribarren values larger than 1.2, no comparison can be made because there is just one result of a breaking wave for the lower geometry. It should be pointed out that the wave heights are not always the same for the lower and higher geometries.

The main conclusion is that, for all different categories, there is almost no difference in the breaker type for the different heights of the geometry. For 0.6 < Ir < 0.8, the higher the submergence and the wave height, the better the tube is for surfing. Figure 4 illustrates the results and shows the breaker type for the lower and higher geometries, for a submergence of 0.16 m.

Based on these results, a first preliminary conclusion can be made, which is that the length of the reef does not have much influence on the breaker type. However, the submergence could have an influence, at least for some values of the Iribarren number.

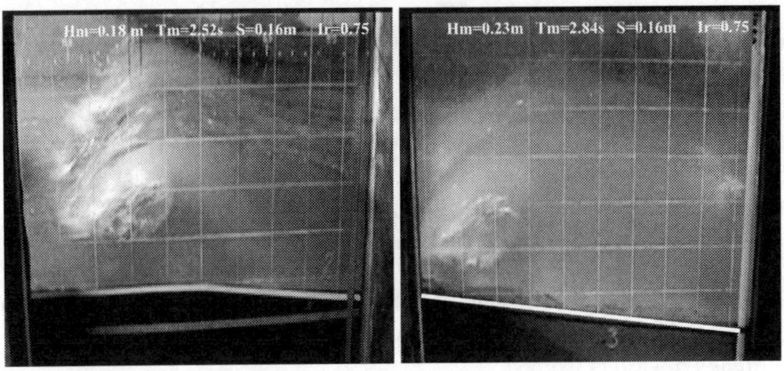

Figure 4: Breaker type for the lower (left) and higher (right) geometries, for 0.6 < Ir < 0.8.

However, on the other hand, with regards to the breaker position, there are differences in the results for different lengths or submergences of the reef. For Ir

< 1.2, the breaker position is more towards the seaward slope of the reef in the case of the higher geometry and for the same submergence. For Ir < 0.8 the difference is between 30 cm and 40 cm and reduces to between 10 cm and 20 cm for 0.8 < Ir < 1.0. This can be observed in Figure 5, for Ir < 0.6, which shows the breaker type for the lower and higher geometries, for a submergence of 0.24 m. For the lower geometry and for the larger wave height tested with large submergences of 0.20 and 0.24 m, most of the waves break before the reef starts. This is due to the fact that the breaker depth for these large wave heights was larger than the depth of the start of the reef. Based on these results, a second preliminary conclusion can be made, which is that the length has an influence on the breaker position and that the shorter the slope of the reef the more the wave breaks towards the crest.

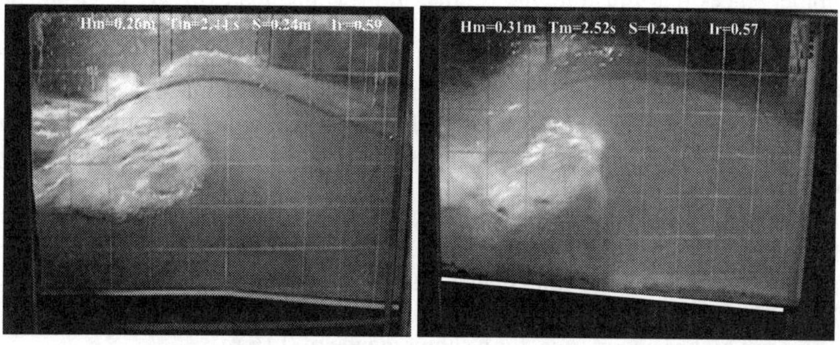

Figure 5: Breaker type for the lower (left) and higher (right) geometries, for Ir < 0.6.

5 Conclusions

Multi-functional artificial reefs (MFARs) are submerged breakwaters that serve several purposes. As well as protecting the local coastline, they enhance surfing possibilities and/or increase the environmental value of the area where they are situated. MFARs have new promising aspects, too: first, they provide an unimpaired visual amenity, and second, they can offer tourist and economic benefits by improving the surfing conditions.

A physical study has been carried out to investigate the influence of the length and the submergence of the reef on the breaker type and on the breaker position. Preliminary conclusions are that: (1) the length of the reef doesn't have much influence on the breaker type, (2) the shorter the slope of the reef the more the wave breaks towards the crest.

More conclusions will be drawn when all results are achieved. As geotextile is an interesting material to construct multi-functional artificial reefs, tests with the same geometries of the reef, using the same wave characteristics, are under way and results will be compared.

Acknowledgements

The authors gratefully acknowledge the Portuguese Foundation for Science and Technology under the project PTDC/ECM/66516/2006 and the financial sponsorship of Ten Voorde's PhD research by the Instituto de Investigação Interdisciplinar, Coimbra, Portugal.

References

[1] Antunes do Carmo, J.S., Schreck Reis, C. and Freitas, H., 2006. Successful rehabilitation of a sand dune system. In: Proc. of the Sixth International Conference on Environmental Problems in Coastal Regions Including Oil and Chemical Spill Studies (Rhodes, Greece), pp. 195–204.

[2] Schreck Reis, C., Antunes do Carmo, J.S. and Freitas, H., 2008. Learning with Nature: A Sand Dune System Case Study (Portugal). *Journal of Coastal Research*, 24(6): 1506-1515.

[3] Mead, S.T. and Black, K.P., 2001. Predicting the breaking intensity of surfing waves. *Journal of Coastal Research*, Special Issue No. 29, pp. 51–65.

[4] Henriquez, M., 2004. Artificial Surf Reefs. Delft, The Netherlands: Delft University of Technology, Master's thesis, 53p. www.waterbouw.tudelft.nl (accessed August 18, 2006).

[5] Battjes, J.A., 1974. Surf similarity. Proc. 14th International Conference on Coastal Engineering, p. 466-479.

[6] Galvin, C.J., 1968. Breaker type classification on three laboratory beaches. *Journal of geophysical research*, 73(12), p. 3651-3659.

[7] Galvin, C.J., 1972. Wave breaking in shallow water. Waves on beaches and resultant sediment transport. Academic Press, New York, N:Y., 413-456.

[8] Smith, E,R. and Kraus, N.C., 1991. Laboratory study of wave breaking over bars and artificial reefs. *Journal of Waterway, Port, and Coastal Engineering*, 117(4): 307-325.

[9] Corbett, B., Tomlinson, R., 2002. Noosa Main Beach Physical Modelling, Research Report No. 17, Griffith Centre for Coastal Management.

[10] Ten Voorde, M., Antunes do Carmo, J.S. and Neves, M.G., 2009. Designing a Preliminary Multi-Functional Artificial Reef to Protect the Portuguese Coast, *Journal of Coastal Research*, 25(1), pp 69-79.

[11] Kaminski, G., and Kraus, N.C., 1993. Evaluation of depth-limited wave breaking criteria. Waves '93, Amer. Soc. Civil Engineers, pp. 180-193.

Beach erosion management in Small Island Developing States: Indian Ocean case studies

V. Duvat
Department of Geography, Institute of Littoral and Environment, La Rochelle University, France

Abstract

In Small Island Developing States (SIDS), the questions of coastal erosion and sea defence structures raise specific issues that this paper will discuss in light of the analysis of the situations in Seychelles and Mauritius. These questions relate back to the role of post-colonial development strategies and have close ties with tourism as beaches have an important economic value. Thus, beach erosion has become a major concern both for the authorities, which lack well-documented analyses as well as the technical and financial capacities for developing appropriate strategies, and for tourism operators. The lack of consistent policies often leads to the systematic use of hard engineering structures without any consideration either for coastal dynamics or socioeconomic factors. Nevertheless, in western Indian Ocean states, beach erosion management has evolved positively for the past 15 years under the influence of internal and external factors. The respective roles of the Regional Environment Programme of the Indian Ocean Commission and of the initiatives of tourism operators in recent progress will be highlighted.

Keywords: beach erosion, coastal protection works, tourism development, Small Island Developing States, Indian Ocean.

1 Introduction

Worldwide beach erosion became apparent during the 1980s due to the works of the International Geographical Union working group on the Dynamics of Coastal Erosion (1972-1976) and the successive Commission on the Coastal Environment (1976-1984). Two hundred participants belonging to 127 countries contributed to a survey that showed that 70 per cent of the world's sandy coasts

WIT Transactions on Ecology and the Environment, Vol 126, © 2009 WIT Press
www.witpress.com, ISSN 1743-3541 (on-line)
doi:10.2495/CP090141

experienced net erosion while 10 per cent sustained progradation and 20 per cent were stable [1]. This survey and the associated works have provided a valuable analysis on the situation of tropical beaches and demonstrated that they were severely affected by shoreline retreat as many coastal areas are located in tropical cyclone tracks [2]. In parallel, the development of coastal tourism and the issues related to climate change and coral reef degradation, which were identified as the main causes of beach erosion, encouraged further scientific studies and expert analyses of beach dynamics and evolution [3, 4].

In Small Island Developing States (SIDS), conducting research into beach erosion has often been made difficult by limited national capacities; regional development programmes have therefore provided useful support. In the 1980s, coastal erosion was identified as a main concern for island states of the western Indian Ocean, however, scientific research, expert analyses and training were only undertaken in the middle of the 1990s when the Regional Environment Programme of the Indian Ocean Commission was launched with the support of the European Union (1995-2003). In such context, on the basis of case studies of the Republic of Seychelles and Mauritius (including the autonomous island of Rodrigues), this paper will highlight and discuss beach erosion management practices. It is based upon the findings of the scientific research carried out by the author from 1996 to 2008 (The author completed her PhD thesis on beach geomorphology and management in Seychelles in 1998. She has carried out research on coastal geomorphology and management in western Indian Ocean islands from 1996. She conducted expert analyses and participated in the launching of regional training programmes on beach dynamics and management, in particular in Seychelles and Mauritius, at the Indian Ocean Commission's demand (see references).). After a brief presentation of the coastal physical settings, we will present the importance and causes of beach erosion and underline the role of coastal development practices in sediment cell destabilization. Then, through the analysis of beach erosion control techniques and policies, we will demonstrate what features the Seychelles and Mauritius have in common as SIDS. Finally, factors of progress and recent trends in beach erosion mitigation will be brought to light, and drivers of erosion control will be discussed.

2 Beach erosion in Seychelles and Mauritius

2.1 The situation of Seychelles

2.1.1 The state of beaches

The granitic islands of Seychelles consist of 41 mountainous islands and islets that rise to a maximum height of 915 m at Morne Seychellois on Mahe. Since colonization times, the steep, rocky and unstable weathered slopes of the four main islands (Mahe, Praslin, Silhouette, la Digue) have encouraged human communities to settle in the narrow and scattered coastal plains which cover 15% (Mahe) to 30% (Praslin and la Digue) of the land area. Most of these plains are less than 700 metres wide and no more than one kilometre long. They are bordered with fringing coral reefs that have played a major role in the formation of sandy beaches as the terrigeneous material brought to the shore by rivers never exceeds 30% of the sediment budget of beaches. The coastline is characterized by an alternation of rocky shores, comprising either gentle continental slopes or masses of granite boulders, and low-lying areas consisting of sandy beaches and mangrove swamps. Sandy beaches represent 34 to 53% of the total length of the shoreline, as shown in table 1, and they vary in length from 60 to 2,500 meters. The main driving forces of the coastal system are swells that are generated by alternating south-eastern trade winds and north-eastern monsoon waves. This climate regime generates the reversal of the longshore drifting from one season to the next. Rarely, tropical cyclones that form in the southeast of Seychelles generate heavy swells that affect the coasts. The three main populated islands (Mahe, Praslin, la Digue) have 62 beaches, half of which are seriously eroded.

Table 1: Characteristics of coasts and beaches in granitic Seychelles.

Island and land area (km^2)	Main characteristics of coasts and beaches			Importance of beach erosion	
	Coastline (km)	Sandy coastline (km and %)	Number of beaches	Number of beaches which are seriously eroded	Number of eroded beaches with hard protection structures
Mahe (154)	105	36 (34%)	23	12	7
Praslin (37)	43	21 (48%)	24	16	11
La Digue (10)	15	8 (53%)	15	7	5
Total	163	65 (40%)	62	33	23

NB The number of eroded beaches is preferred to a figure for the beach length that is affected by erosion because we consider that a beach that is partly eroded is globally threatened, either for natural or human reasons.

2.1.2 Main causes of beach erosion

In granitic Seychelles, 60% of the beach erosion is due to a combination of natural factors and human interferences with physical processes [5].

Firstly, it must be emphasized that the sand budget of beaches has been seriously affected by the diminution of the sediment supply which occurred after the coral reef reached the sea level as it stabilized around 3,000 years BP [5, 6]. More recently, certain changes in monsoon patterns and an increase in storm frequency and intensity have had serious impacts both on sand drifting and on beach budgets. On the north-western coast of Praslin, changes in coastal currents affected the direction and the volume of sediment transport, which resulted in rapid beach erosion that destroyed a coastal road [7]. The analysis of meteorological data indicates a recent increase in the number of storms. The 14 storms that were recorded in the period between 1963 and 1976 represent 50% of the total number of storms that have affected the granitic Seychelles region (5 to 10°S and 55 to 60°E) between 1852 and 1990. El Niño events of 1982-1983, 1987-1988 and 1997-1998 have had serious effects in Seychelles as they affected the climate regime and generated unusual stormy events. The storm that occurred in 1997 had serious impacts on most beaches, as shown in previous works [8]. As tidal data are recent and incomplete, the role of sea level rise in beach evolution is very difficult to estimate [5].

Various human factors have produced or aggravated beach erosion. Until the beginning of the 1990s, sand extraction was freely conducted on beaches and river mouths by residents and building companies. Since then, the reinforcement of sand mining regulations and the production of granitic sand as an alternative material for construction have reduced coastal extraction. However, cleaning has an adverse effect on the sediment budget of beaches as systematic raking removes all coral debris that stabilize beaches and provide them with sand. As a consequence of the narrowness of plains, many residents have reclaimed the inner part of the reef flat in order to extend their properties. Since 1969, coastal reclamation has accelerated as the independence of Seychelles (1976) has stimulated development. On the eastern coast of Mahe, the main fishing harbour of Victoria was extended (1973) and the international airport of Pointe Larue was inaugurated (1971). Since then, the need for flat land for industrial, residential and infrastructural needs has led to the reclamation of some 700 hectares. Coastal reclamation has destabilized sediment cells. The disappearance of mobile sand banks, the disturbance of sand drifting and changes in current patterns have accelerated beach erosion. The induced diminution of sand supply has encouraged residents to build groynes in order to stabilize beaches in front of their properties. As a consequence, destabilization of beaches has extended downdrift. The aggravation of beach erosion has also led to the construction of seawalls. The destruction of coral reefs in reclaimed areas has exacerbated the erosional impacts of storms on back-reef beaches. The narrowness of coastal plains has also encouraged the construction of roads that are very close to the shoreline. Due to their exposure to storm waves, these roads were protected by hard engineering structures such as seawalls and ripraps. Other common causes of beach erosion are channel dredging in reef flats, reclamation of mangroves, destruction of coastal vegetation and construction of harbours and jetties.

2.2 The situation of Mauritius

2.2.1 The state of beaches

The Republic of Mauritius consists of two main volcanic islands, Mauritius (20°S 58°E, 1865 km^2) and Rodrigues (19°S 43°E, 110 km^2), and of about thirty volcanic and coral islets located at a short distance from their coasts. As a consequence of subsidence and erosional processes, the relief is dominated by coastal plains and plateaux. The summits reach 828 m at Piton de la Petite Rivière Noire in Mauritius and 396 m at Mont Limon in Rodrigues.

As shown in table 2, the coastline of Mauritius is composed of sandy beaches bordered with fringing coral reefs on 70% of its length, alternating with mangrove swamps, rocky platforms and cliffs. Most beaches are rectilinear, 20 to 60 m wide and a few hundred to a few thousand kilometres long. About 200 beaches can be distinguished, more than half associated with hotel and tourism residences. Because of late volcanic eruptions that occurred 1.3 to 1.5 million years ago, the coasts of Rodrigues are mainly made of rocky shores. The sixteen sandy beaches only represent 9% of the total length of the shoreline and are located on the eastern and south-eastern coasts. Most are pocket beaches or small crescentic ones. They range from 2 to 25 m wide, some being very narrow as a consequence of the receding shoreline. In both islands, the coastal system is driven by oblique swells produced either by the south-eastern trade winds or by southern storms coming from the southwest, both of which generate a northward longshore current. As they are situated in the cyclone belt, Mauritius and Rodrigues can be affected by hurricanes from the east from November to April.

Table 2: Characteristics of coasts and beaches in Mauritius and Rodrigues.

Island and land area (km^2)	Main characteristics of coasts and beaches			Importance of beach erosion	
	Coastline (km)	Sandy coastline (km and %)	Number of beaches	Number of beaches which are seriously eroded	Number of eroded beaches with hard protection structures
Mauritius (1865)	322	210 (70%)	200*	95*	65*
Rodrigues (110)	67	6 (9%)	16	8	3
Total	389	216 (55%)	216*	103*	68*

NB Asterisked figures are approximate estimations reflecting the complexity of the situation in Mauritius due to the high number of beach sections, which is a consequence of the high level of fragmentation of sediment cells.

2.2.2 Main causes of beach erosion

In Mauritius and Rodrigues, the rapid retreat of isolated beaches indicates a marked decrease in sand supply at the scale of geological ages. In some areas, beachrock slabs have replaced sandy beaches and dunes are now bordered with rocky platforms. Storm waves remove beach material that is entirely lost to sediment cells when it is evacuated by rip currents through reef pass. Beach resilience is generally limited in low energy environments where regular waves are not strong enough to take back to the coast the sand that storm waves have deposited on the top of reef flats [9].

In Rodrigues, human-induced beach erosion is very limited due to the low level of development of the coastal zone. Villages are located inland and only three hotels were built on the seashore. The most impacted sector is that of Port Mathurin, the main town, on the northern coast, where land reclamation and coastal dredging were carried out to extend the harbour and to create flat land for future development.

In Mauritius, in most cases, beach erosion is due to a combination of natural and anthropogenic factors. Human-induced disturbances have accelerated sand loss and coastline recession. The degradation of coastal dunes by sand extraction, flattening for construction requirements, or building on the fore dune, has played a major role in beach destabilization on the western, south-western and eastern coasts. The illegal building of around 200 jetties and groynes has disrupted sand drifting and exacerbated erosion on downdrift beaches, in particular on the western (Flic en Flac), south-western (Morne Brabant), northern (Grand Baie) and south-eastern (Mahebourg) coasts. Many changes have occurred since tourism developed in the 1970s. The building of more than 100 hotels and of numerous coastal tourism residences and villas has had many impacts as most of the regulations concerning sand mining, building and coastal works, which although agreed over the past two decades, have still not been implemented. Owners of coastal plots and tenants of leases often let land either to private individuals or to tourism companies for speculation. There is much confusion on the status of coastal residents and tourism operators, which makes it difficult to ensure the respect of the law. Furthermore, the rapid development of high standard tourism has encouraged the development of many types of coastal works that lack consideration for their local impacts on sensitive ecosystems or for the potential disturbances caused on adjacent sectors. Boat channels have been dredged in reef flats and passes were widened in reef fronts. In touristic areas, coral sands were mined from public beaches, dunes and reef flats without any official authorization in various circumstances, such as the building of a hotel, for beach reprofiling works or post-storm beach nourishment. Therefore, it can be stated that the main cause of beach erosion is the destabilization of sediment cells, which is caused by the multiplication of uncontrolled coastal works.

3 The predominance of hard engineering structures in beach erosion management practices

In Seychelles and Mauritius, beach erosion control is characterized by a resistance strategy that aims at protecting exposed constructions and at fixing the coastline. Over the decades, various types of hard structures have been built, among which seawalls, groynes, ripraps and gabion walls are the most common.

3.1 Beach erosion control in granitic Seychelles

By the beginning of the 19th century, seawalls had already been built along the north-eastern coast of Mahe to protect the main town of Victoria, its headquarters and the harbour area. Over the years, coastal residents developed simple techniques for protecting their houses and properties. The first method, still in usage today, consisted of building vertical seawalls. The second technique was the reclamation of a strip of reef flat in front of one's property. This defensive strategy was highly beneficial as the creation of a buffer zone provided efficient protection for the land, while also extending the property on the seaward side. This traditional practice has declined since the passing of the reclamation law in 1961. Over the past two decades, coastal residents facing severe coastal erosion have tried out a third technique. This involves the building and installation of parallel rows of gabions that limit land loss by fixing the coastline. The other two advantages of this solution are its low impacts on both the quality of the landscape, due to the fact that they are buried under the beach, and also on sediment cells dynamics.

The protection of transport facilities such as roads, harbours and airports, which comes within the jurisdiction of the Division of Environment, has been achieved by the construction of concrete seawalls and cemented groynes until the

1980s. In most sites, groynes failed to induce sand accretion because of the high level of fragmentation of sediment cells or because they were installed in the quiet waters of sheltered bays. Some of the seawalls collapsed under the pressure of storm waves, which led to significant damage to coastal roads. Until the 1990s, seawalls were repaired. Then, the opening of quarries in the granites of Mahe and Praslin sustained the development of a new technical solution. The extraction of massive blocks of granite permitted the construction of solid ripraps that were installed along coastal roads, in front of threatened facilities and around reclaimed plots. With the exception of la Digue Island, where there is no quarry for granite extraction, ripraps have replaced seawalls. The efficiency of these boulder ramparts has encouraged the Division of Environment to systematically use such protections. On the western coast of Praslin, which is affected by rapid beach erosion, it was decided to extend ripraps to non-eroded beaches as a preventative measure. As a result, the amount of artificial shoreline reaches 30 to 40% depending on the island. As most artificial shores were initially sandy coasts, hard defences have reduced the attractiveness of islands. Broken seawalls and groynes and useless structures which were built hastily after storm events are still lying on beaches, spoiling the scenery and disturbing human activities.

The setback distance for construction on the shore is only 15 metres, so most hotels, guesthouses and restaurants were built very close to the sea. Their exposure to storm waves has led to the systematic use of seawalls that have accelerated beach erosion and caused the disappearance of some beaches. As a result, many hotels now only have very degraded beaches to offer their guests.

Due to the weakness of the law and the lack of control measures, both hotel operators and coastal residents have made endless mistakes and damaged the environment by trying out numerous hard protection measures. The lack of a consistent management of sediment cells has led to the gradual destabilization of rectilinear beaches that are hydrodynamically driven by the longshore current. In contrast, the best preserved beaches are isolated pocket or cove beaches.

3.2 Beach erosion control in Mauritius and Rodrigues

In Mauritius, beach erosion control is under the jurisdiction of the Ministry of Environment and Quality of Life (MEQL), which is represented by the Division of Environment (DoE) in Rodrigues.

The intense storms of the 1990s, and in particular tropical cyclone Hollanda which occurred in 1994, caused severe damages to cemeteries (Saint Félix, Cap Malheureux), coastal roads (Grand Baie, Saint Félix) and public beaches (Pomponnette, Flic en Flac, Grand Baie, Morne Brabant) in all areas. In order to limit the destructive effects of such events in future, the Government of Mauritius carried out protection works on the most affected sites and adopted in 1999 what can be considered as the first beach erosion control plan. Massive gabion walls up to 7 meters high in the north were installed at the foot of coastal dunes. Traditional protection structures such as cemented seawalls were abandoned and systematically replaced by gabions. Like in Seychelles, the opening of quarries – here, in volcanic rocks – offered new possibilities for coastal protection. In some places (Grand Baie, Saint Félix), groynes had also been trialled, which failed to trap sand. Thus, in the 1990s, gabions spread to all coastal regions. They were installed in seaside resorts, urban areas and natural sites without any consideration either for the importance and value of the constructions that they defended, or for their environmental impacts. Continuous gabion walls were erected along public beaches even where buildings or roads were not exposed. The policy which became apparent during this decade consists in a strategy of systematic resistance to wave attack by fixing the coastline. An official list of the sites to be defended with gabions was drawn up, detailing 15 kilometres of beach that required urgent treatment; of which, 5 kilometres had been completed by 2000. Within a few years the government plan resulted in the construction of gabions along 50% of the total length of public beaches. On the western coast (Flic en Flac), as gabions were destroyed by waves, the chicken wire and volcanic rocks they contained spread onto the beach and caused injuries to visitors. In other cases (Pomponnette), massive gabion walls proved to be useless because of the resilience of the beach. Such errors were due to lack of national expertise. The University of Mauritius (UoM) and the Mauritius Oceanography Institute (MOI) do not have skills in the field of coastal dynamics. In this context, some erroneous views prevail, such as the belief that maintaining the gabions is favourable to sand deposition and therefore to beach progradation.

Where hotels were built too close to the sea, hard structures were constructed to protect them. This situation refers mainly to the small to medium size hotels of the 1970s and 1980s that offer rooms overlooking the sea (Grand Baie, Belle Mare, Flic en Flac). Wave reflection on the walls accelerated beach erosion and caused damage to hotels.

In Rodrigues, massive ripraps were built at Port Mathurin to protect the reclaimed area from wave attack. As few houses and hotels were built on dunes, hard structures are scarce. Seaside hotels are protected by seawalls and groynes. The expected development of tourism over the coming decades might change this situation as the Division of Environment often takes an interventionist approach. After erosion events in the late 1990s, truck tyres were installed on the southern coast of Ile aux Sables, a nature reserve.

3.3 Beach erosion management in Small Island Developing States (SIDS)

The situations of Seychelles and Mauritius share many common features with other SIDS. Firstly, the ineffectiveness of environmental policies and coastal planning is partly due to a complex heritage that encompasses a mixture of French and English laws inherited from the successive colonial periods. Since independence (1968 in Mauritius, 1976 in Seychelles), new regulations have been adopted ad hoc which do not form a consistent framework.

The background of economic development (from the 1970s) in relation to the adoption of coastal policies, laws and planning (in the 1990s), explains the low level of implementation of regulations and the increase in the number of uncoordinated technical interventions that aimed at controlling beach erosion.

Additionally, both countries lack human, financial and technical capacities for improving coastal management. Coastal managers lack proper diagnostic data to make the right choices for mitigating beach erosion. The same techniques are employed whatever the situation may be. Presently, ripraps are systematically installed on the eroding coasts of Seychelles and gabions on those of Mauritius. Due to their remoteness from developed countries, these countries do not benefit from the advancements made in western states over the past decades.

Moreover, hard structures are considered as progress in coastal management because they provide security to people and protection of possessions. A public survey carried out by the author in 2005 indicated that visitors were satisfied with the installation of gabions for safety reasons. Landscapes have no specific value in Seychellois and Mauritian contexts.

4 New trends in beach erosion management

4.1 General factors that have led to progress

Over the past ten years, national policies for beach erosion control have evolved under the influence of various internal and external factors. One of the main reasons for progress was the establishment of the Regional Environment Programme of the Indian Ocean Commission (PRE-COI 1995-2003) supported by the European Union. This programme gave priority to coastal issues and in particular to the mitigation of beach erosion and coral reef degradation. Thereby, scientific studies were encouraged and supported, and national capacities and policies were assessed. The results of expert reports were presented at the congress of Mahe (2000), where a regional plan of action was developed. In 2002-2003, a practical guide dealing with beach erosion management was edited [10] and a training programme was organized to support capacity building. It took place at the University of Mauritius where it gathered coastal managers from diverse administrations, members of NGOs and tourism operators. Field excursions encouraged case study analysis and experience sharing. This training had practical consequences: the development of bilateral collaboration between the French territory of Reunion, which possesses expertise in the field of coastal geomorphology and beach management, and the neighbouring SIDS; the creation of beach monitoring programmes, in particular in Seychelles (2003), Rodrigues (2003) and Mauritius (2005); and the carrying out of complementary studies aiming at supporting coastal development and management, such as the beach vulnerability assessment of Rodrigues in 2003 [11].

The PRE-COI supported the development of political consciousness and thus encouraged the progress of national policies. In Seychelles, the setback distance was increased from 15 to 25 meters wherever this measure can be applied without obstructing development projects. In Mauritius, in 2003, a new procedure for beach erosion management was discussed with representatives of the MEQL and set up with the purpose of putting an end to hurried technical interventions [12].

Progress was also supported by the building of a new generation of high standard hotels 80 to 100 meters from the high water mark, with tropical gardens separating the private space of rooms from coastal leisure facilities. In a country where the setback distance is only 15 meters, this new hotel concept plays a major role in the reduction of the adverse effects of buildings on beach dynamics and evolution. It is presently being developed in Seychelles by tourism companies from Mauritius.

4.2 The role of the private sector in the diversification of erosion control techniques

In Mauritius, much progress in beach erosion management is due to hotel managers that have experimented with new techniques for controlling beach erosion without reducing the high economic value of beaches. Thus, hotel beaches have acted as field laboratories for testing alternative soft engineering strategies. The pragmatism of tourism operators who look for solutions that are adapted to local situations is favourable to innovation and therefore to the diversification of interventions. New kinds of hard structures were tested, such as underwater (Legends hotel) and submerged breakwaters (Saint-Géran and Sands hotels), buried gabions (Saint-Géran hotel) and sand bags (Palmar hotel). Soft techniques were introduced, which have now become standard, such as mechanical and manual beach reprofiling and the stabilisation of sand dunes with indigenous plants and beach restoration. Artificial beach nourishment became more common after erosion peaks due to storm waves dispersed beach sand on reef flats. The technique consists of pumping sediments that were taken away by the waves onto reef flats. Generally, this operation is completed by the provision of extra sand collected either in coastal dunes or on reef flats. In 2002, after tropical cyclone Dina, about 25% of hotel managers admitted that they have had recourse to beach nourishment to accelerate beach recovery. This figure is definitely less than the true amount. Extraction zones are unknown because sand mining is prohibited, so an effort should be made to regulate this practice in order to limit its potential adverse impacts on coastal dynamics. Its regulation would make it possible to encourage sustainable beach nourishment with sand taken only from authorized sites, thus promoting environment-friendly solutions [12, 13]. Prospecting campaigns should be carried out on the extended submarine platforms that surround Mauritius, Rodrigues and Seychelles for evaluating underwater sand stocks [12].

In Seychelles, adaptive measures were initiated by resort island managers. In Bird (1 km^2) and Desroches (4 km^2) coral islands, the recession of the coastline destroyed hotel bungalows in 1988 and 1991 respectively. In both cases, all buildings were relocated 100 meters inland. The creation of a natural buffer zone was critical as no damage has occurred since then.

4.3 The role of politics in coastal management progress

In Mauritius, the context of the free market and high tourism performance has been favourable to the improvement of beach erosion control practices as the pragmatism and initiative of hotel managers has led to experimentation in new protection and prevention techniques. Another advantage of the tourism system is that most investors are locals who have a long-term view of tourism investment. This situation is favourable to the pursuit of sustainable solutions to problems. Furthermore, local investors have an empirical knowledge of coastal dynamics and therefore a good idea of what or what not to do in coastal development. As 70% of tourism operators belong to the association of hotels and restaurants of Mauritius (ARHIM), they are also aware of the many experiments made at the national scale, and of their findings. This context is favourable to performance and progress.

The situation in the Seychelles is different. Politically, after independence (1976) and a coup d'état (1977), Seychelles became a socialist country opposed to liberalism, which discouraged local initiatives and limited capital formation. At the beginning of the 1980s, internal political problems discouraged foreign investment. As a result, the tourism sector stagnated and financial and monetary problems appeared. Despite the settlement of political issues and the adoption of multipartism (1991), tourism investments are still limited and arise mainly from foreign companies [14]. This global context is not favourable for progress.

5 Conclusion

In Small Island Developing States such as Seychelles and Mauritius, beach erosion raises important economic and environmental issues. Both natural and anthropogenic factors have accelerated coastline recession over the past decades and scientific predictions are globally pessimistic regarding the impacts of climate change on the evolution of coral reefs and back-reef beaches.

In such context, there is an urgent need for coastal managers to work with tourism operators to develop capacity building and to promote consistent policies in the field of beach erosion control. As shown in this paper, regional environment programmes supported by international organizations and NGOs can play a major role in the creation of well-founded regional and national strategies for beach erosion management and their implementation. Currently, the main priorities are to support scientific studies and international expertise in order to put a stop to the systematic installation of hard engineering structures on eroding beaches, and to support the development of techniques that work with coastal processes. This is necessary primarily to improve the efficiency of technical interventions, both in financial and technical terms, and also for integrating environmental and socioeconomic issues in beach erosion control. An integrated beach management strategy is needed to address all threats to coastal environment degradation because it can seriously affect the attractiveness of beaches and therefore the success of insular tourism destinations. As an example, some resort islands of the Maldives and sandy beaches of the island of Saint-Martin (lesser Antilles) have lost their tourism function because of inadequate coastal management.

Acknowledgements

I am grateful to the Indian Ocean Commission, MEQL (Mauritius) and Divisions of Environment in Seychelles and Rodrigues for financial and technical support.

References

[1] Bird, E.C.F., *Coastline changes,* Wiley & Sons: Chichester, New York, Brisbane, Toronto, Singapore, 1985.

[2] Bird, E.C.F. & Schwartz M.L, *The world's coastline,* Van Nostrand Reinhold Company: New York, 1985.

[3] Philips, M.R. & Jones A.L., Erosion and tourism infrastructure in the coastal zone: Problems, consequences and management. *Tourism Management,* **27**, pp. 517-524, 2006.

[4] Wong, P.P., Where have all the beaches gone? *Singapore Journal of Tropical Geography,* **24(1)**, 2003.

[5] Cazes-Duvat, V., *Les littoraux des Iles Seychelles,* L'Harmattan: Paris, 1999.

[6] Camoin, G.F., Montaggioni, L.F., Braithwaite, C.J.R., Late glacial to post-glacial sea levels in western Indian Ocean. *Marine Geology,* **206**, pp. 119-146, 2004.

[7] Shah, N.J., Coastal zone management in Seychelles. *Proc. of the National Workshop on Integrated coastal zone management in Seychelles,* eds. C.G. Lundin & O. Linden, pp. 14-125, 1995.

[8] Cazes-Duvat, V., *Atlas de l'environnement côtier des îles granitiques de l'archipel des Seychelles*, Louis Jean: Gap, 2001.

[9] Cazes-Duvat, V., Les impacts du cyclone Kalunde sur les plages de l'Île Rodrigues (océan Indien occidental). *Zeitschrift Für Geomorphologie*, **49 (3)**, pp. 293-308, 2005.

[10] Cazes-Duvat, V. (coord.), Delmas-Ferré, M., Troadec, R., *Manuel de suivi et de traitement de l'érosion côtière. Pays de la Commission de l'océan Indien.* PRE-COI/7e FED-UE, Le Printemps: Quatre Bornes, 2002.

[11] Duvat, V., *Les littoraux coralliens des petites îles de l'océan Indien (Mascareignes, Seychelles, Maldives). Volume 2 – Aménagement et gestion,* Institut Océanographique: Paris, 2007.

[12] Cazes-Duvat, V., Paskoff, R., *Les littoraux des Mascareignes entre nature et aménagement,* L'Harmattan: Paris, 2004.

[13] Bird, E.C.F., *Beach management,* Wiley & Sons: Chichester, New York, Brisbane, Toronto, Singapore, 1996.

[14] Gay, J.-Ch., Tourisme, politique et environnement aux Seychelles ». *Revue Tiers-Monde,* t. XLV, **178**, pp. 319-339, 2004.

Section 5
Hydrodynamic forces and sediment transport

A numerical study on near-bed flow mechanisms around a marine pipeline close to a flat seabed including estimation of bedload sediment transport

M. C. Ong[1], T. Utnes[2], L. E. Holmedal[1], D. Myrhaug[1] & B. Pettersen[1]
[1]*Department of Marine Technology, Norwegian University of Science and Technology, Norway*
[2]*SINTEF IKT Applied Mathematics, Norway*

Abstract

Near-bed flow mechanisms of high Reynolds number flows around a marine pipeline close to a flat seabed have been studied using a two-dimensional standard high Reynolds number k-ε model. The effects of gap to diameter ratio and seabed roughness for a given boundary layer thickness of the inlet flow upstream of the cylinder have been investigated. The vortex shedding mechanisms have been investigated. Mean pressure, mean friction velocity and the resulting mean bedload sediment transport along the bed have been predicted. Overall it appears that for engineering design purposes the present numerical model is suitable for predicting high Reynolds number flows, which are present near the seabed in the real ocean.

Keywords: numerical model, pipeline, flat seabed, high Reynolds number.

1 Introduction

Marine pipelines are widely used for transporting oil and gas from offshore fields. They are often subject to high Reynolds numbers flow with typical values of $O(10^4) - O(10^7)$, covering subcritical ($300 < Re < 3\times10^5$) to trancritical ($Re > 4\times10^6$) flow regimes. Here $Re = U_\infty D/\nu$ where D is the cylinder diameter; U_∞ is the free stream velocity; and ν is the kinematic viscosity. The hydrodynamic

WIT Transactions on Ecology and the Environment, Vol 126, © 2009 WIT Press
www.witpress.com, ISSN 1743-3541 (on-line)
doi:10.2495/CP090151

characteristics of steady flow around a horizontal smooth circular cylinder near a fixed horizontal boundary represent an idealized situation of a pipeline near the seabed. The proximity of the pipeline to the seabed affects the flows around the pipeline and along the seabed.

Several experimental studies have been carried out to investigate flows at high Reynolds numbers ranging from $O(10^4)$ to $O(10^5)$ in the subcritical flow regime (see, e.g., Bearman and Zdravkovich [1], Lei *et al.* [2] and Wang and Tan [3]). Bearman and Zdravkovich [1] investigated the influence of G/D on the vortex shedding and its spectral behaviour with an upstream flow of $\delta/D = 0.8$ at Re ranging from 2.5×10^4 to 4.8×10^4. Here G is the distance between the bottom of the cylinder and the bed, and δ is the boundary layer thickness of the inlet flow upstream of the cylinder (see fig. 1 for definitions). They measured the distributions of mean pressure around the cylinder and along the bed at Re = 4.8×10^4. They also showed that the vortex shedding motion behind a circular cylinder close to a flat bed is suppressed at $G/D < 0.3$. Here the G/D corresponding to the onset of vortex shedding is defined as the critical ratio, G/D_c. Lei *et al.* [2] studied the flow around a smooth circular cylinder immersed in different boundary layer thicknesses ($\delta/D = 0.14$ - 2.89) at Re ranging from 1.31×10^4 to 1.45×10^4. Their experimental results showed that both drag and lift coefficients strongly depend on G/D, and are affected by δ/D. They found that the variation of the root-mean-square fluctuating lift coefficient (C_{Lrms}) can be used to determine the suppression and onset of the vortex shedding. Their observations also showed that the vortex shedding is suppressed at G/D of 0.2-0.3, depending on different δ/D. Wang and Tan [3] studied the near-wake flow characteristics of a circular cylinder close to a flat bed for Re = 1.2×10^4 and $\delta/D = 0.4$. Their results showed that instantaneous flow fields depend strongly on G/D, and that the flow is characterized by a periodic vortex shedding for $G/D \geq 0.3$.

Only a few numerical studies have been performed for such high Reynolds number flows ($Re > 10^4$) around a circular cylinder near a plane wall. Brørs [4] and Zhao *et al.* [5] applied a standard high Reynolds number k-ε model at Re = 1.5×10^4 and a k-ω model at Re = 2×10^4, respectively. Their results yielded a good qualitative agreement with the published experimental data. However, detailed comparisons with experimental results for $G/D < 0.4$ are not made. Recently Ong *et al.* [6] applied the standard high Reynolds number k-ε model at Re = 1×10^4 - 4.8×10^4 with $\delta/D = 0.14$ - 2. Comparisons of numerical results with the published experimental data were provided for the cases of $G/D < 0.4$. They found that under-predictions of the essential hydrodynamic quantities of the cylinder (such as C_D, C_L, St, C_{Lrms} and C_p) were observed in the subcritical flow regime due to the well-known limited capacity of the k-ε model (and similar two-equation turbulence closures) to capture the vortex shedding correctly. Here C_D is the time-averaged drag coefficient, C_L is the time-averaged lift coefficient, $St = fD/U_\infty$ is the Strouhal number (here f is the vortex-shedding frequency), and C_p is the mean pressure coefficient around the cylinder. C_D and C_L are calculated based on the definitions $F_D = 0.5\rho DC_DU_\infty^2$ and $F_L = 0.5\rho DC_L U_\infty^2$, where F_D and F_L are the time-averaged integrated horizontal and vertical forces per unit length, respectively, acting on the cylinder; and ρ is the fluid density. There is also a limitation of using two-dimensional (2D) models for three-dimensional (3D) flow, as effects from the spanwise secondary flow are not considered in the 2D simulation (see Mittal and Balachandar [7]). However, the mean pressure and the friction velocity along the bed were predicted reasonably well as compared with the published experimental and numerical results in the subcritical flow regime.

Ong *et al.* [8] and Catalano *et al.* [9] presented numerical results on flow around an isolated smooth circular cylinder subject to a steady current at Re ranging from 0.5×10^6 to 4×10^6 by using the standard high Reynolds number k-ε model. Overall, their results are in satisfactory agreement with published experimental data. To our knowledge, neither numerical nor experimental studies are available in the open literature for flows around a circular cylinder close to a flat seabed beyond the supercritical flow regime ($Re > 1\times10^6$).

In the present study, the flows at $Re = 3.6\times10^6$ and $\delta/D = 0.48$ with two different seabed roughnesses ($z_w = 1\times10^{-6}$m and 2×10^{-5}m) are investigated numerically by using 2D Unsteady Reynolds-Averaged Navier-Stokes (URANS) equations with a standard high Reynolds number k-ε model. Here $z_w = d_{50}/12$ where d_{50} is the median grain size diameter. Effects of gap to diameter ratio and seabed roughness are investigated. Mechanisms of vortex shedding are investigated. Near-bed hydrodynamic quantities and the resulting bedload sediment transport are also predicted.

2 Mathematical formulation

2.1 Flow model and numerical solution procedure

The 2D URANS equations are solved using a standard high Reynolds number *k*-*ε* model (see Launder and Spalding [10]) and a Galerkin finite element method with a Segregated Implicit Projection (SIP) solution algorithm proposed by Utnes [11]. This numerical method is 2nd order both in time and space.

2.2 Computational domain, boundary conditions and convergence studies

The computational domain and the boundary conditions imposed for the present simulations are shown in fig. 1. The size of the whole computational domain is 30D by 10D. The upper boundary is located at a distance varying from 8.5D to 9.4D from the centre of the cylinder depending on the corresponding gap ratio; this ensures that the boundary has no effect on the flow around the cylinder. The flow inlet is located 10D upstream from the centre of the cylinder and the flow outlet is located 20D downstream from the centre of the cylinder. These distances are sufficient to eliminate the far field effects from the flow upstream and downstream of the cylinder.

The boundary conditions used for the numerical simulations are as follows:

1. A boundary layer flow is specified at the inlet (see fig. 1)

$$u_1(Y) = \min\{u_* \ln(Y/z_w)/\kappa, U_\infty\};\ u2(Y) = 0 \qquad (1)$$

$$k(Y) = \max\{C_\mu^{-1/2}(1-Y/\delta)^2 u_*^2, 0.0001U_\infty^2\};\ \varepsilon(Y) = C_\mu^{3/4}k(Y)^{3/2}/\ell \qquad (2)$$

$$\ell(Y) = \min\{\kappa Y(1+3.5Y/\delta)^{-1}, C_\mu \delta\} \quad (3)$$

Here Y denotes the wall normal direction starting from the seabed (see fig. 1). k is the turbulent kinetic energy. ε is the rate of viscous dissipation. u_1 and u_2 are the horizontal and vertical velocities, respectively. $C_\mu = 0.09$ is one of the standard coefficients in the k-ε model. The friction velocity is evaluated as $u_* = \kappa U_\infty/\ln(\delta/z_w)$, where z_w is the roughness of the flat bed, and $\kappa = 0.41$ is the von Kármán constant. ℓ is an estimate of the turbulent length scale (see e.g. Brørs [4]).

2. Along the outflow boundary, u_1, u_2, k and ε are specified as free boundary conditions in a finite element context. This means that a traction-free velocity-pressure boundary condition is applied for u_1, u_2 and P (see Gresho and Sani [12] for details), while the flux is set equal to zero for k and ε. Along the upper boundary, u_1, k and ε are free, while u_2 is set equal to zero.
3. No-slip condition is applied on the cylinder surface and the seabed with $u_1 = u_2 = 0$.
4. Standard near-wall conditions are applied for k and ε near the cylinder wall and the bed (see e.g. Rodi [13]) as

$$k = u_*^2/(C_\mu)^{1/2};\ \varepsilon = C_\mu^{3/4} k^{3/2}/(\kappa h_p) \quad (4)$$

where h_p is the normal distance between the first node and the wall, and u_* is the wall friction velocity obtained from the logarithmic (log) law.

$$u_{tan}/u_* = (1/\kappa)\ln(h_p/z_*) \qquad \text{where } z_* = (z_0, z_w) \quad (5)$$

Here u_{tan} is tangential velocity to the wall, z_0 is the roughness parameter of the cylinder surface and z_* is a switch parameter for the wall roughness. A small roughness with $z_0 = 1\times10^{-6}$m (i.e. $d_{50} = 12z_0 = 0.012$mm) is used for the cylinder for all the present simulations. This small roughness leads to almost the same results as a smooth logarithmic wall function, but is preferred because of enhanced numerical stability of the simulations.

Stretching of the mesh is performed to achieve a fine resolution of the region close to the cylinder surface and the seabed. When the grid is refined, the symmetrical grid elements nearest to the cylinder surface are kept constant. The

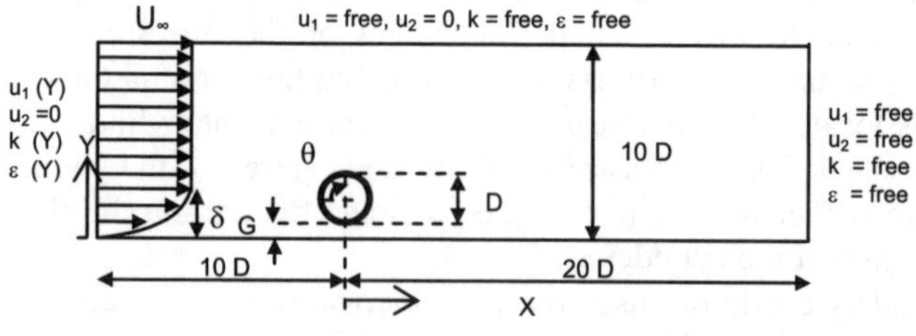

Figure 1: Definition sketch for flow around a circular cylinder close to a flat seabed.

seabeds with small roughness $z_w = 1\times10^{-6}$m and higher roughness $z_w = 2\times10^{-5}$m (i.e. $d_{50} = 12z_w = 0.24$mm) are used.

Both grid and time-step convergence studies have been performed for flows at Re = 3.6×10^6 for cases of $\delta/D = 0.48$, $z_w = (1\times10^{-6}$m, 2×10^{-5}m) and G/D = (0.1, 0.2, 0.3, 0.4, 0.6, 0.8, 1). The variations of C_D and C_L are considered in both grid and time-step convergence studies with a converged deviation of less than 5%. The meshes with approximately 40000 elements are considered to give a sufficient grid resolution, see fig. 2 for Re = 3.6×10^6 with $\delta/D = 0.48$ and $z_w = 1\times10^{-6}$m. The radial distance to the first node from the cylinder surface is 0.0005D. A non-dimensional time step (Δt) of $0.001D/U_\infty$ is found to be sufficient, and the simulations are run for 200 non-dimensional time units (D/U_∞).

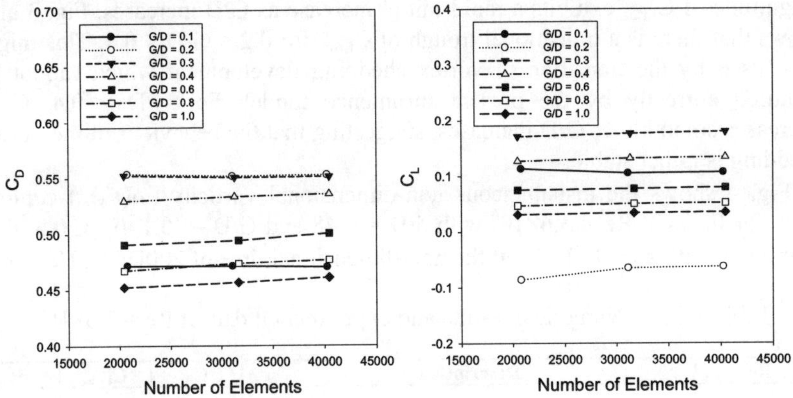

Figure 2: Grid convergence study for C_D and C_L with respect to the number of elements in the computational domain for Re = 3.6×10^6 with $\delta/D = 0.48$ and $z_w = 1\times10^{-6}$m.

3 Results and discussion

3.1 Validation study

High Reynolds number flows at Re = 3.6×10^6 with $\delta/D = 0.48$ and $z_w = (1\times 10^{-6}$m, 2×10^{-5}m) are investigated numerically in the present study. In this flow regime, there are neither experimental nor numerical results available in the open literature. However, the present results are validated by comparing the present numerical results for G/D = 1 with both published experimental data and numerical results for an isolated cylinder subject to a steady current in the same flow regime, since the effect of the seabed on the flow around the cylinder is insignificant for G/D = 1. The values of C_D, C_{Lrms} and St for G/D = 1 and $z_w = 1\times10^{-6}$m are within the range of the published experimental data and numerical results for steady flow around an isolated circular cylinder, see table 1.

3.2 Vortex shedding and suppression

Fig. 3 shows C_{Lrms} versus G/D for δ/D = 0.48 and the seabeds with $z_w = 1\times10^{-6}$m and 2×10^{-5}m. It appears that C_{Lrms} versus G/D has the same qualitative behaviour for both cases, but C_{Lrms} is generally lower for $z_w = 2\times10^{-5}$m (rougher bed) than for $z_w = 1\times10^{-6}$m for 0.15 < G/D < 0.8. The critical value for onset of vortex shedding, G/D_c (i.e. where the curve will intersect the horizontal axis), is between 0.1 and 0.15 in both cases, but it has not been calculated exactly here. It is observed that G/D_c decreases when Re increases by comparing the present results with the lower Re (Re~O(10^4)) results (i.e. $G/D_c \sim 0.3$) reported by Lei *et al.* [2], Ong *et al.* [6] and Wang and Tan [3]. Lei *et al.* [14] found a similar relation between G/D_c and Re for their simulations at Re = 80 - 1000. In fig. 3, $C_{Lrms} = 0$ at G/D = 0.1, suggesting no vortex shedding. For G/D > G/D_c, the magnitude of C_{Lrms} exhibits a rapid initial increase as G/D increases. Fig. 3 also shows that there is a transitional trough of C_{Lrms} for 0.2 < G/D < 0.4. This might be caused by the transition of vortex shedding development which cannot be captured correctly by the present turbulence model. For G/D > 0.4, C_{Lrms} decreases smoothly as G/D increases, suggesting that the behaviour of the vortex shedding is rather stable.

Fig. 4 shows the instantaneous non-dimensional vorticity ($\omega D/U_\infty$) contour plots for flows at Re = 3.6×10^6 with δ/D = 0.48 and G/D = (0.1, 0.15, 0.3, 0.8) near a bed with $z_w = 1\times10^{-6}$m at the non-dimensional time of $200D/U_\infty$. Here ω is

Table 1: Numerical results and experimental data at Re = 3.6×10^6.

Re	Description		C_D	C_{Lrms}	St
3.6×10^6 (Upper-transition regime)	G/D = 1	Present simulation with $z_w = 1\times10^{-6}$m	0.4608	0.0857	0.3052
	Flow around an isolated cylinder	Ong et al. [8]	0.4573	0.0766	0.3052
		Catalano et al. [9] URANS Re = 4×10^6	0.46	-	-
		Published experimental data (summarized by Zdravkovich [15])	0.36-0.75	0.06-0.14	0.17-0.29

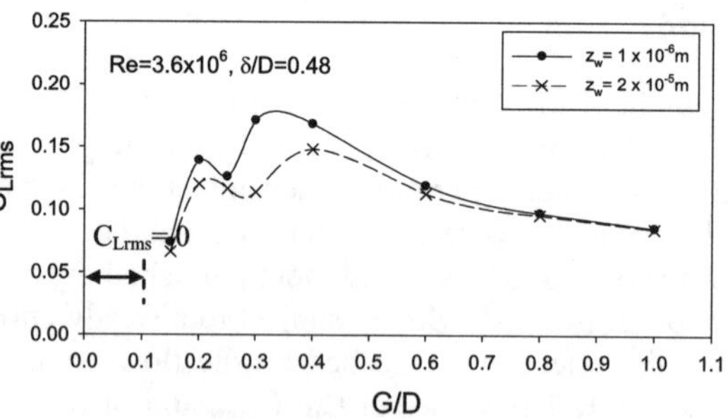

Figure 3: RMS value of the fluctuating lift coefficient versus gap to diameter ratio for the given values of Re, δ/D and z_w.

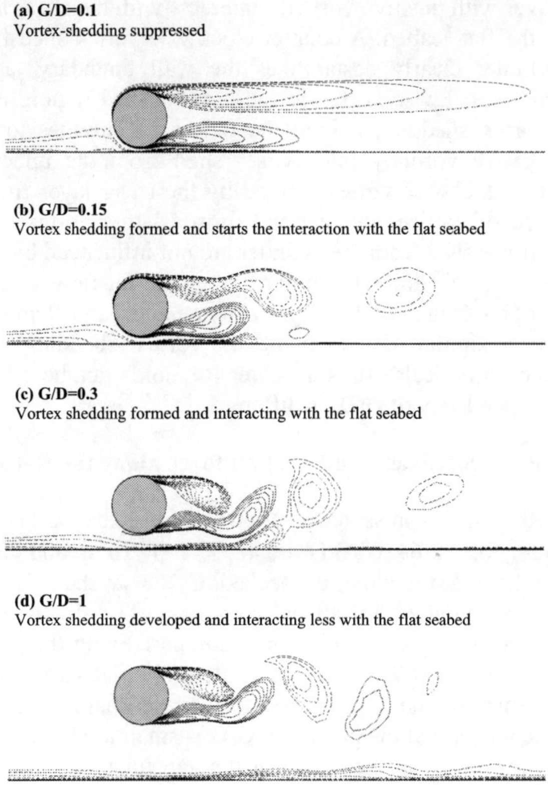

Figure 4: The development of vortex shedding shown by instantaneous non-dimensional vorticity contour plots for Re = 3.6×10^6, $\delta/D = 0.48$ and $z_w = 1\times10^{-6}$m at the non-dimensional time of $200D/U_\infty$. 46 contour levels of $\omega D/U_\infty$ from -540 to 540 are plotted.

the vorticity. The solid contour lines indicate the positive vorticity (counter-clockwise) and the dashed lines indicate the negative vorticity (clockwise). There are three shear layers; two in the vicinity of the cylinder and one at the bed. The suppression and formation of the vortex shedding are also influenced by the interaction between these three shear layers.

It appears that there is no mutual interaction between the two shear layers from the cylinder to form any Kármán-like vortex shedding for G/D = 0.1 (fig. 4a). Both shear layers continue to grow and advect downstream without forming any vortices in the near wake of the cylinder. The flow pattern remains steady. For G/D = 0.15 (fig. 4b), the two shear layers have begun to interact with each other and form Kármán-like vortices in the near wake of the cylinder. The bottom shear layer with positive vorticity interacts with the shear layer (negative vorticity) from the flat seabed. A counter-clockwise vortex shed from the lower side of the cylinder clearly destabilizes the wall boundary layer, and it is accompanied by a clockwise vortex in the near-flat-bed region. For G/D = 0.3 (fig. 4c), the vortex shedding behind the cylinder continues to develop. The vortex with negative vorticity (clockwise) shed from the upper shear layer, interacts with the clockwise vortex formed by the shear layer from the seabed. These two groups of vortices interact and form a larger vortex. For G/D = 1.0 (fig. 4d), the vortices shed from the cylinder are not influenced by the shear layer at the bed. The vortex shedding is similar to the case for flow around an isolated circular cylinder (see Ong *et al.* [8], fig. 6). Wang and Tan [3] and Lei et al. [14] have observed a similar development of vortex shedding in both their experimental and numerical results at lower Reynolds numbers (i.e. $Re < 10^5$), except that the dependency of G/D_c is different.

3.3 Mean pressure coefficient and friction force along the flat seabed

Fig. 5 shows the mean pressure coefficient along the seabed ($C_{pw} = [p_w - p_\infty]/[0.5\rho U_\infty^2]$) for $Re = 3.6\times10^6$, $\delta/D = 0.48$, $z_w = 1\times10^{-6}$m and G/D = (0.1, 0.4, 0.8). Here p_w is the pressure along the seabed. C_{pw} is substantially influenced by the existence of the cylinder. For a small gap, i.e. G/D = 0.1, it appears that the pressure suction at the gap (X = 0) is large compared with those for G/D = 0.4 and 0.8. Here X is the horizontal coordinate along the flat seabed where X = 0 is located at the centre of the gap, see fig. 1. This is mainly due to the higher magnitude of the velocity at the gap when G/D is small as shown in fig. 6 (which shows the velocity profile at the centre of the gap for G/D = 0.1, 0.4 and 0.8). This feature is similar to the lower Re results ($Re = 4.8\times10^4$) reported by Bearman and Zdravkovich [1] and Ong *et al.* [6]. Fig. 7 shows that the effect of the seabed roughness (with $z_w = 2\times10^{-5}$m) on C_{pw} is insignificant as compared with the results for $z_w = 1\times10^{-6}$m.

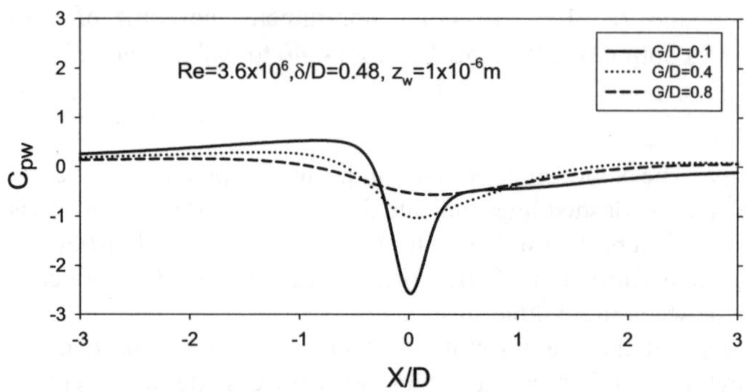

Figure 5: Mean pressure coefficient along the flat seabed for the given values of Re, δ/D and G/D.

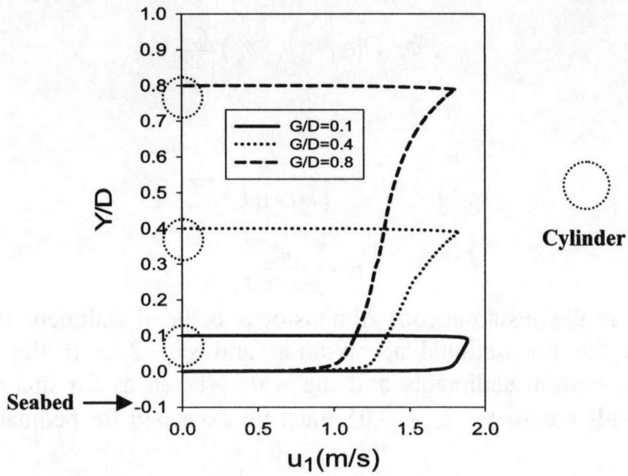

Figure 6: Instantaneous horizontal velocity profile in the gap for Re = $3.6{\times}10^6$, $\delta/D = 0.48$, $z_w = 1{\times}10^{-6}$m and G/D = (0.1, 0.4, 0.8) at the non-dimensional time of $200D/U_\infty$.

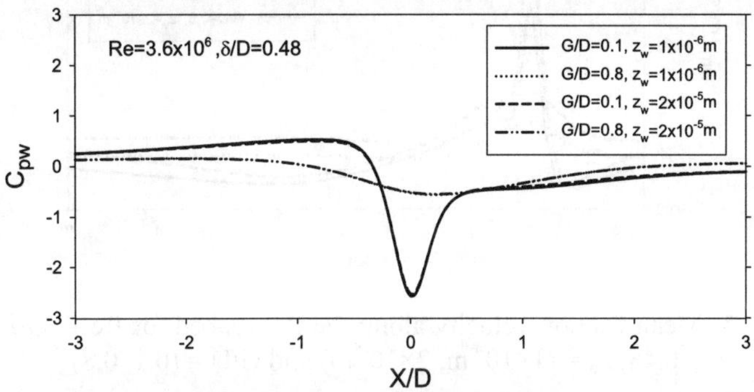

Figure 7: Mean pressure coefficient along the flat seabed for the given values of Re, δ/D, z_w and G/D.

Fig. 8 shows the mean friction velocity (u_{*wm}) for Re = $3.6{\times}10^6$, $\delta/D = 0.48$, G/D = (0.1, 0.8) and $z_w = (1{\times}10^{-6}$m, $2{\times}10^{-5}$m). It is observed that u_{*wm} is higher for the rougher seabed ($z_w = 2{\times}10^{-5}$m) than that for the less rough seabed ($z_w = 1{\times}10^{-6}$m), as expected. Fig. 8 also shows that u_{*wm} at the gap is much higher for G/D = 0.1 than that for G/D = 0.8. This is due to the higher velocity at the gap when G/D is small as shown in fig. 6.

3.4 An example of bedload sediment transport calculation

The calculation of the bedload sediment transport along the flat seabed is demonstrated in this section. The instantaneous non-dimension bedload sediment transport Φ is a function of the instantaneous non-dimensional seabed shear stress (Shields parameter) θ_s and is given by (Nielsen [16])

$$\Phi = 12|\theta_s|^{1/2}(\theta_s - \theta_{sc})\frac{\theta_s}{|\theta_s|} \tag{6}$$

where

$$\Phi = \frac{q_b}{\left(g(s-1)d_{50}{}^3\right)^{1/2}} \tag{7}$$

$$\theta_s = \frac{u_{*w}^2}{g(s-1)d_{50}} \tag{8}$$

Here q_b is the instantaneous dimensional bedload sediment transport, $g = 9.81 m/s^2$ is the gravitational acceleration and $s = 2.65$ is the density ratio between the bottom sediments and the water (taken as for quartz sand). The critical Shields parameter $\theta_{sc} = 0.05$ must be exceeded for bedload transport to occur.

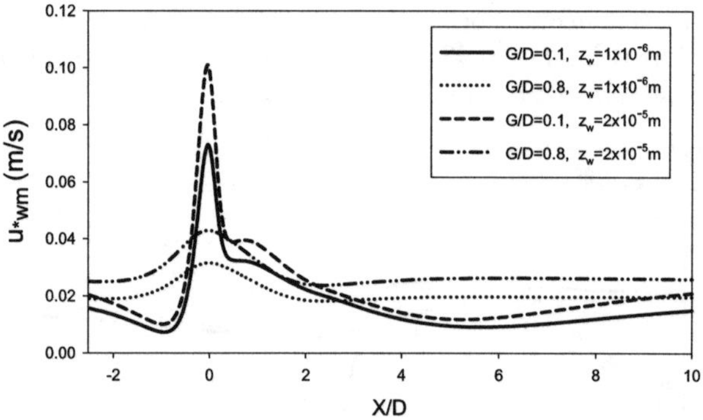

Figure 8: Mean friction velocity along the flat seabed for $Re = 3.6\times10^6$, $\delta/D = 0.48$, $z_w = (1\times10^{-6}m, 2\times10^{-5}m)$ and $G/D = (0.1, 0.8)$.

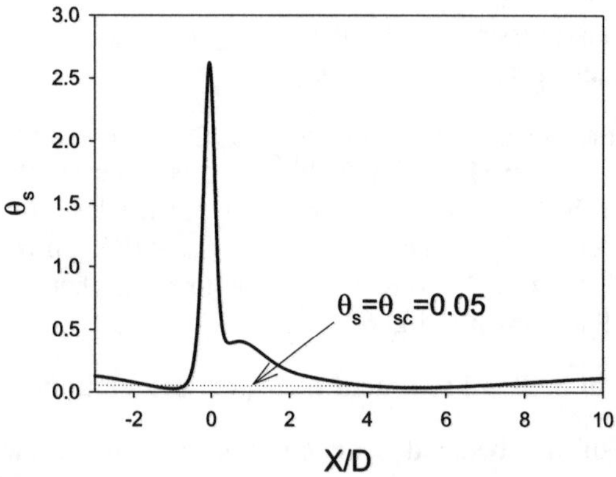

Figure 9: Instantaneous Shields parameter along the seabed for $Re = 3.6\times10^6$, $\delta/D = 0.48$, $z_w = 2\times10^{-5}m$ and $G/D = 0.1$.

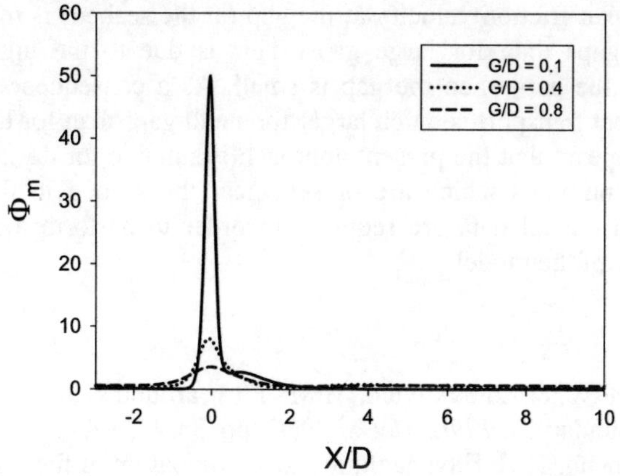

Figure 10: Mean non-dimensional bedload sediment transport along the bed for Re = 3.6×10^6, $\delta/D = 0.48$, $z_w = 2\times10^{-5}$m and G/D = (0.1, 0.4, 0.8).

Fig. 9 shows θ_s along the flat seabed for Re = 3.6×10^6, $\delta/D = 0.48$, $z_w = 2\times10^{-5}$m (i.e. $d_{50} = 12z_w = 0.24$mm, fine sand) and G/D = 0.1. The locations where the sediment transport takes place for $\theta_s > \theta_{sc}$ can be determined from the figure. Fig. 10 shows Φ_m (the mean non-dimensional bedload transport) along the bed for Re = 3.6×10^6, $\delta/D = 0.48$, $z_w = 2\times10^{-5}$m and G/D = (0.1, 0.4, 0.8). It is observed that the bedload sediment transport is significantly amplified at the location of the gap (X/D = 0) for G/D = 0.1 compared with those for G/D = 0.4 and 0.8. If the flat seabed is movable, scouring around the cylinder will take place. The scouring process will not be investigated here. Detailed explanations of the flow mechanisms and the development of the scour can be found in Sumer and Fredsøe [17].

4 Conclusions

Near-bed flow mechanisms of high Reynolds number flows around a marine pipeline close to a flat seabed have been studied using a 2D standard high Reynolds number k-ε model. The main results are summarized as follows:

1. Suppression and formation of the vortex shedding are influenced by the interaction between three shear layers; two from the top and the bottom of the cylinder and one at the seabed. The vortex shedding is suppressed when the gap is smaller than the critical gap (i.e. corresponding to the onset of vortex shedding). Beyond the critical gap, vortex shedding develops as the gap increases, and becomes fully developed as the influence of the bed diminishes.
2. For the same Reynolds number, inlet boundary layer thickness, seabed roughness and cylinder, the magnitude of negative pressure coefficient at the seabed at the location of the gap increases as the gap becomes smaller.

3. The mean friction velocity at the gap (at the seabed) is much larger for small gaps than for large gaps. This is due to the higher velocities within the gap when the gap is small. As a consequence, the bedload sediment transport is much larger for small gaps than for large gaps.

Overall it appears that the present approach is suitable for design purposes at high Reynolds numbers which are present near the seabed in the real ocean. However, experimental data are required in order to perform a more detailed validation study of the model.

References

[1] Bearman, P.W. & Zdravkovich, M.M., Flow around a circular cylinder near a plane boundary. *J. Fluid Mech.*, 89(1), pp. 33-47, 1978.

[2] Lei, C., Cheng, L. & Kavanagh, K., Re-examination of the effect of a plane boundary on force and vortex shedding of a circular cylinder. *J. Wind Eng. & Ind. Aerodynamics*, 80(3), pp. 263-286, 1999.

[3] Wang, X. & Tan, S.K., Near-wake flow characteristics of a circular cylinder close to a wall. *J. Fluids & Struc.*, 24(5), pp. 605-627, 2008.

[4] Brørs, B., Numerical modelling of flow and scour at pipelines. *J. Hydraulic Eng.*, 125(5), pp. 511-523, 1999.

[5] Zhao, M., Cheng, L. & Teng, B., Numerical modelling of flow and hydrodynamics forces around a piggyback pipeline near the seabed. *J. Waterway, Port, Coast. & Ocean Eng.*, 133(4), pp. 286-295, 2007.

[6] Ong, M.C., Utnes, T., Holmedal, L.E., Myrhaug, D. & Pettersen, B., Numerical simulation of flow around a marine pipeline close to the seabed. *Proc. 31st Int. Conf. Coast. Eng.*, Hamburg, Germany, 2008. (In press).

[7] Mittal, R. & Balachandar, S., Effect of three-dimensionality on the lift and drag of nominally two-dimensional cylinders. *Phys. Fluids*, 7, pp. 1841-1865, 1995.

[8] Ong, M.C., Utnes, T., Holmedal, L.E., Myrhaug, D. & Pettersen, B., Numerical simulation of flow around a smooth circular cylinder at very high Reynolds numbers. *Marine Struc.*, 22, pp. 142-153, 2009.

[9] Catalano, P., Wang, M., Iaccarino, G. & Moin, P., Numerical simulation of the flow around a circular cylinder at high Reynolds numbers. *Int. J. Heat & Fluid Flow*, 24, pp. 463-469, 2003.

[10] Launder, B.E. & Spalding, D.B., *Mathematical Models of Turbulence*, Academic Press, London, 1972.

[11] Utnes, T., A segregated implicit pressure projection method for incompressible flows. *J. Comp. Phys.*, 227, pp. 2198-2211, 2008.

[12] Gresho, P.M. & Sani, R.L., *Incompressible flow and the finite element method,* John Wiley & Sons Ltd, West Sussex, England, 1999.

[13] Rodi, W., *Turbulence models and their application in hydraulics. A state-of-the-art review. IAHR Monograph Series, 3rd Ed.,* A.A. Balkema, Rotterdam, The Netherlands, 1993.

[14] Lei, C., Cheng, L., Armfield, S.W. & Kavanagh, K., Vortex shedding suppression for flow over a circular cylinder near a plane boundary. *Ocean Eng.*, 27, pp.1109-1127, 2000.

[15] Zdravkovich, M.M., *Flow around Circular Cylinders, Vol. 1: Fundamentals*, Oxford University Press, New York, 1997.
[16] Nielsen, P., *Coastal Bottom Boundary Layers and Sediment Transport*, World Scientific, Singapore, 1992.
[17] Sumer, B.M. & Fredsøe, J., *The Mechanics of Scour in the Marine Environment: Advanced series on ocean engineering- Vol. 17*, World Scientific, Singapore, 2002.

Wave-induced steady streaming and net sediment transport in ocean bottom boundary layers

L. E. Holmedal & D. Myrhaug
Department of Marine Technology,
The Norwegian University of Science and Technology, Norway

Abstract

The sediment transport resulting from the interaction between two important streaming-generating mechanisms has been investigated by numerical simulations of the seabed boundary layer beneath both sinusoidal waves and Stokes second order waves. These two mechanisms are streaming caused by turbulence asymmetry in successive wave half-cycles (beneath asymmetric forcing), and streaming caused by the presence of a vertical wave velocity within the seabed boundary layer.

Keywords: sea bed boundary layers, streaming, sediment transport.

1 Introduction

In coastal waters of intermediate or shallow water depths the surface waves induce water particle trajectories from the free surface to the bottom, dominating the flow in the water column. Near the bottom an oscillating boundary layer is formed because of the bottom friction. Inside this boundary layer, the wave-induced forcing is responsible for the transport of sea bed material either as bedload or as suspended load. This material includes sediments, chemical compounds, as well as biological material such as fish larvae.

Ocean surface waves are progressive, and for finite water depths the near-bottom water particle trajectories are ellipses where the horizontal axis is much larger than the vertical axis. Thus a small vertical wave velocity exists in the flow, and the existence of this vertical wave velocity gives rise to a weak mass transport within the oscillatory bottom boundary layer. This happens because the vorticity and turbu-

WIT Transactions on Ecology and the Environment, Vol 126, © 2009 WIT Press
www.witpress.com, ISSN 1743-3541 (on-line)
doi:10.2495/CP090161

lence created within the oscillating boundary layer are transported upwards from the bottom with time. Hence a weak vorticity exists in a layer much thicker than the oscillating boundary layer (Batchelor [3]). As a result, the vertical and horizontal velocity components are not 90 degrees out of phase within this layer (as they are in potential flow), and the vertical wave velocity combines with the horizontal wave velocity through the convective terms in the governing boundary layer equations, giving rise to a non-zero wave-averaged drift within the oscillatory boundary layer. This effect is caused by the bottom friction and wave action, and is commonly referred to as steady streaming. This streaming phenomenon for oscillating bottom boundary layers beneath gravity waves was first explained by Longuet-Higgins [5]. However, steady streaming also arises because of wave asymmetry, as described in detail by Scandura [4] and Davies and Li [13] for flows in the transitional laminar to turbulent regime and flows in the rough turbulent regime, respectively. This phenomenon was first measured in an oscillating water tunnel by Ribberink and Al-Salem [1]. As pointed out by Scandura [4] the effect of wave asymmetry is particularly important in shallow waters. However, as explained by Longuet-Higgins [5], the steady streaming velocity will also be present in realistic near bottom flows beneath symmetric waves.

This work will focus on the sediment transport resulting from the interaction between two important streaming-generating mechanisms: The first is streaming caused by turbulence asymmetry in successive wave half-cycles (beneath asymmetric boundary layer forcing); the second is streaming caused by the presence of the vertical wave velocity within the seabed boundary layer as explained by Longuet-Higgins [5]. A more complete description is given by Holmedal and Myrhaug [12] including a detailed discussion of streaming-generating mechanisms and their physical implications.

2 Model formulation

2.1 Governing equations

Here the main governing equations and boundary conditions are given; for a more detailed description the reader is referred to Holmedal and Myrhaug [12]. Wave-induced mass transport in bottom boundary layers over an infinitely long flat bottom is considered. The horizontal coordinate at the bottom is given as x, whilst the vertical coordinate z gives the distance from the bottom. The bottom is fixed at $z = z_0 = k_N/30$, where k_N is the equivalent Nikuradse roughness. The limits of the horizontal coordinate x is such that $x = 0$ at the start of the wave length, and $x = \lambda$ at the end of the wave length. For intermediate and shallow water depths, the water particle trajectories are ellipses where the horizontal axis is much larger than the vertical axis. Hence the boundary layer approximation applies, and the simplified Reynolds-averaged equations for conservation of the mean momentum and mass become

$$\frac{\partial u}{\partial t} + \frac{\partial (u^2)}{\partial x} + \frac{\partial (uw)}{\partial z} = -\frac{1}{\rho}\frac{\partial p}{\partial x} + \frac{\partial}{\partial z}(\nu_T \frac{\partial u}{\partial z}), \tag{1}$$

$$\frac{\partial u}{\partial x} + \frac{\partial w}{\partial z} = 0, \tag{2}$$

where u is the horizontal velocity component, w is the vertical velocity component, p is the pressure, ρ is the density of the water, and ν_T is the kinematic eddy viscosity. The turbulence closure is provided by a k-ϵ model. Subjected to the boundary layer approximation, these transport equations are given by (see e.g. Rodi [7])

$$\frac{\partial k}{\partial t} + \frac{\partial (uk)}{\partial x} + \frac{\partial (wk)}{\partial z} = \frac{\partial}{\partial z}(\frac{\nu_T}{\sigma_k}\frac{\partial k}{\partial z}) + \nu_T (\frac{\partial u}{\partial z})^2 - \epsilon, \tag{3}$$

$$\frac{\partial \epsilon}{\partial t} + \frac{\partial (u\epsilon)}{\partial x} + \frac{\partial (w\epsilon)}{\partial z} = \frac{\partial}{\partial z}(\frac{\nu_T}{\sigma_\epsilon}\frac{\partial \epsilon}{\partial z}) + c_{\epsilon 1}\frac{\epsilon}{k}\nu_T (\frac{\partial u}{\partial z})^2 - c_{\epsilon 2}\frac{\epsilon^2}{k}. \tag{4}$$

where k is the turbulent kinetic energy and ϵ is the turbulent dissipation rate. Here Eq.(2) has been applied to write Eqs.(1), (3) and (4) in conservative form. The kinematic eddy viscosity is given by

$$\nu_T = c_1 \frac{k^2}{\epsilon}. \tag{5}$$

The standard values of the model constants have been adopted, i.e. (c_1, $c_{\epsilon 1}$, $c_{\epsilon 2}$, σ_k, σ_ϵ) = (0.09, 1.44, 1.92, 1.00, 1.30).

The instantaneous dimensionless bedload transport Φ is a function of the instantaneous dimensionless sea bed shear stress (Shields parameter) θ and is given by a formula by Nielsen [2]

$$\Phi = 12\theta^{\frac{1}{2}}(\theta - \theta_c)\frac{\theta}{|\theta|}. \tag{6}$$

where

$$\Phi = \frac{q_b}{(g(s-1)d_{50}^3)^{\frac{1}{2}}}, \tag{7}$$

$$\theta = \frac{\tau_b}{\rho g(s-1)d_{50}}. \tag{8}$$

Here q_b is the instantaneous dimensional bedload transport, τ_b is the dimensional instantaneous sea bed shear stress, g is the gravity acceleration, $s = 2.65$ is the density ratio between the bottom sediments and the water, ρ is the water density, and d_{50} is the median grain size diameter. The critical Shields parameter $\theta_c = 0.05$ must be exceeded for bedload transport to take place.

By using the boundary layer approximation, the equation for the sediment concentration c is written

$$\frac{\partial c}{\partial t} + u\frac{\partial c}{\partial x} + (w - \mathrm{w}_s)\frac{\partial c}{\partial z} = \frac{\partial}{\partial z}(\epsilon_s \frac{\partial c}{\partial z}), \tag{9}$$

$$\epsilon_s = \nu_T + \nu. \tag{10}$$

Here ϵ_s is the sediment diffusivity, w_s is the settling velocity of the median sand grains in still water, and ν is the laminar kinematic viscosity of water. Here the laminar viscosity has been included in the sediment diffusivity in order to stabilize the numerical scheme; this model is described in more detail in Holmedal et al. [6].

2.2 Simplification of equations

In order to simplify the mathematical solution of Eqs.(1)-(4) and (9) the relation $\partial/\partial x = -(1/c_p)\,\partial/\partial t$ is applied. This is an approximation which is only valid for weakly decreasing waves (i.e. the wave height decay over a wave length due to the energy dissipation is small); this will be discussed further in conjunction with Eq.(14). This approximation leads to the two-dimensional boundary layer equations (i.e. Eqs.(1), (3), (4) and (9)) reducing to spatially one-dimensional equations. Physically this transformation implies a mapping from two spatial dimensions to one spatial dimension. The length of the physical two-dimensional space is one wave length, and the height is $z_{max} - z_0$; in one dimension the height $z_{max} - z_0$. The results obtained in one dimension can be mapped back to the physical two-dimensional space.

As a consequence of this simplification the vertical velocity component is found from the continuity equation and is evaluated as

$$w = -\int_{z=z_0}^{z} \frac{\partial u}{\partial x} dz = \frac{1}{c_p}\int_{z=z_0}^{z} \frac{\partial u}{\partial t} dz. \tag{11}$$

and inserted into Eqs. (1), (3), (4) and (9). Here c_p is the phase velocity which will be defined further below. Furthermore, $w = 0$ at $z = z_0$ has been utilized (see Eq.(15)).

2.3 Boundary conditions and numerical solution

The sea bed is assumed to be hydraulically rough. At the bed ($z = z_0$) no-slip conditions are imposed on the velocity. The $k - \epsilon$ model is coupled in a standard way with the logarithmic wall law near the bottom. Neumann conditions are imposed at the upper edge of the flow domain. This hydrostatic boundary layer flow is driven by progressive sinusoidal and second order Stokes waves. The boundary conditions for the sediment concentration are given by a specified reference concentration above the bed (depending on the Shields number) given by Zyserman and Fredsøe [8]. On top of the flow domain Neumann conditions are specified for the sediment concentration.

A finite difference method was used to solve the governing equations, and the equations were integrated in time using the integrator VODE (Brown et al. [9]).

3 Results and discussion

This paper presents the wave-induced mass transport within the ocean bottom boundary layer for realistic wave conditions, bottom roughnesses, and water depth. Ocean surface waves with an amplitude of $a = 1.22$m and a period of 6s propagate over a flat rough bottom. The water depth is 8m, the resulting wave length is 45m, and the bottom roughnesses are, $z_0 = 1 \cdot 10^{-5}, 3 \cdot 10^{-5}, 6 \cdot 10^{-5}, 1 \cdot 10^{-4}$ and $2.3 \cdot 10^{-4}$m. By using the empirical formula $k_{\mathrm{N}} = 2.5\, d_{50}$, these roughnesses correspond to fine sand, medium sand, coarse sand, very coarse sand and gravel, respectively (Soulsby [10], Chapter 2). This wave condition represents intermediate water depth ($k_p h = 1.11$) with wave steepness $ak_p = 0.17$. The near-bottom potential flow is approximated by second order Stokes theory.

3.1 Streaming-induced sediment transport

If the wave-induced forcing is strong enough to move the sea bed material (for example sediments and/or pollutants), or to bring it into suspension, then the weak streaming-induced boundary layer drift and non-zero wave-averaged bottom shear stress will cause a net transport of this material over time. This transport may take place either as net transport of suspended sediments or bedload.

Table 1: Mean bedload transport, suspended sediment transport and total sediment transport ($\overline{q}_{\text{total}}$) for linear waves and Stokes second order waves. L.W denotes sinusoidal progressive waves; S.W denotes second order Stokes waves.

	d_{50} (mm)	$\overline{q_b}$ (mm^2/s)	$\int_{2\,d_{50}}^{z_{max}} \overline{uc}\, dz$ (mm^2/s)	$\overline{q}_{\text{total}}$ (mm^2/s)
L.W	0.13	8.3	69.8	78.1
L.W.	0.32	10.6	11.9	22.5
S.W	0.13	10.2	57.9	68.1
S.W.	0.32	13.3	12.4	25.7

Table 1 shows the mean suspended sediment transport $\int_{2d_{50}}^{z_{max}} \overline{uc}\, dz$, the mean bedload transport $\overline{q_b}$ and the total sediment transport (the mean suspended sediment transport plus the mean bedload transport). It appears that the total sediment transport beneath second order Stokes waves is not very different from the total sediment transport beneath sinusoidal waves, given the uncertainty which is inherent in these sediment models. However, for $d_{50} = 0.13$ mm the total sediment transport is larger for sinusoidal waves than for second order Stokes waves, while for $d_{50} = 0.32$ mm the total sediment transport is larger for second order Stokes waves than for sinusoidal waves. In these simulations the settling velocities for the suspended sediments are $w_s = 0.0119$ m/s and 0.0429 m/s, for $d_{50} = 0.13$ mm and $d_{50} = 0.32$ mm, respectively; these are taken from Dohmen-Janssen et al. [11].

Holmedal and Myrhaug [12] showed that both the Eulerian and the Lagrangian wave averaged seabed boundary layer velocity is larger beneath sinusoidal progressive waves than beneath Stokes second order waves. This has some implications for the resulting sediment transport. The trend seems to be that for fine sediments the suspended sediment flux is larger beneath sinusoidal waves than beneath second order Stokes waves, since fine sediments tend to follow the particle trajectories. Since the major part of the sediment transport is taking place as suspension for fine sediments, the total sediment transport is larger beneath sinusoidal progressive waves than beneath Stokes second order waves for fine sediments. For coarser sediments the trend is opposite: the total sediment transport is larger beneath Stokes second order waves than beneath sinusoidal progressive waves. For these coarser sediments the bedload yields a larger contribution to the sediment transport, and this contribution appears to be slightly larger for second order Stokes waves than for sinusoidal progressive waves. This might be due to that the maximum bottom shear stress during a wave cycle is larger for second order Stokes waves is larger than beneath sinusoidal progressive waves.

4 Summary and conclusions

The sediment transport resulting from the interaction between two important streaming-generating mechanisms has been investigated by numerical simulations of the seabed boundary layer beneath both sinusoidal waves and Stokes second order waves. These two mechanisms are streaming caused by turbulence asymmetry in successive wave half-cycles (beneath asymmetric forcing), and streaming caused by the presence of a vertical wave velocity within the seabed boundary layer.

It appears that the total sediment transport beneath second order Stokes waves is not very different from the total sediment transport beneath sinusoidal waves, given the uncertainty which is inherent in these sediment models. However, there is a trend showing that the total sediment transport is larger beneath sinusoidal progressive waves than beneath Stokes second order waves for fine sediments while as for coarser sediments this trend is opposite.

References

[1] Ribberink, J.S. & Al-Salem, A.A., Sheet flow and suspension of sand in oscillatory boundary layers. *Coastal Eng*, **25**, pp. 205–225, 1995.

[2] Nielsen, P., *Coastal bottom boundary layers and sediment transport*. World Scientific Publishing Co. Pte. Ltd., Singapore, 1992.

[3] Batchelor, G.K., *An Introduction to Fluid Dynamics*. Cambridge University Press, 1967.

[4] Scandura, P., Steady streaming in a turbulent oscillating boundary layer. *Journal of Fluid Mechanics*, **571**, pp. 265–280, 2007.

[5] Longuet-Higgins, M.S., Mass transport in water waves. *Phil Trans R Soc Lond A*, **245**, pp. 535–581, 1953.

[6] Holmedal, L.E., Myrhaug, D. & Eidsvik, K.J., Sediment suspension under sheet flow conditions beneath random waves plus current. *Cont Shelf Res*, **24**, pp. 2065–2091, 2004.

[7] Rodi, W., *Turbulence Models and Their Application in Hydraulics, A state-of-the-art review.* IAHR Monograph series, A. A. Balkema, Rotterdam, Netherlands, 3rd edition, 1993.

[8] Zyserman, J.A. & Fredsøe, J., Data analysis of bed concentration of suspended sediment. *J Hydr Res*, **120(9)**, pp. 1021–1042, 1994.

[9] Brown, P.N., Byrne, G.D. & Hindmarsh, A.C., VODE: A Variable Coefficient ODE Solver. *SIAM J Sci Stat Comput*, **10**, pp. 1038–1051, 1989.

[10] Soulsby, R.L., *Dynamics of marine sands*. Thomas Telford Publications, 1997.

[11] Dohmen-Janssen, C.M., Hassan, W.N. & Ribberink, J.S., Mobile-bed effects in oscillatory sheet flow. *J Geophys Res*, **106(C11)**, pp. 27103–27115, 2001.

[12] Holmedal, L.E. & Myrhaug, D., Wave-induced steady streaming, mass transport and net sediment transport in rough turbulent ocean bottom boundary layers. *Continental Shelf Research*, **29**, pp. 911–926, 2009.

[13] Davies, A.G. & Li, Z., Modelling sediment transport beneath regular symmetrical and asymmetrical waves above a plane bed. *Cont Shelf Res*, **17(5)**, pp. 555–582, 1997.

Measuring suspended sand transport using a pulse-coherent acoustic Doppler profiler

T. Aagaard[1] & B. Greenwood[2]
[1]*Institute of Geography and Geology, University of Copenhagen, Denmark*
[2]*Department of Physical and Environmental Sciences, University of Toronto at Scarborough, Canada*

Abstract

A simple method to derive sediment transport rates in the nearshore from Pulse-Coherent Acoustic Doppler Profilers (PC-ADP) is described. Measured sediment concentration profiles are compared with theoretically estimated concentration profiles and exhibit good agreement. Cross-shore and longshore sediment transport rates obtained during 5-6 week long deployments on the lower shoreface in 6-8 m water depth are compared with predictions from an energetics transport model. The two estimates agree to within 20% in the longshore direction while the cumulated cross-shore wave-dominated transport component is trending significantly less offshore for the measurements than for the model predictions. The measurements indicate that significant onshore sediment transport from the lower to the upper shoreface only occurs when waves are close to breaking.

Keywords: PC-ADP, sediment concentrations, sediment transport, shoreface.

1 Introduction

Field measurements of suspended sediment transport in the nearshore can be obtained using either optic or acoustic devices (Osborne et al. [1]). While standard procedures exist for estimating sediment concentrations from optical backscatter, it is considerably more difficult to obtain reliable estimates from acoustic backscatter sensors (ABS). One reason is that acoustic backscatter needs to be corrected for signal attenuation due to water absorption and sediment scattering. Significant effort has been invested in developing algorithms to invert

WIT Transactions on Ecology and the Environment, Vol 126, © 2009 WIT Press
www.witpress.com, ISSN 1743-3541 (on-line)
doi:10.2495/CP090171

the backscattered acoustic signal to provide sediment concentration (e.g. Betteridge et al. [2]; Sheng and Hay [3]; Thorne and Hanes [4]) but while reliable results from ABS can now be obtained, procedures for signal correction are complicated and time-consuming when long-term sediment transport rates need to be assessed. In addition, unless multi-frequency sensors are used (e.g. Crawford and Hay [5]), these techniques can only be used for single grain sizes and are difficult to apply for sand when water clarity is significantly affected by for example more or less permanently suspended silts and/or organics.

Here, we test a simple method for obtaining long-term sediment transport rates from a Pulse-Coherent Acoustic Doppler Profiler (PC-ADP). The PC-ADP is similar to ABS sensors in the sense that backscattered sound pressure is proportional to the amount of scatterers (sediment particles) in the water column. In contrast to ABS, PC-ADP's have only rarely been used to study sediment dynamics (Kostachuk et al. [6]; Wren and Leonard [7]), possibly because of the relatively large footprint of the sensor (approx. 0.13 m at a distance of 0.50 m above the bed). Hence, while the reliability of velocity estimates from PC-ADP's have been established recently (Lacy and Sherwood [8]), the capability of the instrument to resolve sediment transport is not well understood.

However, when estimating sediment transport in the bottom boundary layer under waves, it is important that velocity and sediment concentration measurements are collocated. This is the case for PC-ADP sensors because velocity and backscatter is recorded at the same point in the water column, while ABS and paired velocity sensors are separated by some finite distance at least in the horizontal. In addition, the large sensor footprint of the PC-ADP may be an advantage in some respects because the effect of bedforms on sediment suspension and concentration will be averaged over an entire bedform. In this paper, mean sediment concentration profiles and concentration time series estimated using a simple methodology are compared to outputs determined from more elaborate techniques. Cross-shore and longshore sediment transport rates are then compared with predictions from an often-used energetics sediment transport model to assess sediment transport rates on the lower shoreface under a range of wave conditions.

2 Study site

The field site at Vejers Beach is located on the exposed Danish North Sea coast. The mean annual offshore significant wave height (H_s) is 1.3 m with an average zero-crossing wave period (T_z) of 4.3 s. Waves from the northwest are dominant and offshore significant wave heights are up to 7 m during storms. Persistent low ($H_s < 1$ m) background swell impinges from the northwest. The site is micro-tidal with a mean spring tidal range of 1.2 m. Tidal current speeds on the lower shoreface are in the order of 0.2-0.3 m s^{-1} with a northward directed residual current.

The cross-shore profile and the mean sand grain size at the study site are illustrated in Figure 1. The shoreface at Vejers has a mean slope of $\beta = 0.006$ and it exhibits 3-4 longshore bars. The PC-ADP was deployed seaward of the bars in mean water depths of 6 and 8 m, respectively, for periods of 5-6 weeks during the fall seasons of 2005 and 2006. The shallower deployment depth corresponds to the boundary between the upper and the lower shoreface where bathymetric change over time pinches out, see Figure 2. The mean sand grain size on the lower shoreface is ≈ 120 microns and it coarsens to ≈ 180 microns on the upper shoreface (Figure 1). In addition, significant amounts of permanently suspended silts and organics (wash load) are present in the area; this sediment originates from the Wadden Sea region further south and is transported northward by the residual tidal currents.

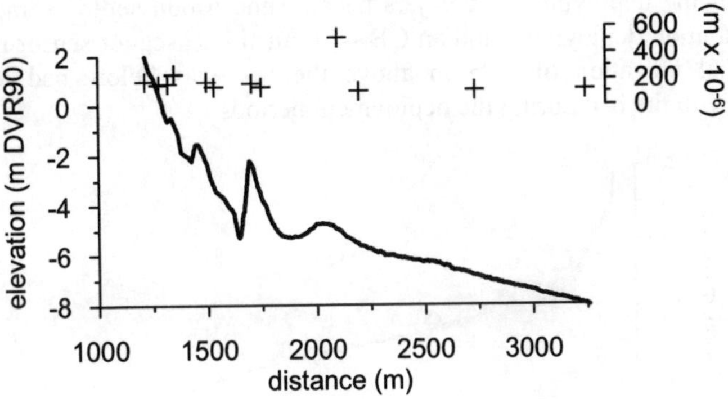

Figure 1: The cross-shore bathymetry of the field site at Vejers at the time of the first instrument deployment. The right-hand axis shows mean sediment grain size across the profile. DVR90 is Danish Ordnance datum, corresponding to mean sea level.

3 Methodology

At both deployment positions, a 1.5 MHz Sontek PC-ADP was mounted 0.51 m above the bed and recorded fluid velocities and acoustic backscatter intensity in 1.6 cm vertical bins from the sensor blanking distance (0.05 m from the transducer head) to the seabed. The sampling mode was one burst per hour with a record duration of 8.5 minutes and a sampling rate of 2 Hz. A Druck pressure sensor was installed in the transducer head to record wave heights and mean water depths. Velocity outputs were rotated with respect to shoreline orientation to provide cross-shore (u) and alongshore (v) velocities, they were corrected for ambient sound speed using data from a CTD-probe, smoothed using a horizontal 3-point Box filter and velocity ambiguity resolution was applied when required. When true velocities exceed the maximum velocity resolvable by the instrument, the velocity signal is wrapped resulting in erroneous readings. A second set of acoustic pulse pairs allows the signal to become unwrapped. For the particular instrument and the given profiling lag, velocity ambiguities occurred when orbital velocity (predicted from the local wave height measurements) exceeded 0.65 m s^{-1} near the bed. Records containing velocities theoretically exceeding this threshold were corrected using the software supplied by the manufacturer and subsequently despiked and smoothed.

Acoustic backscatter intensity was used to estimate suspended sediment concentrations (c) and subsequently cross-shore (q_x) and longshore (q_y) transport rates. Methods to compute c directly from backscattered pressure, accounting for signal loss due to water absorption and sediment scattering (e.g. Betteridge et al. [2]; Thorne and Hanes [4]; Crawford and Hay [5]) are difficult to apply for single-frequency sensors when the sediment is inhomogeneous, consisting of a mixture of sand and wash load, and they usually further require an independent estimate of c at some point within the water column. The latter could not be achieved for the deployments at Vejers because the tripod settled some 15 cm into the bed upon deployment and an OBS-3 optical backscatter sensor mounted at a nominal elevation of 0.05 m above the bed was below bed level, or interfering with the bed during the deployment periods.

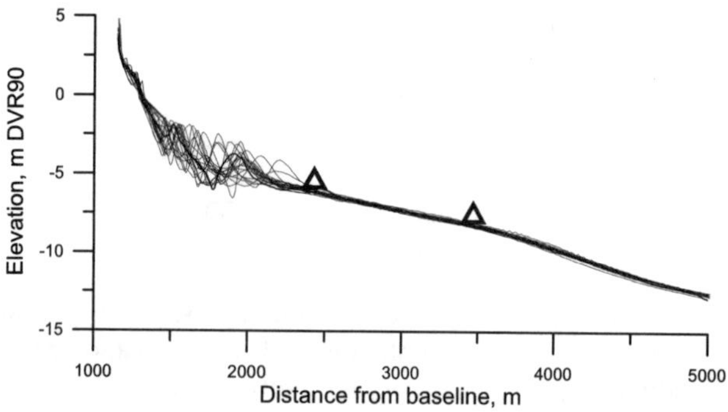

Figure 2: Cross-shore profile envelope comprising surveys 1969-2006 at Line 6170 off Vejers. The two triangles indicate the PC-ADP deployment positions.

Instead, a simple method of obtaining the concentration of sand in the water was developed: Acoustic backscatter was corrected for signal loss due to sound attenuation by water absorption and by the permanently suspended wash load. This attenuation was determined from instrument bursts with negligible wave activity and a lack of sediment suspension from the bed, and it was estimated as -18 dB m^{-1}. Field offsets induced by inherent system characteristics and wash load were then removed from the individual instrument records. They were taken as the average over the uppermost five measurement bins of the 5th percentile of output signal values and subtracted from the signals in all bins assuming that the fine-grained sediment fractions were uniformly distributed in the vertical. Acoustic pressure was then converted to sediment concentration using a transfer function that was developed from a short-term experiment during which measurements of optical backscatter were available. This experiment was done at a location with comparable sand characteristics ($\overline{d} = 148$ microns) and the OBS-sensor was subsequently calibrated in a recirculating tank to obtain a cross-calibration between backscattered acoustic pressure and sediment concentration.

The resulting transfer function was applied to all acoustic bins after offset removal. Hence, backscatter intensity was not corrected for sound attenuation due to sediment scattering. However, failure to do so may not a large problem, as reported errors are typically within ± 20% when the sensor operating frequency is less than 3MHz (Vincent [9]).

Finally, net cross-shore and longshore suspended sediment transport rates were estimated using:

$$q = \sum_{n=1}^{N} \sum_{t=0}^{T} u(t)c(t)dndt \qquad (1)$$

where u is fluid velocity, c is (estimated) sediment concentration, N is the number of measurement bins (N = 22-26) and T is the duration of the time series.

4 Mean concentration profiles

To compare the predictions of c from the simple method with estimates using a standard algorithm and to assess the quality of the predicted sediment concentrations, measurements from the short-term deployment were used. In this case, an OBS-sensor was located at an elevation of 8-10 cm above the bed. The instruments were deployed in a trough landward of a longshore bar, and the seabed was covered with wave ripples having a wavelength of 11 cm and a height of 2 cm. The significant wave height for the particular example shown here was 0.33 m with a period of 5.6 s and the mean water depth was 1.31 m yielding a relative wave height of 0.25; the waves were non-breaking.

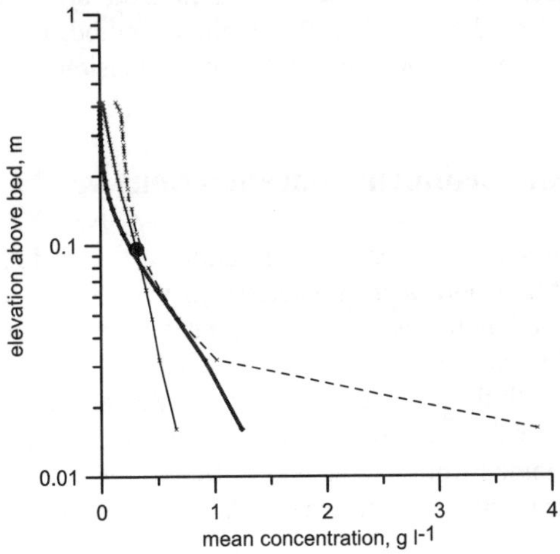

Figure 3: Mean concentration profiles determined from the simple method (dashed line) and from the explicit method of Thorne and Hanes [4; solid line]. The dot indicates the mean concentration recorded by the OBS. The thick blue line is the mean concentration profile determined from eqn. (2).

The mean concentration profile was determined from the simple method outlined above and from the explicit method of Lee and Hanes [10] and Thorne and Hanes [4; their eqn.21]. The explicit method requires an independent estimate of the sediment concentration at a given elevation, which is obtained from the OBS and it assumes a constant particle diameter with elevation above the bed. The attenuation due to water was taken as 0.067 N m^{-1} (Thorne et al. [11]) and the sediment attenuation was computed from Thorne and Hanes [4; their eqn.9]. The output values were finally corrected for near-field effects. Comparing the two estimates (Figure 3), the mean concentration profile obtained from the explicit method is considerably steeper, while measured concentrations increase more rapidly towards the bed.

The results were also compared with the independent model of Nielsen [12]:

$$C(z) = C_0 \exp\left(-\frac{z}{l_s}\right) \tag{2}$$

where C_0 is a reference concentration near the bed, given as $C_0 = 0.005\theta_r^3$ with θ_r being the effective Shields parameter in the presence of ripples (Nielsen [13]). z is elevation above the bed, and the vertical mixing length, l_s, is computed from Aagaard [14]:

$$\frac{l_s}{h} = 0.012 \exp\left(4.78 \frac{H_s}{h}\right) \tag{3}$$

Except for the lowermost measurement point which may be affected by proximity to the bed (with the sensor in fact intermittently seeing the bed), the concentrations determined from the simple method are much closer to the independent model predictions than the explicit method, probably because the explicit method assumes a homogeneous sand suspension and does not account for signal dampening by the wash load.

5 Instantaneous sediment concentrations

Instantaneous values of sediment concentration, $c(t)$, are required in order to calculate sediment transport rates in wave-dominated settings and $c(t)$ can not be obtained from the explicit method which considers only mean concentrations. Instantaneous sediment concentrations obtained from the simple method were instead compared with the time series from the OBS, and with predictions using the algorithm provided by Osborne et al. [1; their eqn.1]; the required system calibration was obtained from the method of Betteridge et al. [2] and attenuation due to sediment scattering was determined from the high-pass model of Sheng and Hay [3]).

Figure 4 compares the three sediment concentration time series at an elevation of 9.6 cm above the bed. The measured time series from the PC-ADP captures the basic characteristics of the concentration time series as seen by the OBS, although the two time series signals are not identical. The two instruments were separated in the horizontal by approximately 0.5 m and the measurement volume of the ADP is far larger than the measurement volume of the OBS. The latter is conceptually a point measurement of sediment concentrations, while the former provides an estimate which is averaged over approximately 80 cm^2, at this particular distance from the sensor head. Hence, the diameter of the sensing volume roughly corresponded to the ripple wavelength and the concentration estimate from the ADP is a measure of the average concentration over one ripple form.

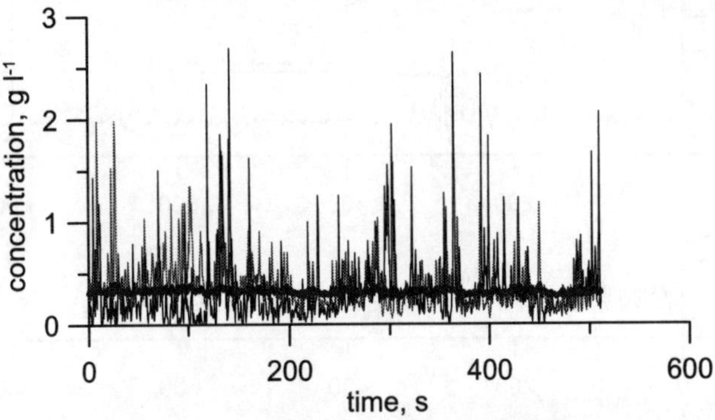

Figure 4: Time series of sediment concentrations at 9.6 cm above the bed measured by the OBS (black), by the PC-ADP (red) and estimated from the method of Osborne et al. [1; thick blue line].

In contrast to the time series from the ADP, the time series predicted from the method of Osborne et al. [1] bears little resemblance to the actually occurring concentrations and completely lacks the spiky nature of the observed concentration signal. The reason is probably again that the wash load dampens the amplitude of the backscattered signal and this effect is not taken into account by the model. In contrast, the simple method first removes the offset induced by the fine-grained sediment and then uses the cross-correlation with the OBS-signal from which the offset has also been removed.

6 Net sediment transport

The acoustic backscatter calibration obtained from the trial experiment was applied to estimate sediment concentrations and transports during the experiments at Vejers. Significant wave height, mean current speeds and longshore and cross-shore instantaneous and cumulated sediment transport from the experiment at 6 m water depth are plotted in Figure 5. The figure also plots suspended sediment transports rates estimated from the Bailard energetics model:

$$< q_t(t) >= K_s(< |\vec{u}|^3(t)u(t) >) - K_{sg} < |\vec{u}|^5(t) > \quad (4)$$

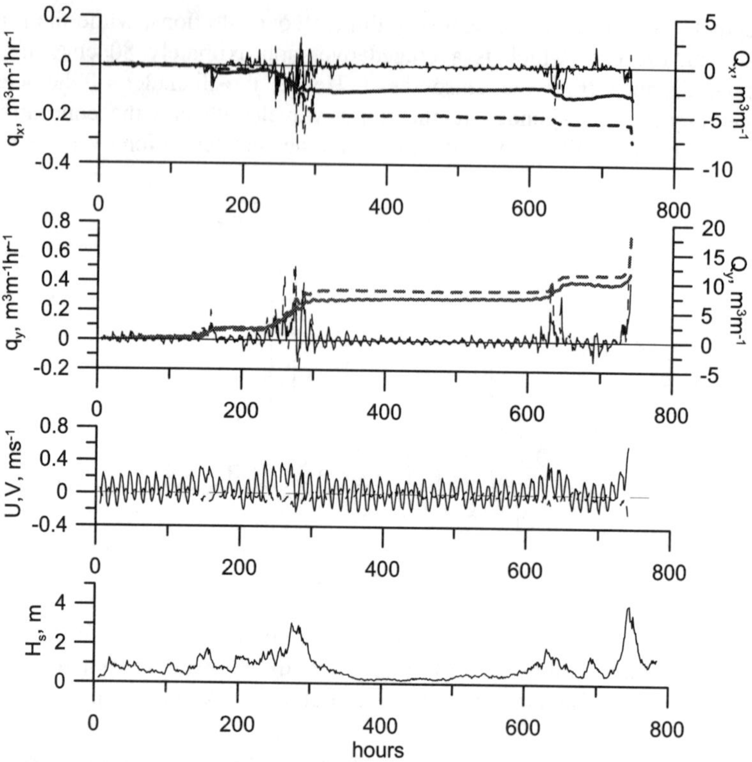

Figure 5: Measurements from the PC-ADP deployment in 6 m water depth at Vejers. The lower panel shows significant wave height (H_s) and the second panel shows cross-shore (U; dashed) and longshore (V; solid) mean current speeds. The third panel from the bottom illustrates measured (solid line) and model-predicted (dashed line) suspended longshore sediment transport rates (q_y) and the upper panel shows the cross-shore transport rates (q_x). The thick coloured lines illustrate the cumulated transport rates with the solid lines representing measurements and dashed lines representing model predictions. Positive transports are directed onshore and northward.

where the suspended load term is:

$$K_s = \left(\frac{\rho}{\rho_s - \rho}\right) c_f g a' \left(\frac{e_s}{w_s}\right)$$

and the gravity term is

$$K_{sg} = K_s \left(\frac{e_s}{w_s}\right) \tan\beta$$

In eqn. (4), $u(t)$ is the instantaneous cross-shore velocity, $\vec{u}\,(t)$ is the magnitude of the instantaneous horizontal velocity vector, ρ_s and ρ are sediment and water density, g is gravitational acceleration, a' is the pore space factor (0.6), $\tan\beta$ is the local slope of the bed, $\tan\varphi$ is angle of sediment repose, w_s is sediment fall velocity, taken as 0.012 m s^{-1}, and the transport efficiency factors were set at the model default values $e_b = 0.21$ and $e_s = 0.025$ (Bailard [15]). c_f is a drag coefficient = $f_w/2$, where f_w is the wave friction factor. The longshore transport component was calculated accordingly. Velocities recorded at 0.25 m above the bed were used as input to the model

In the longshore direction, measured cumulated longshore sediment transport agrees with model predictions to within 20%, and time-dependent transport rates were in good agreement, except for situations when wave heights exceeded 3 m. In such cases, the energetics model provides larger transport rates which may be because the measurements were not corrected for sound attenuation due to sediment scattering; this is considered to be more significant when waves (and therefore sediment concentrations) are large.

In the cross-shore direction where sediment transport due to wave motions is much more important, measurements trended significantly more onshore than model predictions. While the energetics model can be expected to perform well when mean currents are dominant, which was the case for the longshore transport, it generally provides imprecise results when wave motions are dominant (Thornton et al. [16]). Reasons for the model inaccuracies include the assumption of in-phase relationships between velocity and concentration. In reality, phase shifts between velocity and concentration tend to develop when bedforms are present. For the present measurements, cumulated (offshore-directed) model-predicted transport was a factor of 2 larger than the measured transports, and the model does not replicate the onshore transport that is indicated by the measurements during low- and moderate energy situations (Figure 5; hours 350-600 and 650-700).

7 Sediment supply from the lower to the upper shoreface

The supply of sand from the lower to the upper shoreface is an important factor to the long-term development of the cross-shore profile. Figure 6 plots the measured cross-shore sediment transport rates obtained at 6 and 8 m depth as a function of relative wave height and the mobility number:

$$\psi = \frac{\vec{u}^{\,2}}{(s-1)g\bar{d}} \tag{5}$$

where $(s\text{-}1)$ is relative sediment density and $\bar{d}$ is mean sediment grain size. According to Dingler and Inman [17], ripples occur for $\psi<240$, while flat bed and sheet flow prevail for $\psi>240$.

Because tidal currents set obliquely to the depth contours at Vejers, the mean currents introduce a site-specific bias which is not representative in a general assessment of cross-shore sediment exchange between the upper and lower shoreface. Therefore, only the wave-induced oscillatory transport rates are considered in Figure 6 where mean (current-induced) transports rates have been subtracted from the total transport rate. Oscillatory transports were mainly onshore directed while the total net transports at 6 m depth (Figure 5) were mainly offshore directed, because of the dominance of the mean transport component at this depth at Vejers.

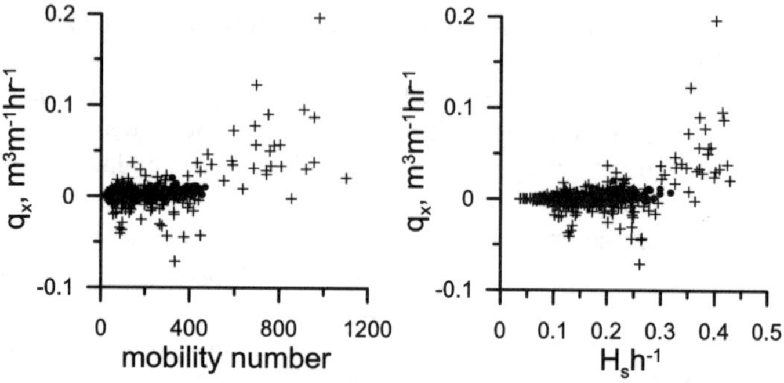

Figure 6: Net wave-induced cross-shore suspended sediment transport rates at 6 m (crosses) and 8 m (dots) water depth, as a function of mobility number and relative wave height. Positive transport rates are directed onshore.

The measurements suggest that significant onshore transport rates due to waves on the lower shoreface at occur only when $\psi > \approx 400$ and the relative wave height $H_s/h > \approx 0.3$. This corresponds to flat bed conditions when waves are strongly shoaling and close to breaking. For smaller relative wave heights and when bedforms occur, net cross-shore wave-induced transport rates were small and inconsistent at Vejers. This indicates that sediment supply from the lower to the upper shoreface may be limited and occurs from a rather narrow zone seaward of the longshore bars.

8 Conclusions

Acoustic backscatter sensors are increasingly used to estimate sediment concentrations and transports on the shoreface, and algorithms exist for inversion of the backscattered signal to provide sediment concentrations. Due to their complexity, these algorithms may not be favourable for using with large data sets and they normally assume uniform sediment grain sizes. Here, we have derived a simple method to obtain sediment concentrations and transport rates from PC-ADP sensors in settings with a coexistence of sand and wash load. The method is based on a transfer function between acoustic backscatter and concentration, and using a heuristic estimate of sound attenuation caused by water absorption and wash load. Tests suggest that this method performs better than accepted methods under the particular conditions sampled. Data comparison with predictions from a frequently used sediment transport model suggest that a PC-ADP can produce reasonable estimates of sediment transport rates on the shoreface, even when integrated over relatively long time scales. The results obtained further indicate that onshore transport rates from the lower to the upper shoreface are insignificant, unless waves are close to breaking.

References

[1] Osborne, P.D., Vincent, C.E. & Greenwood, B. Measurement of suspended sand concentrations in the nearshore: field intercomparison of optical and acoustic backscatter sensors. *Continental Shelf Research*, **14**, pp. 159-174, 1994.

[2] Betteridge, K.F.E., Thorne, P.D. & Cooke, R.D. Calibrating multi frequency acoustic backscatter systems for studying near-bed suspended sediment transport processes. *Continental Shelf Research*, **28**, pp. 227-235, 2008.

[3] Sheng, J. & Hay, A.E. An examination of the spherical scatterer approximation in aqueous suspension of sand. *Journal Acoustic Society of America*, **83**, pp. 598-610, 1988.

[4] Thorne, P.D. & Hanes, D.M. A review of acoustic measurement of small-scale sediment processes. *Continental Shelf Research*, **22**, pp. 603-632, 2002.

[5] Crawford, A.M. & Hay, A.E. Determining suspended sand size and concentration from multifrequency acoustic backscatter. *Journal Acoustic Society of America*, **94**, pp. 3312-3324, 1993.

[6] Kostachuk, R., Best, J., Villard, P., Peakall, J. & Franklin, M. Measuring flow velocity and sediment transport with an acoustic Doppler current profiler. *Geomorphology*, **68**, pp. 25-37, 2005.

[7] Wren, P.A. & Leonard, L.A. Sediment transport on the mid-continental shelf, Onslow Bay, North Carolina during Hurricane Isabel. *Estuarine, Coastal and Shelf Science*, **63**, pp. 43-56, 2005.

[8] Lacy, J.R. & Sherwood, C.R. Accuracy of a pulse-coherent acoustic Doppler profiler in a wave-dominated flow. *Journal Atmospheric and Oceanic Technology*, **21**, pp. 1448-1461, 2004.

[9] Vincent, C.E. Measuring suspended sand concentration using acoustic backscatter: a critical look at the errors and uncertainties. *Coastal and Shelf Sediment Transport*, ed. Balson, P.E. & Collins, M.B., Geological Society of London, Special Publication **274**, pp. 7-15, 2007.

[10] Lee, T.H. & Hanes, D.M. Direct inversion method to measure the concentration profile of suspended particles using backscattered sound. *Journal of Geophysical Research*, **100**, pp. 2649-2657, 1995.

[11] Thorne, P.D., Vincent, C.E., Hardcastle, P.J., Rehman, S. & Pearson, N. Measuring suspended sediment concentrations using acoustic backscatter devices. *Marine Geology*, **98**, pp. 7-16, 1991.

[12] Nielsen, P. *Coastal Bottom Boundary Layers and Sediment Transport*. World Scientific, Singapore, 324 pp, 1992.

[13] Nielsen, P. Suspended sediment concentration under waves. *Coastal Engineering*, **10**, pp. 23-31, 1986.

[14] Aagaard, T. *Beach Morphodynamics: two Danish case studies*. Medd. Skalling-Laboratoriet, 38, 160 pp, 2002.

[15] Bailard, J.A. An energetics total load sediment transport model for a plane sloping beach. *Journal of Geophysical Research*, **86**, pp.10938-10954, 1981.

[16] Thornton, E.B., Humiston, R.T. & Birkemeier, W.A. Bar/through generation on a natural beach. *Journal of Geophysical Research*, **101**, pp. 12097-12110, 1996.

[17] Dingler, J.R. & Inman, D.L. Wave-formed ripples in nearshore sands. *Proc.15th Conf. on Coastal Eng.*, ASCE, New York, pp. 2109-2126, 1977.

Sediment flux in a rip channel on a barred intermediate beach under low wave energy

B. Greenwood[1], R. W. Brander[2], E. Joseph[2], M. G. Hughes[3], T. E. Baldock[4] & T. Aagaard[5]
[1]*Department of Physical & Environmental Sciences, University of Toronto Scarborough, Canada*
[2]*School of Biological, Earth & Environmental Sciences, University of New South Wales, Australia*
[3]*Geoscience Australia, Marine and Coastal Environment Group, Australia*
[4]*Department of Civil Engineering, University of Queensland, Australia*
[5]*Institute of Geography & Geology, University of Copenhagen, Denmark*

Abstract

The classic model of water and sediment flux in barred surf zones is a net flux landward across a *nearshore bar*, alongshore in a *feeder channel*, and offshore in narrow jet-like flows in a *rip neck* cut through the bar; this circulation is frequently modulated by the tide, even under micro-tidal conditions. Water levels, waves, currents and suspended sediment transport (SST) were recorded at elevations of z = 0.13, 0.26 and 0.39 m in a *rip neck* on an *intermediate bar-rip* beach at a micro-tidal site, Bennett's Beach, NSW, Australia. Measurements revealed SST was driven by *quasi-steady* rip currents and by gravity and infragravity waves. The balance between these components determined the magnitude and direction of the overall net SST. Tentative conclusions are that:

(i) the direction of the overall net SST rate in the rip neck was dominated, as expected, by offshore-directed mean cross-shore currents, especially around high tide; at this time the SST rates due to gravity and infragravity waves were relatively small and somewhat variable in direction.

(ii) as the tide fell, relatively large SST rates were directed onshore by shoaling gravity waves propagating through the neck, which opposed and even exceeded the rip current transport. At mid-tide, the transport by infragravity waves complemented the gravity waves such that the overall net flux of

WIT Transactions on Ecology and the Environment, Vol 126, © 2009 WIT Press
www.witpress.com, ISSN 1743-3541 (on-line)
doi:10.2495/CP090181

suspended sediment was directed onshore into the rip cell at the two lowest elevations.

(iii) whenever the SST by the *rip current* and by *shoaling waves* was close to a balance, it was the infragravity waves that controlled the rate and direction of the overall net suspended sediment flux.

Keywords: rip currents, sediment flux, wave and tide modulation.

1 Introduction

Rip currents are an integral component of the cellular water and sediment circulation in surf zones, both marine and lacustrine. On barred coasts, rips are directed offshore as jet-like flows in constricted channels cut through the bar, and often fed by longshore-directed feeder currents. The classic circulation model is a net flux of water landward across the bar followed by a longshore and seaward flux through the feeder channel and rip neck (Komar [1]). This flux is often modulated by the tide, even under micro-tidal conditions (Aagaard et al. [2]; Brander [3, 4]; MacMahan et al. [5, 6]). Rip currents may play an important role in maintaining the nearshore sediment balance, although this is far from proven, since measurements are limited. In this paper, the flux of suspended sediment in a rip channel neck under breaking swell and wind waves is documented, and the role of currents, waves at a range of frequencies, and the tide is assessed.

2 Location and methods

Mean water levels, waves, horizontal currents, and suspended sediment concentrations were recorded in a rip neck at Bennett's Beach, NSW, Australia (Figs 1, 2 and 3) in February, 2004. The beach is micro-tidal, with a mixed semi-diurnal regime, and was in an "*intermediate bar-rip*" state (Wright and Short [7]) during the measurement period. The beach consists of medium sand, and samples from the bar were well-sorted, but bimodal, with a mean grain size of 1.79 ø (300 μm), a standard deviation of 0.37 ø (91 μm) and a negative skewness, -0.61 ø, typical of wave-winnowed sand (Greenwood and Davidson-Arnott [8]). The beach was subject to long period swell from the SE during the experiment, coupled with a variable wind-wave field forced by a sea breeze. The beach is oriented to a SSE exposure, with fetch restricted to the north by the Little Gibber headland (Dark Point) and Broughton Island, and to the south by Yacaaba Head, and Cabbage Tree and Boondelbah Islands. The study rip cell was asymmetric with a dominant feeder current flowing from the southwest. Figure 2 illustrates the rip cell on Feb. 19th with tracer injected at the head of the feeder channel; the narrow rip neck and diffused rip head are shown. The nearshore bathymetry (Fig. 3) consisted of a dominant, oblique nearshore bar welded to a shoal to the south and extending alongshore in a north-easterly direction to end at the rip neck; a trough separated the bar from the beach face, varying in depth from near zero where the bar was welded to the shoal, to ~ 1 m deep before it turned offshore into the rip neck, which at high tide was ~1.75 m deep. Over the three-day monitoring period, the transverse bar moved both northeastwards and offshore forcing the rip neck to migrate slightly northward also, but causing a distinct constriction and increase in depth of the channel (Figure 3).

The study cell was instrumented with sensors deployed each day at high tide in the centre of the rip neck (P1, Fig. 3) to measure: (a) cross-shore and alongshore (relative to the average local shoreline orientation) horizontal currents, at elevations of z = 0.13, 0.39, 0.50 m, using biaxial electromagnetic urrent meters (Marsh-McBirney, OEM 512); (b) sediment concentrations at z = 0.13, 0.26, 0.39 m, using optical backscatter sensors (D&A Instruments, OBS-1P and OBS-3); and (c) mean water surface elevation and waves, using a strain-gauge pressure sensor (Druck-1830). Waves incident to the surf zone were recorded by a pressure sensor deployed ~130 m directly offshore of the bar (P3). Data were recorded continuously at 4 Hz, and stored in consecutive ~17-minute "bursts". Records were taken during falling spring tides from high water to low water and back to mid-tide.

Figure 1: Bennett's Beach, NSW, Australia; the arrow (upper right panel) marks the study site. Also shown is an oblique aerial view of the study site. Note: the shoals, nearshore bars and the rip cells are not identical to those during the study period on this intermediate bar-rip beach.

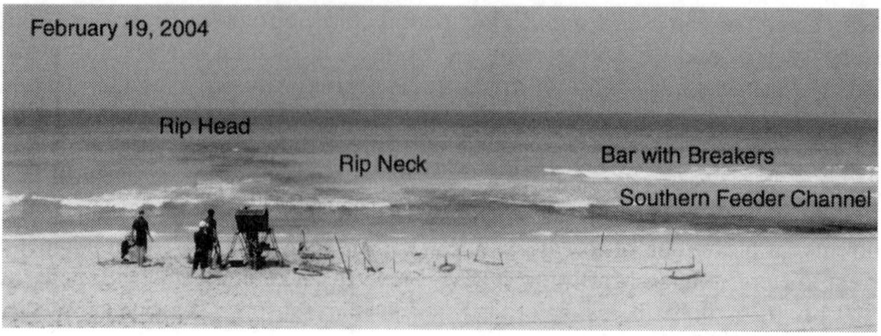

Figure 2: Study rip cell at Bennett's Beach with tracer, Feb. 19th, 2004. Note the breaking waves on the bar, the feeder current, the rip neck and the rip head.

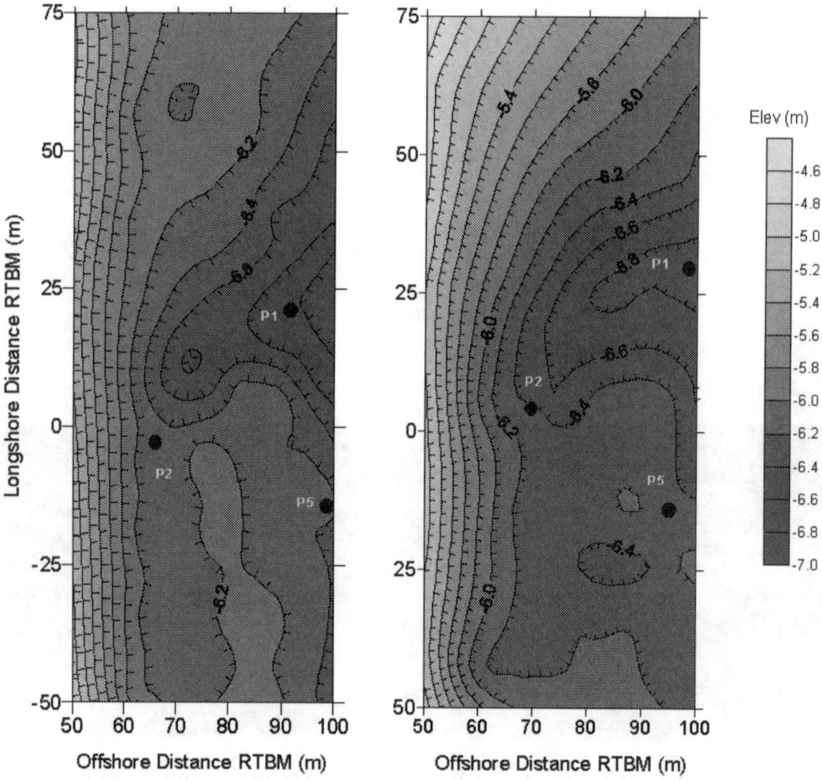

Figure 3: Nearshore bathymetry, Bennett's Beach, Feb. 19th (left panel) and Feb. 21st (right panel). P1 marks the location of the instrument pod in the rip neck and P2 and P5 locates the pods in the feeder channel and on the bar crest. Pods were relocated each day as the morphology changed. Note: RBTM is relative to the bench mark.

Time series of instantaneous horizontal fluid velocities and suspended sediment concentrations were used to compute the SST rates for each "burst". Concentrations measured at 0.13 and 0.26 m were coupled with velocities measured at 0.13 m; concentrations measured at z = 0.39 m were coupled with velocities measured z = 0.39 m. There is potential error here (see Austin and Masselink [9]); however, in all cases the measured currents at 0.13 m were only very slightly different from those measured at 0.39 and 0.50 m; the latter two were essentially identical. The sensors were supported on a solid heavy base, which did not allow significant shifts in elevation once the support settled, and in any case the transports were averaged over 17-minute blocks.

Time series of "collocated" velocities and concentrations were used to compute the time-averaged net, $<q_s>_{net}$, mean, $<q_s>_{mean}$ and oscillatory $<q_s>_{osc}$ components of the SST rate at each elevation (see Jaffe et al. [10]; Osborne and Greenwood [11]):

$$<q_s>_{net} = \frac{1}{T}\int_0^T \int_0^h U_{(z,t)} \; C_{(z,t)} \; dzdt$$

$$<q_s>_{mean} = \frac{1}{T}\left[\int_0^T \int_0^h U_{(z,t)} dzdt\right] * \frac{1}{T}\left[\int_0^T \int_0^h C_{(z,t)} dzdt\right]$$

$$<q_s>_{osc} = \frac{\Delta f}{F}\int_0^T \int_0^h C_{UC(f)} dzdt$$

where U = instantaneous velocity, which can be disaggregated into cross-shore and alongshore components; C = instantaneous sediment concentration; z = elevation; t = time; $C_{UC\,(f)}$ = cospectrum of U and C; Δf = unit bandwidth; F = frequency range; h = water depth; and T = time. The cospectrum identifies SST due to oscillatory motions at different frequencies (Huntley and Hanes [12]; Davidson et al. [13]); gravity and infragravity waves were separated at 0.04 or 0.05 Hz, at pronounced reductions in variance at these values on each day.

3 Results: February 19th and 21st, 2004

3.1 Water levels

The range in water levels recorded in the rip neck (P1) were larger than the tidal range recorded offshore at P3; there was also a slight shift in the timing of the low water recorded in the rip neck again relative to the tide offshore (Fig. 4). The maximum water level range recorded over the measurement period offshore at P3 was 0.49 m on both Feb. 19th and 21st; in contrast, in the rip neck the average range in water level for the two days was 0.61 m, an amplification of ~30%.

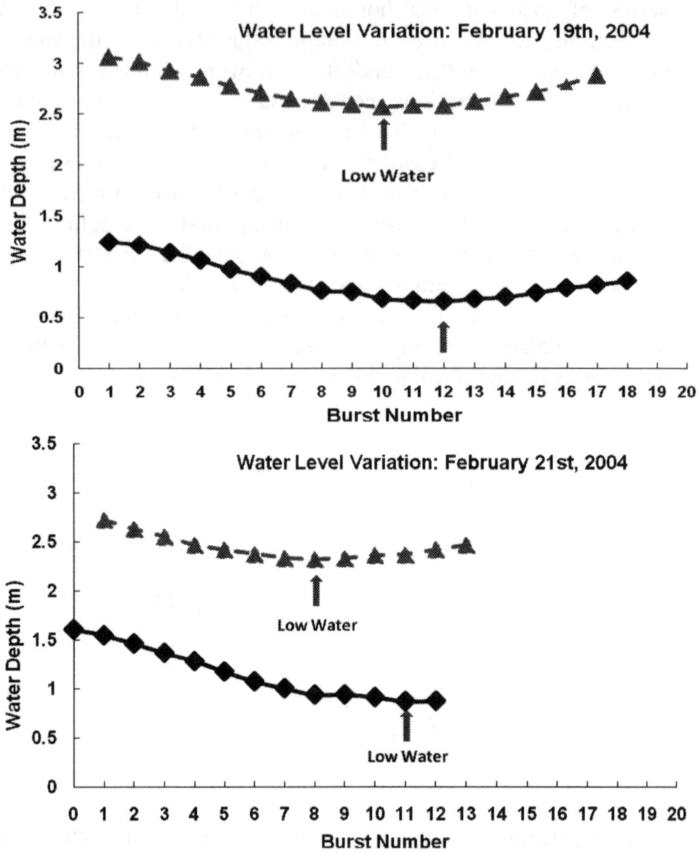

Figure 4: Water levels outside the surf zone at P3 (triangles) and in the rip neck at P1 (diamonds). Note the small differences in amplitude and phase between the two measurement locations.

3.2 Waves

The primary energy source driving rip circulation on both days was an incident gravity wave field consisting of both long period swell (0.100-0.065 Hz; 10–16s) and locally generated wind waves (0.17–0.12 Hz; 6–8 s; Fig. 5). Feb. 21st was more energetic than the 19th, but spectra on both days were dominated by narrow banded, long-crested swell. Wind waves grew in the afternoons with the sea-breeze, broadening the incident spectra (Fig 5). Outside the surf zone, little energy was recorded at infragravity frequencies, although significant peaks at 0.013 Hz (73 s) and 0.010 Hz (100 s) occurred late in the tidal cycle on the 21st (Rip 21_4 and Rip 21_11; see Fig. 5). There was a reduction in the overall variance at incident frequencies in the rip neck (P1) compared to offshore, as well as a significant "red shift" in the current spectra, with distinct peaks at infragravity frequencies, 0.015 – 0.019 Hz (53-67 s; Fig. 6).

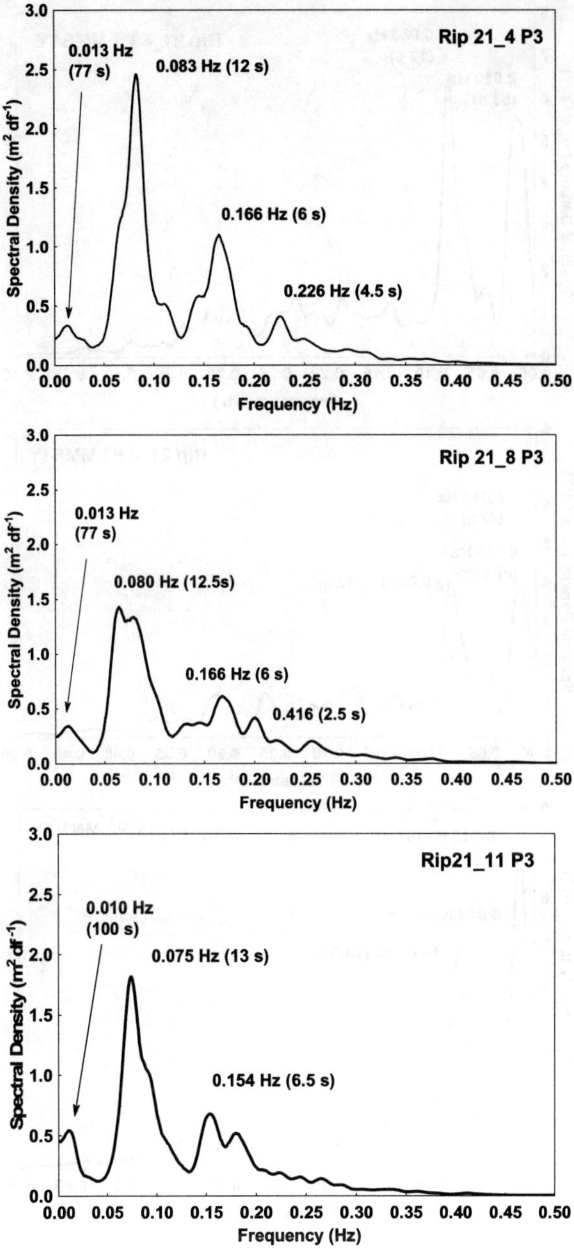

Figure 5: Wave spectra recorded offshore at P3 from mid tide (Rip 21_4; 1458 h) to low tide (Rip 21_11; 1658 h) on Feb. 21st, 2004.

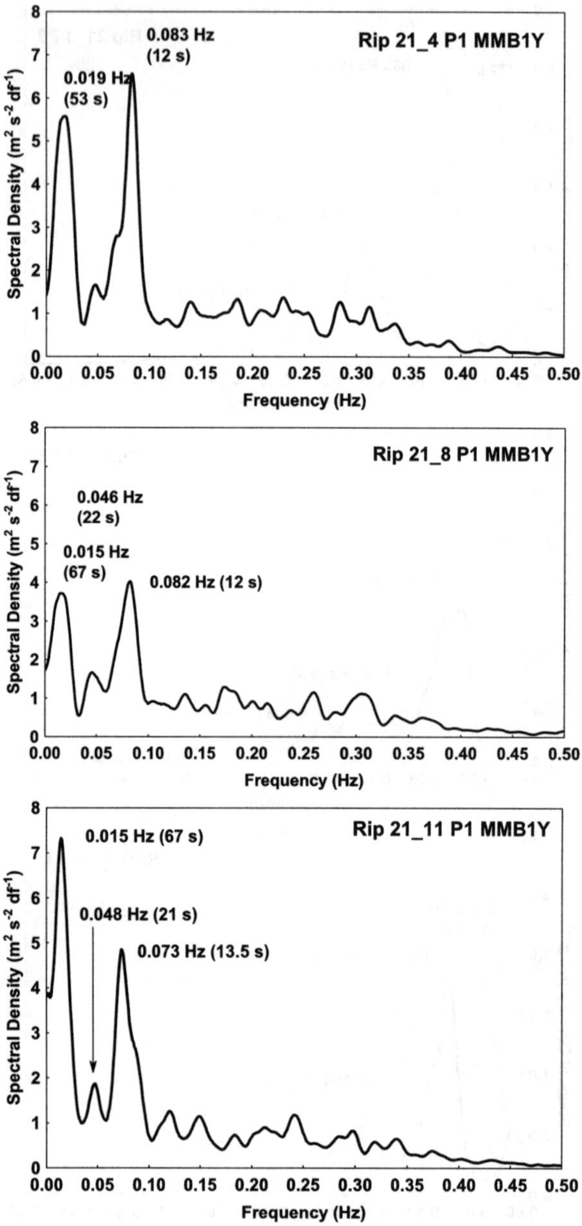

Figure 6: Cross-shore current spectra recorded in the rip neck from mid tide (Rip 21_4; 1458 h) to low tide (Rip 21_11; 1658 h) on Feb. 21st, 2004.

3.3 Mean current speed in the rip neck and tidal levels

Rip current speeds have been shown to increase as the tide falls under both low and high energy conditions, as a result of increased wave breaking and also topographic forcing as the cross-sectional area of a rip channel is reduced by falling water levels in barred systems (e.g. Aagaard et al. [2]; Brander [3]; Brander [4]; MacMahan et al. [5]). Fig. 7 illustrates the relationship between water level in the rip neck and the magnitude of the mean cross-shore current during the falling stage of the tide at Bennett's Beach. A positive relationship existed between rip current speed and water level (see also Castelle and Bonneton [14], Castelle et al. [15]).

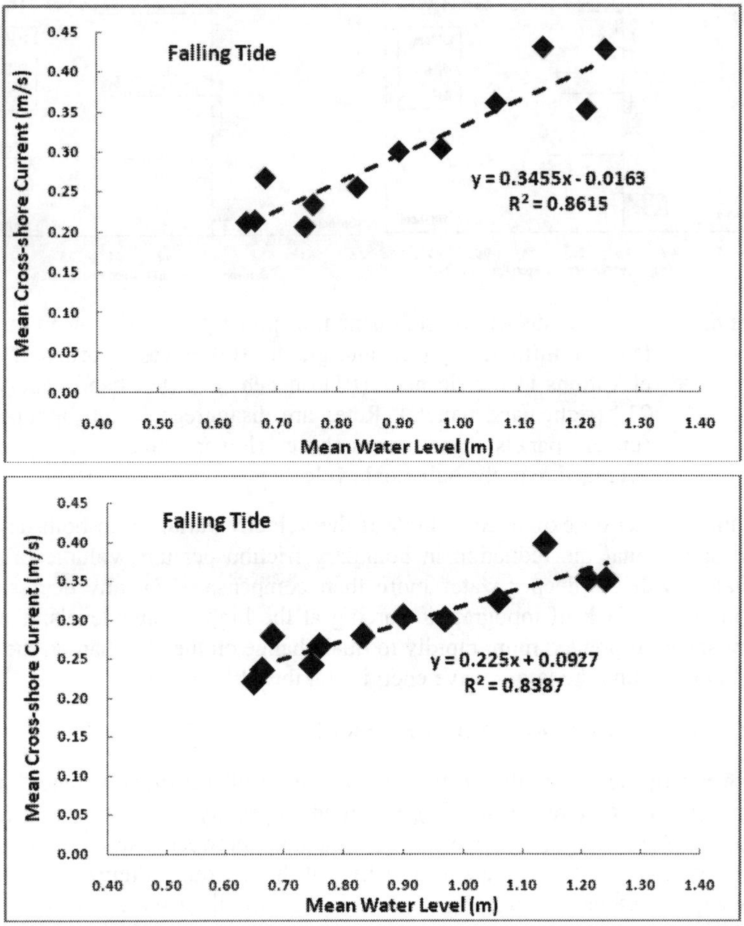

Figure 7: Mean cross-shore current speeds measured at z = 0.15 m and the meawater levels recorded in the rip neck during falling water levels (Feb. 19th (upper panel) and Feb. 21st, 2004 (lower panel).

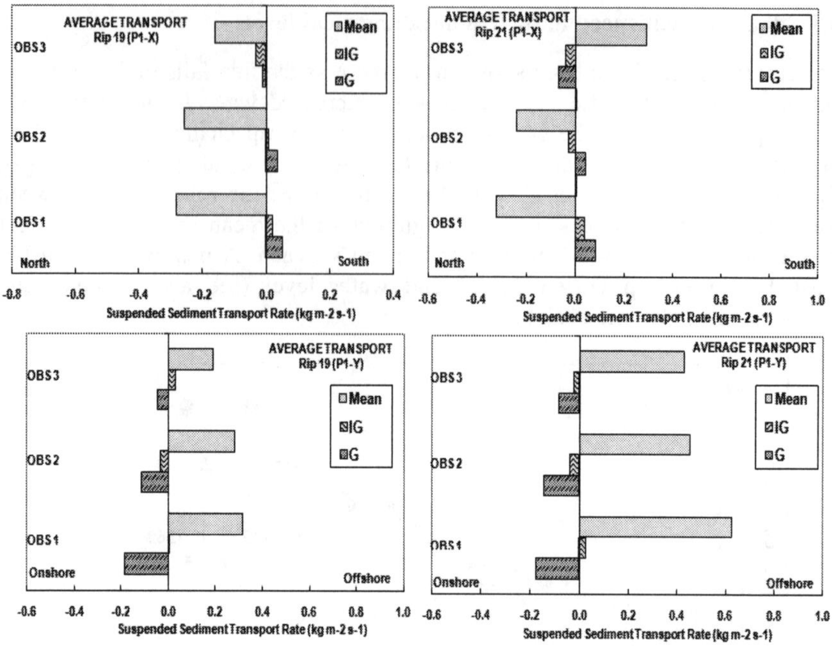

Figure 8: Average suspended sediment transport rates by the mean current (Mean), infragravity (IG) and gravity (G) waves recorded at three elevations in the rip neck (P1) on Feb. 19th (left hand panels)and 21st (right hand panels). Rates are disaggregated into alongshore (upper panels) and cross-shore (lower panels) components averaged over the half-tidal cycle.

A linear function explained >80 % of the velocity variance on both days. It would appear that the reduction in boundary friction per unit volume of flow associated with the deeper water more than compensates for any decrease in breaking or the lack of topographic forcing at the higher water levels. The rip current speed responded more rapidly to tidal change on the 19th than on the 21st, probably reflecting the larger wave energies on the 21st.

3.4 Suspended sediment flux in the rip neck

SST in the rip neck was driven by a combination of: (a) time-averaged, mean currents (the rip current essentially), (b) incident gravity waves and, to a lesser extent, by (c) infragravity waves. The balance between these components controlled both the magnitude and direction of the net flux. Figure 8 illustrates the average cross-shore and alongshore SST rates for all bursts recorded over the half-tidal cycle at each of the three elevations in the rip neck (P1) on Feb. 19th and 21st. As expected, the average net flux was seaward and to the north, driven dominantly by the mean rip current following the NE-SW axis of the rip channel. However, this pattern was not consistent throughout the tidal cycle. At high tide on both days, the net SST was indeed dominated by the mean current and

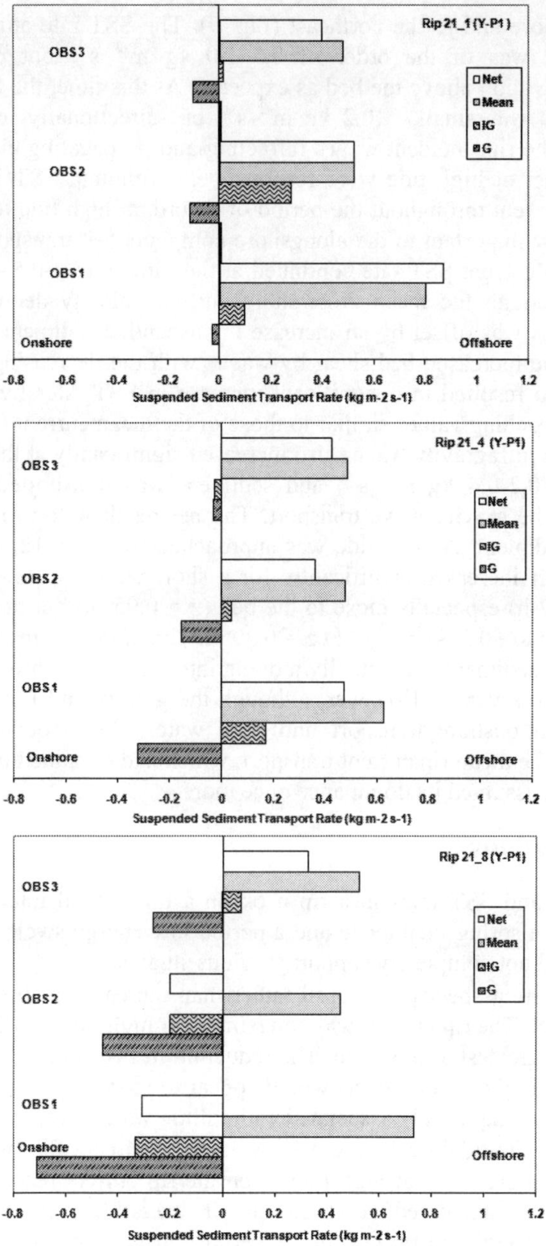

Figure 9: Cross-shore SST rates in the rip neck (P1) at high tide (Rip 21_1), mid-tide (Rip 21_4 and Rip 21_8). The mean current (Mean), infragravity (IG) and gravity (G) wave transports are shown as well as the 17-minute total net (Net) flux.

directed offshore and to the northeast (Fig. 9). The SST rate attributable to the mean current was of the order of 0.5–1.0 kg m^{-2} s^{-1}, but decreased with increasing elevation above the bed as expected. At this time, the SST rate of the gravity waves was small, ≤ 0.2 kg m^{-2} s^{-1}, but directionally in opposition to transport by the rip; incident waves refracting and propagating along the axis of the rip channel at high tide were responsible. Although SST by infragravity waves was present throughout the period of record, at high tide it was relatively small and only important to the alongshore component of transport. As the tide fell, the mean current SST rate continued at the same rate (~0.5–1.0 kg m^{-2} s^{-1}). Thus, even though the mean cross-shore current velocity decreased as water levels fell, this was offset by an increase in suspended sediment concentrations induced by the increased bed shear by waves with the decreasing water depths. The latter also resulted in a significant increase in SST rates by gravity waves themselves, reaching values similar to those of the mean current (~0.5–1.0 kg m^{-2} s^{-1}). SST by infragravity waves also increased significantly at this time (now of the order of 0.2-0.4 kg m^{-2} s^{-1}) and sediment was transported shoreward to complement the gravity wave transport. The net result was a *landward flux* of suspended sediment. As low tide was approached (bursts 8-12; Fig. 4), gravity wave transport increased significantly for a short while, up to ~0.7 kg m^{-2} s^{-1} (burst 8, 1607 h) especially close to the bed ($z = 0.05$ m) but decreased rapidly with elevation to ~0.3 kg m^{-2} s^{-1} at $z = 0.39$ m. This transport reversal meant that at these times sediment was actually moving into the rip cell through the rip neck rather than the reverse. However, although the gravity and infragravity waves maintained an onshore transport until low water, the associated magnitudes dropped and the mean rip current transport, which had remained between 0.5 and 0.8 kg m^{-2} s^{-1}, assumed its dominance once more.

4 Conclusions

The currents and SST rates in a rip neck on a micro-tidal intermediate barred beach during a spring tidal cycle and a period low energy swell and wind wave conditions did not completely support previous studies:

- Rip current speeds decreased rather than increased as tidal levels fell in the rip channel. The rip current was a maximum at high tide and thus the net SST rate was also greatest at this time. The reduction in bed friction per unit volume of water with higher water levels would appear to more than compensate for the bathymetric forcing usually associated with falling water levels in the rip neck.
- During the tidal cycle, the net suspended sediment flux was *not* directed uniformly offshore; only at high tide when the rip current reached a maximum was sediment transported offshore at all elevations. At mid-tide, swell propagating through the rip channel forced a transport reversal and an onshore net transport of suspended sediment at least near the bed.
- The net flux of suspended sediment did not increase with the falling tidal levels as expected. Although at mid- and low-tide overall transport rates increased significantly, some of this transport was directed onshore by both gravity and infragravity waves in opposition to the mean current, which still transported sediment offshore.

Acknowledgements

Funding is acknowledged from NSERC Canada (BG) and an FRGP grant from UNSW (RWB). Mark Daly and Felicity Weir provided valuable field assistance; we are indebted to Dave Mitchell (University of Sydney) for his technical, field and culinary skills. Tom Meulendyk and Heena Dhawan (UTSC) assisted with data analysis, supported by funding from the Government of Canada.

References

[1] Komar, P.D. *Beach Processes and Sedimentation.* 2nd Edition, Prentice Hall, New Jersey, p.338, 1998.

[2] Aagaard, T., Greenwood, B., Nielsen, J. Mean currents and sediment transport in a rip channel. *Marine Geol.*, 140: 25-45, 1997.

[3] Brander, R.W. Field observations on the morphodynamic evolution of low-energy rip systems. *Marine Geol.*, 157: 199-217, 1999.

[4] Brander, R.W. Morphodynamics of a large-scale rip current system at Muriwai Beach, New Zealand. *Marine Geol.*, 165: 27-39, 2000.

[5] MacMahan, J.H., Thornton, E.B., Stanton, T.P., Reniers, A.J.H.M. RIPEX: observations of a rip current system. *Marine Geol.*, 218: 113-134, 2005.

[6] MacMahan, J.H., Thornton, E.B., Reniers, A.J.H.M. Rip Current Review. *Coastal Eng.*, 53: 191-208, 2006.

[7] Wright, L.D., Short, A.D. Morphodynamic variability of surf zones and beaches: a synthesis. *Marine Geol.*, 56: 93-118, 1984.

[8] Greenwood, B, Davidson-Arnott, R.G.D Textural variation in the sub-environments of the shallow-water wave zone, Kouchibouguac Bay, New Brunswick. *Canadian J. Earth Sciences*, 9: 679-688, 1972.

[9] Austin, M.J., Masselink, G. The effect of bedform dynamics on computing suspended sediment fluxes using optical backscatter sensors and current meters. *Coastal Eng.* 55: 251-260, 2008.

[10] Jaffe, B.E., Sternberg, R.W., Sallenger, A.H. The role of suspended sediment in shore-normal beach profile changes. *Proceedings of the 21st Coastal Engineering Conference*, American Society Civil Engineers, p. 1725-1743, 1984.

[11] Osborne, P.D., Greenwood, B. Frequency dependent cross-shore suspended sediment transport 1: a non-barred shoreface, Queensland Beach, Nova Scotia, Canada. *Marine Geol.*, 106: 1-24, 1992.

[12] Huntley, D.A., Hanes, D.M. Direct measurement of suspended sediment transport. *Proceedings of Coastal Sediments '87*, American Society Civil Engineers, New York, p. 723-737, 1987.

[13] Davidson, M.A., Russell, P.E., Huntley, D.A., Hardisty, J. Tidal asymmetry on a macro-tidal intermediate beach. *Marine Geol.*, 110: 333-353, 1993.

[14] Castello, B., Bonneton, P. Nearshore waves and currents over crescentic bars. *Journal Coastal Res.*, SI39: 687-691, 2004.

[15] Castello, B., Bonneton, P., Sénéchal, N., Dupuis, H., Butel, R., Michel, D. Dynamics of wave-induced currents over a multiple barred sandy beach on the Aquitaine coast. *Continental Shelf Res.*, 26: 113-131, 2006.

Section 6
Pollution and dispersion

Environmental impact assessment and HazOp study of the drilling cuttings confinement process into non-productive wells in marine platforms in Campeche, Mexico

M. Muriel-García[1], J. G. Cerón[2] & R. M. Cerón[2]
[1]*Instituto Mexicano del Petróleo, México*
[2]*Universidad Autónoma del Carmen, México*

Abstract

Potential risk and environmental impacts associated with the cuttings re-injection process as an alternative method for drilling wastes disposal were identified and assessed in two marine platforms facilities located in Campeche, Mexico. Environmental impacts were identified and assessed on physicochemical, biological-ecological, socio-cultural and economical elements using the Rapid Impact Assessment Matrix (RIAM). Drilling cuttings re-injection can cause negative effects with light changes. Most of negative impacts were on the physicochemical and biological-ecological elements, whereas positive impacts were on the socio-cultural and economical elements. The most critical negative impacts were: effects due to extreme events, persistent substances presence, effects on aquatic fauna, and changes in biodiversity. The most important positive impacts were those that generated changes in the regional and local economy. By applying the correct mitigation measures it would be possible to decrease the impacts, and in some cases, to eliminate them. Risks were evaluated using the HazOp methodology, deviations in the process were analyzed, recommendations were provided and each risk analyzed was categorized as tolerable or non-tolerable. Consequences analysis from an accidental spill of slurry and/or cuttings was carried out by the YAXUM-3D mathematical model. The results of the consequences analysis showed that even the concentrations in the discharge point exceed the recommended criteria for protection of marine life and marine water quality; the spill is rapidly dispersed complying with the permitted levels in a period of 8 h as a maximum.

Keywords: environmental assessment, HazOp, drilling cuttings disposal, re-injection, platforms.

doi:10.2495/CP090191

1 Introduction

Most industries generate pollutants; their volumes and risk level depend on the nature of activities, and on the interactions of these pollutants with the environment. Until 2004, drilling wastes from offshore drilling platforms were sent to Dos Bocas Port located in Tabasco, Mexico and then they were transported within containers by land to a thermal desorption plant. This process constitutes an expensive option to dispose of drilling wastes. A more efficient and economical alternative is the drilling wastes confinement by re-injection in non-productive wells. At beginning of 2004, one offshore facility in the Mexican Gulf implemented this method showing good results: a safer elimination of drilling wastes, a decrease in the current costs related to disposal of wastes, marine logistic activities and holding times in drilling equipments using oil based fluids. The Mexican Oil Industry decided to apply this methodology to other facilities located in the northeast of the Campeche Sonda.

Cuttings injected by this process only come from wells using drilling fluids that are water, oil or polymeric based, because it is prohibited to inject drilling cuttings or residual mud containing inverse emulsion fluids. To comply with Mexican legislation [1], in order to implement this process it is necessary to carry out an environmental assessment and a risk analysis to identify impacts on environmental factors and risks derived from the drilling cuttings reinjection process into non-productive wells. This study used a Rapid Impact Assessment Matrix (RIAM) to identify and assess environmental impacts. HazOp methodology was used to identify and evaluate risks related to this process.

2 Methodology

2.1 Process description

Drilling cuttings disposal by re-injection involves the following stages: 1) Mechanical grinding of drilling cuttings until one obtains a slurry or a proper semi-liquid phase to be injected; 2) Treatment of this slurry by adding and diluting chemicals to keep the proper characteristics of density, viscosity and rheology; 3) Injection of cuttings slurry into a proper formation through the surfacing piping or drilling piping; and, 4) Ensuring the long term confinement of the injected wastes to avoid a future escape toward the surface. Each one of these stages involves an inherent risk related to operation of any mechanical equipment, possible human errors and environmental impacts. An undesirable event is a possible accidental spill of cuttings slurry and/or drilling cuttings, for this reason, in this study spill transport in marine water was modeled. The confinement stage was not considered due to the following reasons: 1) the injection process is not carried out within the annular section and 2) the injection is carried out into a non-productive well, which has been previously characterized so that its stability over the long term can be assured.

Table 1: Specific coordinates of both Offshore Platforms (A and B).

Platform	Geographic Coordinates
Platform A	19º12′0” N; 92º13′9” W
Platform B	19º10′6” N; 92º12′1’’W

2.2 Study site location

The offshore facilities considered in this study are located in the Campeche Sonda, 65 km NW from the Carmen City coast in Campeche, Mexico. The specific location of these facilities is shown in table 1.

2.3 Environmental impact assessment

To define the environmental impacts generated from the Project we used the Rapid Impact Assessment Matrix (RIAM) proposed by Pastakia [2]. This methodology uses a scoring system in a matrix format. Generated impacts from the activities of the project are assessed against the environmental components, which are designed with a specific punctuation that provides a measure of the expected impact for each component. The analysis is carried out based on criteria that are divided into two groups (A) and (B), and four environmental areas (physicochemical, biological-ecological, socio-cultural and economic/ operational).

2.3.1 Criteria

(A) criteria are related to the importance of the condition (they can cause changes in the obtained score in an individual level), and (B) criteria are related to the importance of the situation (they cannot change the obtained score). Both groups, (A) and (B), are constituted by different criteria, which are defined as shown in table 2. Values assigned to any of these groups are determined by the following equations, where: (a1) and (a2) are the criteria individual score for group (A); (b1) and (b3) are the criteria individual score for group (B); aT is the result of multiplying the criteria score of group (A); bT is the result of adding the criteria score of group (B); and VA is the final score for the condition analysis.

$$(a1) \times (a2) = aT \qquad \text{for Group (A)} \tag{1}$$

$$(b1) + (b2) + (b3) = bT \qquad \text{for Group (B)} \tag{2}$$

$$(aT) \times (bT) = VA \qquad \text{Condition Final Result} \tag{3}$$

2.3.2 Environmental components

In table 3 the different environmental categories used in this method are shown. A matrix is generated where each of the environmental components and its criterion are listed. In table 4 the individual criterion (VA) is shown, grouped into categories (VC), which can be compared to each other. Categories are

Table 2: Criteria description used in RIAM matrix.

Criteria	Description	Values Scale
A1	*Condition importance.* - This criterion is evaluated against a space limit or human interests that could be affected	(4) Importance at an international/national level; (3) importance at a national/regional level; (2) importance at immediate areas outside of the local condition; (1) importance only at the local condition; and (0) not important.
A2	*Change magnitude/effect.* - Magnitude is defined as a measure of the benefit/loss caused by an impact or condition	(+3) greater positive benefit; (+2) significant improvement of the current status; (+1) improvement of the current status; (0) no change/current status; (-1) negative change of the current status; (-2) loss or significant negative change; and (-3) greater change/loss.
B1	*Permanent.* - Defines if a condition is temporal or permanent and it must only be a measure of the condition temporal status	(1) no change/it does not apply; (2) temporal; and (3) permanent.
B2	*Reversibility.* - Defines if the condition can be changed and it is a measure of the control over the condition effect	(1) no change/it does not apply; (2) reversible; and (3) irreversible.
B3	*Cumulative.* - A measure of either the simple direct impact or cumulative effect over time	(1) no change/it does not apply; (2) no cumulative/simple; and (3) cumulative/ synergistic.

Note: Positive and negative impacts can be assessed by a values scale considering negative and positive values for group (A). A zero value means that "there is no change" or "it is not important". For group (B), a zero value is not considered because if all criteria were zero, the final result of VA would be zero. This condition cannot occur even if the group (A) criteria show a condition importance that could be considered; to avoid this, a value of "1" is used to define the situation "without change/or not important".

defined by conditions that act as markers showing a change in the (A) group score combined with the highest or lowest (B) group score. These conditions have been defined in an interval of $\pm$ 5 and each value describes a generated impact derived from the project. Limits of the categories are showed in table 4 with numerical and alphabetic values.

2.4 HazOp analysis

HazOp analysis involves two stages: 1) identification of the risk involved, and 2) ranking of these risks. The most feasible event is the accidental spill of drilling cuttings and/or slurry to the ocean. Risks were evaluated using the procedure to determine the tolerable risk level in the northeast marine region facilities [3]. A hazard is identified when a deviation in the normal operation state exists. The Risk Analysis Group identified the existent protections and if these were not enough to neutralize the hazard or mitigate its consequences, they proposed specific actions to accomplish that. First, study nodes are selected (process lines, vessels and/or process equipment) according to the process flow and considering auxiliary services; then the design intention of each equipment is defined; after, a guide word is selected in combination with a process parameter to develop the

Table 3: Environmental components used in the RIAM matrix.

Physicochemical components: all physical and chemical aspects of the environment, including finite natural resources (not biological) and degradation of the physical environment by pollution.
Biological-ecological components: all biological aspects of the environment, including renewable natural resources, biodiversity conservation, interactions between species and biosphere pollution.
Socio-cultural components: all human aspects of the environment, including social issues affecting individuals and communities, cultural aspects, such as heritage conservation, and human development.
Economical/operational components: economical consequences of the temporal and permanent environmental changes and complexity in project management in terms of project activities.

Table 4: Categories used in the RIAM matrix.

RIAM Environmental Value (VA)	Alphabetic values of the category (VC)	Numerical values of the category (VC)	Category Description
108 to 72	E	5	Greater change/positive impact
71 to 36	D	4	Significant change/positive impact
35 to 19	C	3	Moderate change/positive impact
10 to 18	B	2	Change/positive impact
1 to 9	A	1	Simple change/positive impact
0	N	0	There is no change/current status/it does not apply
-1 to -9	-A	-1	Simple change/light negative change
-10 to -18	-B	-2	Change/negative impact
-19 to -35	-C	-3	Moderate change/negative impact
-36 to -71	-D	-4	Significant change/negative impact
-72 to -108	-E	-5	Greater change/negative impact

deviation in a study node; then consequences are assessed (considering that all safeguards have failed); after, causes of the deviation and all existent safeguards are listed; and finally, actions and recommendations needed to mitigate deviations are listed. This procedure was applied to each study node. Finally, this information was documented in a HazOp worksheet for each node.

2.5 Hydrodynamic study

The YAXUM/3D tool [4] was used to model the dispersion on aquatic medium of a slurry/cuttings spill derived from an error in the operation. Three scenarios were modeled in a period of 61 days: the dry period (from February to the middle of June), the rainy period (from the middle of June to the middle of October) and the "norths" period (from the middle of October to January). Two cases were identified: 1) a spill of 250 bbls from retention tank of 500 bbls; and 2) a rupture of a 2" line to the injection well due to corrosion that could spill 160 bbls. The discharge point has the initial coordinates of the corresponding platform location. Bathymetric data base was created from the World Database ETOPO 2 [5, 6] and a numerical mesh of the study area was generated by using the ARGUS ONE program [7]. To carry out two-dimensional simulations in the 10x10 km domain and to generate the initial hydrodynamic parameters in the

3x3 km domain, it was necessary to process the marine current velocities (different harmonics were obtained for different ports of the Gulf of Mexico), wind intensities [5, 8, 9] and variation of tides [8, 10–12]. Velocity vectors for currents in a sequence of 15 days were determined for each climatic period, finding a predominant current direction from the east to the west with a light deviation toward the north-west quadrant for the three periods. The dispersion study was limited to the liquid phase, as the solid phase does not have an important effect on the water column due a short residence time (including metals and total hydrocarbons of petroleum (THPs)).

3 Results and discussion

3.1 Environmental assessment

When the maximum confinement capacity of wastes into a non-productive well is reached, this will be plugged and process equipment will be re-installed in another well/facility. Activities are limited to installing the re-injection process equipment. Construction activities are not required because the project is located in existent facilities, so the environmental impact assessment was only focused on the operation/maintenance stage. Effects on different environmental components are showed in table 5.

In table 6, environmental assessment results are shown for each group of effects. Most of the physicochemical impacts were classified as *Simple Change/ Light Negative Impact* (F/Q1, F/Q3, F/Q7, F/Q8 and F/Q9). F/Q4 causes a negative impact, although even persistent substances can be quickly dispersed. In table 6 it can be observed that B/E2 and B/E4 are classified as *Simple Change/ Light Negative Impact*, whereas *B/E3 and B/E1* are classified as *Negative Impacts.* Socio-cultural impacts were classified as follows: S/C3 and S/C4 were assigned within the *Simple Change/Light Negative Impact* category, and S/C1 and S/C2 were considered as *Simple Change/Positive Impact.* Regarding economical-operational impacts, there was a *Simple Change/Positive Impact* related to O/E2 and there was a *Significant Change/Positive Impact* related to O/E1.

The drilling cuttings re-injection process will produce light negative impacts, most of which affect physicochemical and biological-ecological components, whereas positive impacts are caused on socio-cultural and economical-operational components. During the installation, operation and maintenance stages, there will be positive and negative changes that will affect the regional economy and ecological environment. Twelve negative impacts were identified against four positive impacts and three impacts that do not generate any change. Most critical negative impacts were those that affect F/Q2, F/Q4, B/E1, and B/E3 environmental components. The most important positive impact was that affecting the E/O1 component. Positive impacts with a minor importance were those that cause simple changes on S/C1, S/C2 and O/E2.

Table 5: Environmental components assessment.

Physicochemical	Biological-ecological	Socio-cultural	Economical/ Operational
F/Q1: Water quality (eventual discharges from shipments, sanitary residual water and process effluents)	*B/E1: Fauna* (a spill could have effects on the zooplankton development due toxic substances, and a temporal migration of nektonic and benthic species could happen due to noise emissions)	*S/C1: Employment* (during installation and removal stages of equipment, permanent employment will be minimal, but temporal employment will be important)	*O/E 1: Changes in local/regional economy* (there will be a strengthening of the state economy in Campeche and Tabasco)
F/Q2: Extreme events (tropical storms and norths could affect process operation)			
F/Q3: Navigation (tug boats, boats for transportation of equipment and materials, and launches for the relocation of workers are typical of the area of platforms; fishing boats could be affected in a temporal way on a minor scale due to there being a restricted area for boating activities and fishing)			
F/Q4: Persistent Substances (A spill could impact the marine water, however, changes will depend on meteorological conditions, prevent and control measures applied to contain it, and the presence and concentration of the persistent substances into the spilled material)			
F/Q5: Landscape alteration (this project will be located in existent facilities, for this reason, landscape is not going to be altered)			
F/Q6: Emissions from wastewater (Domestic wastewater will be generated and will be sent to treatment plants. Treated waters will be spilled into the sea water if they comply with the established limits (NOM-001-ECOL-1996)	*B/E2: Marine vegetation* (a spill could have effects on the fitoplankton development due toxic substances)	*S/C2: Auxiliary services* (some services will be temporarily required: rent of machinery and equipment, food, boat, and accommodation services, and logistic support)	*O/E2: Operation and maintenance costs* (some activities related to the installation and operation of the re-injection equipment and services, such as food, accommodation, logistic support, and transport, will be required
F/Q7: Emissions of hazardous and non-hazardous wastes (domestic materials and industrial wastes will be managed according to the corresponding legislation)			
F/Q8: Gas and particles emissions (CO, SO_2, NO_x particles and unburned hydrocarbons will be generated from the internal combustion process of tug boats and cranes, and process equipment located in the platforms, but these emissions do not exceed the permissible limits)			
F/Q9: Noise emissions (Noise emissions from ship and crane engines during the transfer of equipment and during drilling cuttings and containers stowed during operations could affect the marine environment, however, noise sources will be into confined and isolated sites located at 25 or 30 m over sea level, decreasing the impact from these noise emissions. Besides they will comply with the legislation (NOM-080-STPS-1993))	*B/E 3: Changes in biodiversity* (in case of a spill, marine flora and fauna biodiversity could be affected in a punctual and temporal way at the discharge point)	S/C3: *Solid wastes* (solid wastes will be generated during some project stages and a collection and disposal system will be required)	
	B/E 4: Eutrophication (a spill of wastewater and industrial effluents without treatment under some climatic and biological conditions could cause water eutrophication)	S/C4: *Sewage treatment* (sewage and sanitary residual waters will be produced, for this reason a sewage treatment system will be required)	

Table 6: Environmental impact assessment results.

	Criterion (A)		Criterion (B)			Group A	Group B	Condition result	Category numeric Value	Category alphabetic value	
Component	a1	a2	b 1	b2	b3	At	Bt	VA	VC	VCA	Description
F/Q 1	1	-1	2	2	2	-1	6	-6	-1	(-A)	Simple Change/ Light negative impact
F/Q 2	3	-1	2	1	1	-3	4	-12	-2	(-B)	Change/ Negative Impact
F/Q 3	1	0	2	2	2	0	6	0	0	(N)	There is no change/Current Status
F/Q 4	1	-2	2	2	3	-2	7	-14	-2	(-B)	Change /Negative Impact
F/Q 5	0	0	2	2	1	0	5	0	0	(N)	There is no change/ Current Status
F/Q 6	1	0	1	1	1	0	3	0	0	(N)	There is no change/Current Status
F/Q 7	1	-1	2	2	2	-1	6	-6	-1	(-A)	Simple Change/ Light negative impact
F/Q 8	1	-1	2	2	2	-1	6	-6	-1	(-A)	Simple Change/ Light negative impact
F/Q 9	1	-1	3	3	1	-1	7	-7	-1	(-A)	Simple Change/ Light negative impact
B/E 1	1	-1	2	2	3	-1	7	-7	-1	(-A)	Simple Change/ Light negative impact
B/E 2	1	-1	3	3	3	-1	9	-9	-1	(-A)	Simple Change/ Light negative impact
B/E 3	1	-1	3	3	3	-1	9	-9	-1	(-A)	Simple Change/ Light negative impact
B/E 4	1	-1	3	3	1	-1	7	-7	-1	(-A)	Simple Change/ Light negative impact
S/C 1	2	1	2	1	1	2	4	8	1	(A)	Simple Change/Positive impact
S/C 2	2	1	2	1	1	2	4	8	1	(A)	Simple Change/ Positive impact
S/C 3	1	-1	2	2	2	-1	6	-6	-1	(-A)	Simple Change/ Light negative impact
S/C 4	1	-1	2	2	2	-1	6	-6	-1	(-A)	Simple Change/ Light negative impact
O/E 1	3	3	2	2	3	9	7	63	4	(D)	Significant Change/Positive Impact
O/E 2	2	1	2	1	1	2	4	8	1	(A)	Simple Change/ Positive impact

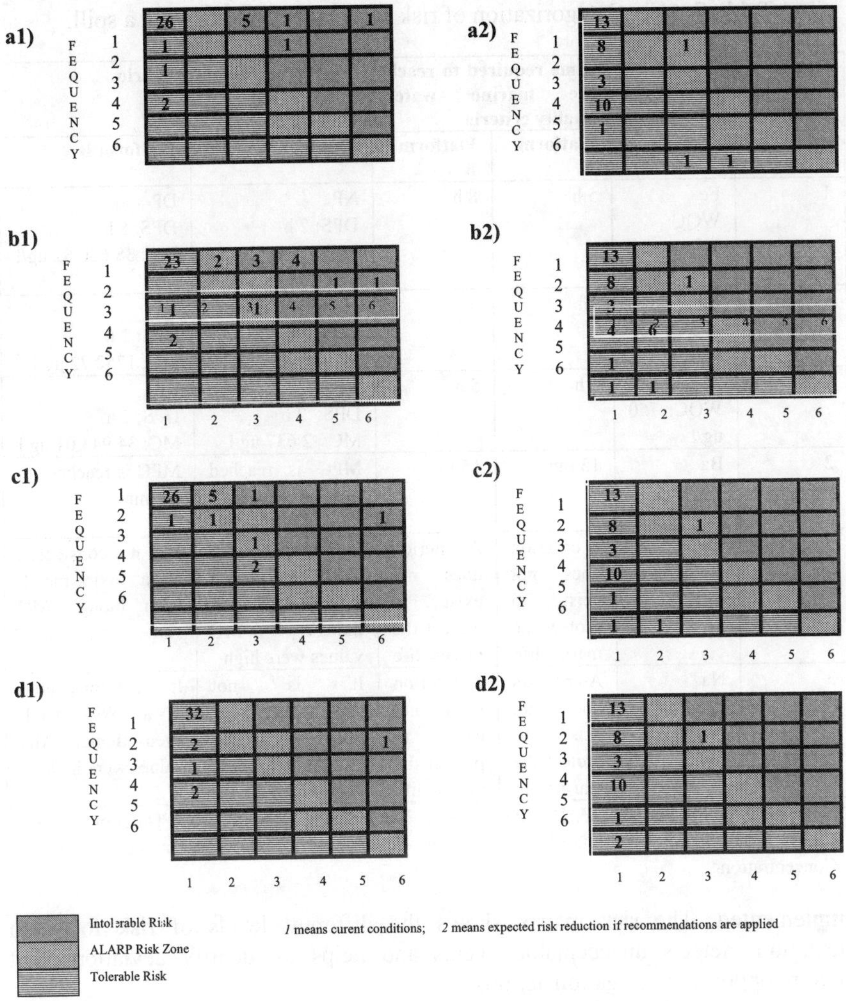

Figure 1: Risk matrixes: a) for damage to personnel, b) for environmental impacts, c) for damage/losses in production, and d) for damage to facilities.

3.2 HazOp results

Each identified risk was evaluated considering damage to personnel and population, environmental impacts, economical and production losses according to the procedure described in the methodology section. These results are used to create the risk matrix, which contains the number of sceneries corresponding to each combination of the frequency/consequence ratio (figure 1). These matrixes show the expected risk reduction as soon as the recommendations are

Table 7: Categorization of risk scenarios resulting from a spill.

Category	Parameter	Time required to reach the marine water quality criteria		Worst scenario	
		Platform A	Platform B	Platform A	Platform B
1	Fe WQC: 1000 ug/l	5 h	8 h	NP DPS: 2 h MC: 8 330.14 ug/l	DP DPS: 5 h MC: 38 680.82 ug/l
1	Be WQC: 130 ug/l	5h	5 h	NP DPS: 2 h MC: 530.16 ug/l	NP DPS: 2 h MC: 1752.72 ug/l
1	Al WQC: 750 ug/l	5 h	5 h	NP DPS: 2 h MC: 2 637 ug/l	RP DPS: 2 h MC: 34 943.01 ug/l
2	Ba WQC: 50 000 ug/l	15 min	15 min	MPC is reached in 15 min	MPC is reached in 15 min
3	Ca	A criterion does not exist to protect the marine life	A criterion does not exist to protect the marine life	It is not considered as a toxic metal, even though MC values were high	It is not considered as a toxic metal, even though MC values were high
3	Na	A criterion does not exist to protect the marine life	A criterion does not exist to protect the marine life	It is not considered as a toxic metal, even though MC values were high	It is not considered as a toxic metal, even though MC values were high

Note: NP: Norths Period; DP: Dry Period; RP: Rainy Period DPS: Scenario Duration Period; WQC: Water Quality Criteria; MC: Maximum Concentration; MPC: Maximum Permissible Concentration.

implemented. The risk matrix shows the different levels of risk for each deviation, detects unacceptable events and helps to identify deviations that require opportune mitigation actions.

These categories were assigned according to toxicity criteria in the aquatic life [13–16], concentrations found in analyzed samples and the time required to disperse and to reach permissible concentrations. For platform *B*, the concentrations of metals and THPs were greater than those found for platform *A*. The *Worst scenario column* shows the climatic period where maximum concentrations exceeded the permissible values to protect marine life and to meet the water quality requirements, indicating the maximum duration of this scenario. Platform *A* for all climatic periods showed concentrations of THPs greater than the value reported for marine water quality (374 016 μg/l) after 30 minutes. After 2 hr, permissible concentrations were reached for three sceneries modeled. Dispersion is not good in the dry period; for this reason this scenario was assigned as *category 1*, sceneries modeled for the rainy and norths periods were assigned as *category 2*, as their effects can be mitigated in 2 h by actuation of the spill control system.

In Platform *B* for all climatic periods, concentrations were about twice the criteria value for marine water quality (202 971.36 μg/l) after 30 minutes. In the rainy and norths periods, permissible concentrations were reached after 2 hr. In the dry season, the concentration remained at 202 971.36 μg/l after 2 hr and permissible concentrations were reached after 8 hr. The worst scenario in platform *B* was obtained in the dry season and it was assigned as *category 1*, the modeled sceneries in the norths and rainy periods were assigned as *category 2* as their effects can be mitigated in 2 hr by actuation of the spill control system.

4 Conclusions

Environmental impact assessment results show that negative impacts identified on the natural environment can be decreased by applying proper mitigation measures, and it is even possible to eliminate some of them. It is possible to infer that the drilling cuttings re-injection project is feasible and viable for carrying out disposal of wastes into non-productive wells in both platforms. From the HazOp Analysis a total of 63 recommendations were made, some of the most important are the following: to use only and exclusively the fire fighting moto-pumps according to the NFPA 2031; to install the firefighting network and cabinet in the housing platform; to provide enough lifeboats according to the maximum number of people allowed on the platforms; to install audible and visible alarms in the presence of fire, smoke and gas in the housing platform and to install toxic gas detection equipment and self-contained breathing equipment in the operations area; and to meet the maintenance programs for the firefighting network and monitors on the platforms and for crane and auxiliary services. The simulation results for a slurry spill scenario showed that the risk level is tolerable, even though concentrations at the discharge point exceeded recommended criteria to protect marine life and marine water quality. The spill is dispersed quickly, reaching permissible levels in a period from 5 to 8 hr; this means that these concentrations are diluted to tolerable levels before spill control equipment arrives on the site.

References

[1] Ley General de Equilibrio Ecológico y Protección al Ambiente. SEMARNAT, 1996.

[2] Jensen, Kurt. *Environmental Impact Assessment using the Rapid Impact Assessment Matrix (RIAM)*. Olsen & Olsen-Fredensborg. ISBN: 87-85215-32-5, 1998.

[3] PEMEX-PEP-RMNE. *Lineamiento para la Determinación del Nivel de Riesgo Tolerable en las Instalaciones de la Región Marina Noreste*, clave 250-22100-SI-212-0001, version 1, Enero 2003.

[4] Casulli, V. & Cheng, R.T; Semi-implicit finite difference methods for three dimensional shallow water flow. *International Journal for numerical methods in fluids,* **(15)**, pp. 629-648, 1992.

[5] World Data Base ETOPO 2 from the National Center for Atmospheric Research, USA.
[6] Instituto Nacional de Estadística, Geografía e Informática (INEGI), http://www.inegi.org.mx
[7] ARGUS ONE, http://www.argusint.com/
[8] González, S. R. *Modelación numérica de circulación de corrientes oceánicas para el Golfo de México. Generación de escenarios hidrodinámicos*, Tesis de Maestría en Ciencias, Instituto Politécnico Nacional, México D. F., 2005.
[9] Barrios, P. H. *Modelación baroclínica y dispersión de partículas en medios marinos. Aplicación al Golfo de México*. Tesis de Maestría en Ciencias, Instituto Politécnico Nacional, México D. F., 2005.
[10] Salas, D. L. D., y Monreal, G. M. *Mareas y circulación residual en el Golfo de México. Contribuciones a la oceanografía física en México*, Monografía No. 3, Unión Geofísica Mexicana, 1997.
[11] Gómez, R. E., y Vélez, M. H. *Medición de corrientes con perfilador acústico doppler, Reporte Análisis de Datos*, Universidad Autónoma Metropolitana, Unidad Iztapalapa, México D. F., 2004.
[12] CICESE, http://oceanografia.cicese.
[13] 3 CFR Ch. 1 (7-1-97-Edition) Document EPA 151.11.
[14] Current National Recommended Water Quality Criteria, http://www.epa.gov/waterscience/criteria/wqcriteria.html.
[15] Quality Criteria for Water, 1986. EPA 440/5-86-001.
[16] NOM-001-Semarnat-1993. Norma Oficial Mexicana que establece los límites máximos permisibles de contaminantes en las descargas de aguas residuales en aguas y bienes nacionales. DOF, México. 23 de abril de 2003.

Operational tools in the Basque Country (south-eastern Bay of Biscay) for water quality management within harbours

A. Del Campo, L. Ferrer, A. Fontán, M. González, J. Mader,
A. Rubio & Ad. Uriarte
Marine Research Division, AZTI-Tecnalia, Spain

Abstract

In this contribution, a pre-operational local oceanic system, which combines data acquisition with numerical modelling in the two main harbours of the Basque Country, Bilbao and Pasaia, is described. The analysis of wind and current measurements showed the adequacy of the data acquisition system to provide open boundary conditions to solve the hydrodynamics in the internal part of the harbour domains. The use of the Finite Element hydrodynamic model, *TRIMODENA*, combined with a Lagrangian Particle Tracking Model (*LPTM*), offers a viable method to simulate the spatio-temporal evolution of pollutant dispersion for water quality management in an operational way. Some applications of the present modelling system are shown.
Keywords: Bay of Biscay, operational oceanography, numerical modelling, harbour, management, pollution.

1 Introduction

Oceanography has progressed rapidly over the past three decades. This has been driven by the need to develop new technologies, issues related to navigation, climate change, marine ecosystems and water quality management. The latest advances in oceanography provide a new challenge to the scientific community: Operational Oceanography, defined by Fischer *et al.* [1] as the activity of systematic and long-term routine measurements of seas, oceans and atmosphere, and their rapid interpretation and dissemination.

Recently, many public administrations have decided to provide financial support for a wide range of projects in Operational Oceanography, such as:

WIT Transactions on Ecology and the Environment, Vol 126, © 2009 WIT Press
www.witpress.com, ISSN 1743-3541 (on-line)
doi:10.2495/CP090201

EuroGOOS (European co-operation on the Global Observing System) and ECOOP (European COastal-shelf sea OPerational observing and forecasting system) in Europe and ESEOO (Establishment of a Spanish Operational Oceanography System) in Spain. In the Basque Country, two projects facilitate the development of local operational systems: ITSASEUS (supported by the Basque Government) and LOREA (Litoral, Océano y Riberas de Euskadi-Aquitania, funded by the European INTERREG-IVA project). All these projects are focused on providing useful tools, for the atmosphere and ocean, through operational systems which combine data acquisition with numerical modelling. The developed tools will enable the creation of products for the monitoring of several unusual phenomena as well as routine activities such as: atmospheric pollution, storms and surge warnings, high waves, sediment transport, oil spills, river plumes, aerial and maritime traffic, design of marine structures, or coastal water quality.

Within the framework of the ITSASEUS and LOREA projects, the Basque pre-operational oceano-meteorological system is made up of different sub-systems that will be highly co-ordinated to provide different resolutions on global, regional and local scales. Preliminary results by Ferrer *et al.* [2] have shown the capacity of the applied models to predict oceano-meteorological phenomena at regional scales. In this paper, the capabilities of the present pre-operational local oceanic sub-system to forecast currents and pollutant dispersion, within the two main harbours of the Basque Country, are described.

2 Study areas

The Basque country is located in the innermost part of the Bay of Biscay, in the Cantabrian Sea, included as part of the North Atlantic Ocean (Figure 1). In the western part of the Basque Country, Bilbao harbour is located at the mouth of the Nervión estuary (Figure 1). This estuary is the largest in this region, with a surface area of about 20 km^2, an average depth of 30 m and a mean annual river flow of about 36 $m^3 \cdot s^{-1}$. The harbour has been growing progressively from the inner part of the Nervión River to the outer estuarine and open sea areas, becoming one of the most important harbours in Europe. In the eastern part of the Basque Country, Pasaia harbour is located within the Oiartzun estuary (Figure 1). This estuary is surrounded by mountains and relatively small with a surface area of about 1 km^2, an average depth of 10 m and a mean annual river flow of about 5 $m^2 \cdot s^{-1}$. The Pasaia harbour is, from a commercial point of view, the second most important harbour in the Basque Country.

3 Operational observation and modelling tools

3.1 Real-time oceano-meteorological data and bathymetric information

The available oceano-meteorological data, including location, instrument type, record length and sampling rate, is presented in Table 1.

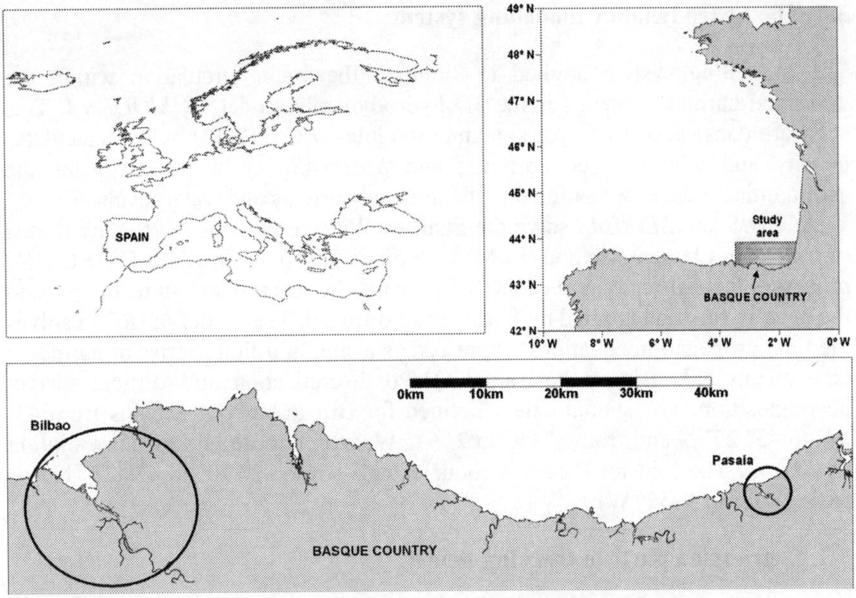

Figure 1: Location of Bilbao and Pasaia harbours in the Basque coast.

Table 1: Description of presented oceano-meteorological stations (EOM), including its location, record length and sampling rate.

Station reference	Location	Instrument type	Station elevation / water depth (above and below sea level) (m)	Mean depth below surface (m)	Start date	Sampling rate (min)
EOM PASAIA	43° 20.3' N 1° 55.5' W	AANDERAA DCM-12 Doppler, Tide gauge and Automatic Weather Station	16 / 24	0 - 4 - 8 - 12 - 16 - 20	Aug. 2001	10
EOM BILBAO	43° 22.7' N 3° 04.9' W		22 / 30	0 - 4 - 9 - 13 - 17 - 22	Aug. 2003	10

Oceanic data (currents, sea surface variations, temperature and turbidity) and atmospheric variables (temperature, wind, pressure and solar radiation) are collected at the stations every ten minutes. At these stations, information on current is provided by an upward-looking ADCP, which measures the current speed and direction at 6 depths within the water column. Atmospheric measurements are collected at 16 and 22 m above mean sea level at the Pasaia and Bilbao stations, respectively.

In both harbour domains, 1 m resolution seafloor Digital Terrain Models from high resolution multibeam system (Seabat 7125) are used for bathymetric information.

3.2 The hydrodynamic modelling system

The modelling system applied to simulate the water circulation within the described harbour domains is the 3D hydrodynamic model *TRIMODENA*. This software consists of two hydrodynamic modules: *ECADIS* [3], which calculates density and wind-induced currents; and *MAREAS* [4], which estimates the astronomical tidal propagation and the induced currents and water levels.

ECADIS and *MAREAS* solve the Shallow Water Equations (SWE), by means of the Finite Element Method (FEM) with quasi-3D approximations [5]. *ECADIS* processes the stationary part of the SWE, using the macro-element technique and the penalty function applied by Fortin and Fortin [6]. The model *MAREAS* solves the tidal propagation equations, expressed as a sum of a finite series of harmonic constituents [7], using a horizontal Q1/P0 discretization and vertical spectra decomposition. The spatial extent defined for Bilbao harbour extends from 43° 14' to 43° 25' N and from 3° 09' to 2° 51' W, with a mean horizontal resolution of 20.4 m. The grid for Pasaia harbour extends from 43° 20' to 43° 23' N and from 1° 58' to 1° 52' W).

3.3 Lagrangian particle-tracking model

A Lagrangian Particle-Tracking Model, *LPTM*, is used to forecast the pollutant dispersion within the harbours of Bilbao and Pasaia. This model is fed by the output of the *TRIMODENA* hydrodynamic model simulations. The current fields computed on the *TRIMODENA* grid are used by the *LPTM*, to estimate the particle velocities where the pollutants are located in each time step. *LPTM* uses random turbulent velocity terms to parameterise unresolved sub-grid phenomena along both, the horizontal and vertical axes. The method used for the dispersion estimation is based upon the 4th order Runge-Kutta scheme.

4 Description of oceano-meteorological conditions

The oceano-meteorological conditions over the study areas are described by means of the available wind and oceanic data from Pasaia and Bilbao stations (Table 1). These also are compared with the model results.

Wind frequency distribution analyses were performed on data acquired from January 2004 to December 2006. At the Bilbao station, the wind direction field showed a seasonal pattern, in agreement with the results of Medina [8] and Usabiaga *et al.* [9]. In accordance with observations of these authors for the Basque coast, the dominant winds are from the south in winter and autumn and from the north in spring and summer. Wind data at the Pasaia station showed a different pattern from that observed at the Bilbao station; the former is dominated by southerlies throughout the year. This observation suggests that seasonal southerlies are reinforced by land breezes at the Pasaia station, due to their channelling along Pasaia Bay as pointed out by Fontán *et al.* [10]. Land effects are important in coastal areas of the Basque Country, especially at Pasaia harbour due to its complex topography.

In order to estimate the contribution of wind to the variability of the surface current at Pasaia station, the squared coherence spectra for wind and sea surface current components, from January 2004 to December 2006, were calculated (Figure 2). The contribution of wind, to surface current fluctuations, occurs over a wide range of frequencies, ranging from: terdiurnal, semidiurnal and diurnal land-sea breezes; to periods of several days, representing the passage of cyclones/anticyclones; to lower frequencies near fortnightly periods, representing changes in the prevailing weather conditions; and, finally, to seasonal variability as pointed out by Fontan *et al.* [10]. Close agreement was observed between sea surface and sub-surface currents and wind measurements, indicating that water circulation is governed mainly by wind forcing fluctuations, over a wide range of meteorological frequencies. This is in accordance with observations of Fontán *et al.* [10].

Surface current data measured at the Pasaia station throughout 2008 and surface currents computed by the hydrodynamic model *TRIMODENA* were compared (Figure 3). For surface current computation within the model, the input wind data were those measurements observed at the Pasaia station. Overall, a high level of agreement is observed between computed and observed surface currents; 57% of the variability in the surface current measurements is explained by model computations.

5 Applications to water quality management

The described tools have been used for some operational applications within the two harbour domains. One of these applications is the analysis of the extension and dilution of industrial effluent discharges in the far-field (i.e. the region where

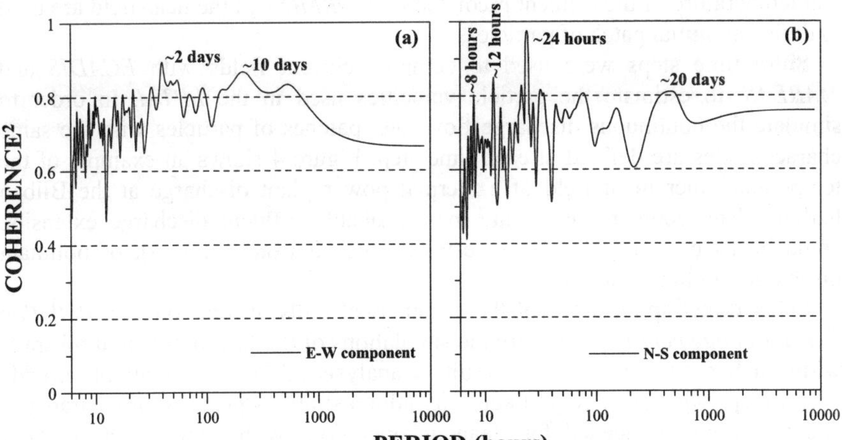

Figure 2: Smoothed squared coherence spectra (99% significance level) of current and wind components, at the Pasaia station, from January 2004 to December 2006 for: eastward (a) and northward (b) components at sea surface.

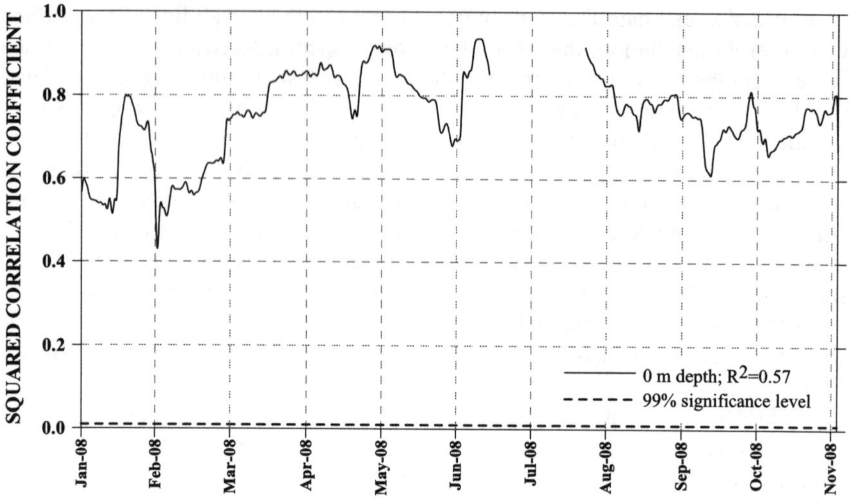

Figure 3: Squared correlation coefficient sequences (99% significance level) between measured and computed surface currents at the Pasaia station in 2008. The sample size in each case is 720 hours.

the mixing is dominated by the ambient flow conditions). Historical wind data from the oceano-meteorological stations were analysed to obtain the local wind probability distributions. Each probable wind is used as an input in *ECADIS* to compute the local wind-induced currents. In these types of studies, the transport processes in the near-field are simulated with *CORMIX1* [11]. The dimensions and temperature of the effluent predicted by *CORMIX1* in the near-field are used to define an initial patch of particles.

Short time steps were used to compute current fields with *ECADIS* and *MAREAS,* to estimate the particle velocities used in the *LPTM.* In order to simulate the continuous discharge flow rate, patches of particles with the same characteristics are defined at each time step. Figure 4 shows an example of the temperature increment field of a thermal power plant discharge at the Bilbao harbour. The computation of the most probable effluent discharge extension could be used as a parameter to estimate the environmental risk of pollutant discharges within a harbour.

Other direct applications of these numerical tools, in relation to a *Pollution Event Management Program,* are the simulations of the fate of potential oil spills within a harbour domain. The resultant analysis of these simulations provide useful maps which enable port authority administrations plan response strategies and the resources needed for clean up operations within the *Local Oil Spill Contingency Plan.* An example of these maps is shown in Figure 5 where the oil fate or retention zones for each study area are represented with a line. The potential pollution points shown in these maps correspond to the locations used by the oil companies as working space on both land and at sea.

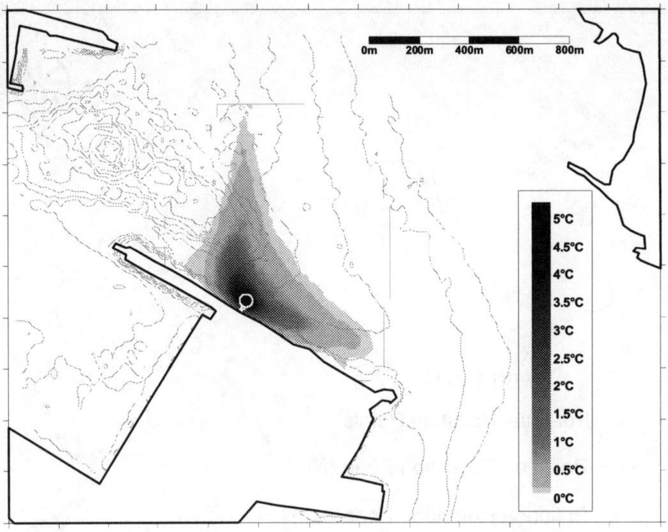

Figure 4: Thermal plume dispersion at the Bilbao harbour.

It is worthwhile to note, that the extrapolation of the prevailing winds observed by these stations, to the entire computational harbour domain may not always be appropriate, particularly for the inner part of the harbour. This may be especially valid in the case of Pasaia, where the presence of mountains almost certainly modifies the wind pattern.

At present, *TRIMODENA* and *LPTM* are being used as prediction tools, in real-time, for a rapid response to a potential oil spills in the Bilbao harbour; AZTI-Tecnalia is providing daily local 72-hour forecasts of particle trajectories to the Port Authority of Bilbao. At the moment, in the absence of a higher resolution local wind model, *TRIMODENA* is being forced by the forecasts obtained from the meteorological model MM5 (Fifth-Generation NCAR / Penn State Mesoscale Model).

6 Conclusions

The meteorological conditions at the study areas have been described by three years of wind measurements from the Pasaia and Bilbao oceano-meteorological stations. Whilst winds at the Bilbao station showed similar seasonal patterns to those described by other authors for the Basque coast, winds at the Pasaia station showed a marked influence of land-sea breezes due to their channelling along the Pasaia Bay. With respect to the marine conditions, a surface current data analysis from the Pasaia station revealed that surface water circulation is governed by wind forcing in the external area of this harbour. The clear agreement between measured and numerically simulated currents with the *TRIMODENA* hydrodynamic model throughout the year at the Pasaia station has shown the adequate capability of the model to predict wind-induced currents outside the

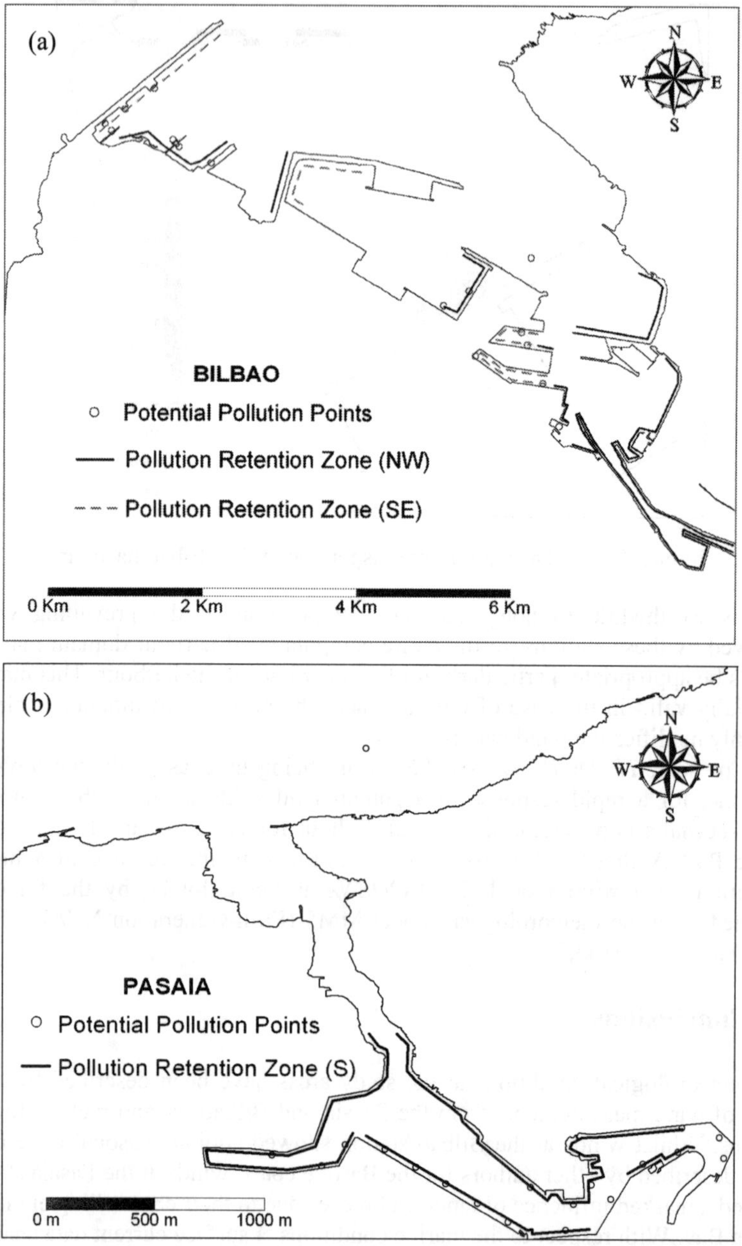

Figure 5: Oil retention zones estimated using prevailing wind conditions: (a) Bilbao harbour (NW and SE winds); and (b) Pasaia harbour (S wind). The location of potential pollution points is also shown.

harbour. Data acquired from these oceano-meteorological stations seem to be adequate for establishing external surface and open boundary conditions. The extrapolation of the wind patterns registered by these stations to the entire computational harbour domain should be improved by using a high resolution local wind model. Additionally, further research is needed in order to simulate currents induced by breezes in localised areas such as harbours. The pre-operational local modelling system described here is being used as a tool for several applications related to harbour management. Some applications, such as environmental risk management of point source pollution and local contingency plans, have been described.

Acknowledgements

This study has been undertaken with the financial support from different sources: Ministry of Education and Science of the Spanish Government together with the Port Authority of Bilbao (PETRI Program, ref.:PET2006_0111), Port Authority of Pasaia, Department of Industry, Trade and Tourism and Department of Transport and Public Works of the Basque Government (ETORTEK Program, ref.: ITSASEUS and; SAIOTEK Program, ref.: MODELTOX) and Aquitania-Euskadi cooperation (INTERREG Program, ref.: LOREA). We would like to acknowledge Department of Transport and Public Works of the Basque Government and Meteorological and Climatology Direction staff for public provision of oceano-meteorological data.

References

[1] Fischer, J., Flemming, N., Holt, M. & R. Rogers, J., 1999. *A profile of operational oceanography*. EuroGOOS Secretariat.

[2] Ferrer, L., Fontán, A., Mader, J., Chust, G., González, M., Valencia, V., Uriarte, Ad., Collins, M.B., 2009. *Low-salinity plumes in the oceanic region of the Basque Country*. Continental Shelf Research, 29 (8): 970-984.

[3] Espino, M., 1994. *Estabilización de la superficie libre en la solución de Shallow-Water por elementos Finitos*. Aplicaciones oceanográficas. Ph. D. thesis, UPC, Barcelona.

[4] González, M., García, M.A., Espino, M., Arcilla, S., 1995. *Un modelo numérico en Elementos Finitos para la corriente inducida por la marea. Aplicaciones al Estrecho de Gibraltar*. Revista Internacional de Métodos Numéricos para Cálculo y Diseño en Ingeniería, 11 (3): 383-400

[5] Zienkiewicz, O.C., Heinrich, J.C., 1979. *A unified treatment of steady state shallow water and two-dimensional Navier-Stokes equations-finite element penalty function approach*. Computer Methods in Applied Mechanics and Engineering, 17-18: 673-698.

[6] Fortin, M., Fortin, A., 1985. *A generalization of Uzawa's Algorithm for the solution of the Navier-Stokes Equations*. Communications in Applied Numerical Methods, 1: 205-208.

[7] Walters, R.A., 1986. *A finite Element Model for tidal and residual circulation*. Communications in Applied Numerical Methods, 2: 393-398.

[8] Medina, M., 1974. *La mar y el tiempo. Meteorología náutica para aficionados, navegación deportiva y pescadores*. Editorial Juventud, 160 pp.

[9] Usabiaga, J.I., Sáenz, J., Valencia, V., Borja, A., 2004. *Climate and Meteorology: variability and its influence on the Ocean*. In: Borja, A. and Collins, M. (eds.). *Oceanography and Marine Environment of the Basque Country*, Elsevier Oceanography Series, 70: 75-95, Elsevier, Amsterdam.

[10] Fontán, A., González, M., Wells, N., Collins, M., Mader, J., Ferrer, L., Esnaola, G., Uriarte, Ad., 2009. *Tidal and wind-induced circulation within the southeastern limit of the Bay of Biscay: Pasaia Bay, Basque coast.* Continental shelf research, 29 (8): 998-1007.

[11] Jirka, G.H., Doneker, R.L., Barnwell, T.O., 1991. *CORMIX: A Comprehensive Expert System for Mixing Zone Analysis of Aqueous Pollutant Discharges*. Water Science and Technology, 24 (6): 267-274.

Bayesian inference for oil spill related Net Environmental Benefit Analysis

R. Aps[1], K. Herkül[1], J. Kotta[1], I. Kotta[1], M. Kopti[1], R. Leiger[1,3], Ü. Mander[2] & Ü. Suursaar[1]
[1]*Estonian Marine Institute, University of Tartu, Estonia*
[2]*Institute of Ecology and Earth Sciences, University of Tartu, Estonia*
[3]*Estonian Maritime Academy, Estonia*

Abstract

This paper investigates the applicability of Bayesian inference to oil spill related situation assessment in order to facilitate the Net Environmental Benefit Analysis (NEBA) based decisions in evaluating the threat or probable overall environmental impact of the spill. Bayesian networks are believed to be useful in integrating the NEBA related information imported from 1) oil spill scene surveillance, 2) simulation results on an oil spill incident with human response, and 3) ecological sensitivity maps. This paper exemplifies the use of Bayesian Belief Networks in answering the questions: can the oil spill be combated at sea, and if it cannot then is the oil threatening a sensitive environment?
Keywords: net environmental benefit analysis, Bayesian inference, oil spill response simulation, Gulf of Finland (Baltic Sea).

1 Introduction

At the International Maritime Organization's (IMO) Marine Environment Protection Committee's 53rd session in July 2005, the Baltic Sea was designated as a Particularly Sensitive Sea Area (PSSA). At the same time, oil transportation is growing significantly in the Baltic Sea area and especially in the Gulf of Finland exceeding 200 million tons a year by 2010. Despite improving navigation measures, there is a growing risk for incidental oil spills and associated oil pollution.

In an actual spill situation, everything possible is done to prevent oil washing ashore. The Net Environmental Benefit Analysis (NEBA) is defined as a method

WIT Transactions on Ecology and the Environment, Vol 126, © 2009 WIT Press
www.witpress.com, ISSN 1743-3541 (on-line)
doi:10.2495/CP090211

to determine the most appropriate response option(s) in order to minimize the overall environmental impact of an oil spill [1, 2]. The NEBA based oil spill response related decision making is considered as essentially a multi stage process. Any stage of decision making starts with the inputs which are oil spill surveillance data collected from diverse sources. With the state of the oil spill appraised, an assessment of the situation is conducted next which, among other aspects, involves assessing 1) expected drift, behaviour and fate of the spilled oil, 2) predicting its future behaviour, and 3) the level of threat it poses to sensitive environment. The oil spill response decision maker is now in a position to weigh the appropriateness of alternate courses of oil combat action and decide upon one – and the cycle starts again.

Bayesian inference is an important statistical tool that is increasingly being used by ecologists in general to evaluate decision making alternatives: in a Bayesian analysis, information available is summarized in a prior probability distribution while posterior probability distributions provide a direct measure of the degree of belief that can be placed on models, hypotheses, or parameter estimates [3–5]. The use of Bayesian techniques in ecological risk assessment has recently attracted considerable attention because (1) they are able to employ subjective interpretations of probability, and (2) they immediately direct the analyst to the full distributional qualities of parameter uncertainty, through the posterior distribution function [6].

This paper investigates the applicability of Bayesian inference to oil spill related situation assessment in order to facilitate the NEBA based decisions in selecting the best available oil spill response alternative, and in evaluating the threat or probable overall environmental impact of the spill.

2 Material and methods

The BBN for situation assessment are constructed using HUGIN RESEARCHER software. General relationships between the variables of interest, in terms of the relevance of one variable to others, are taken into account in a graphical representation capturing the conditional dependencies in a qualitative fashion (parent–child nodes). The links in the graphical representation are then assigned conditional probabilities (Bayesian networks). The BBH constructed for this study is representing the uncertainties in oil spill accident situation assessment. Prior probabilities are obtained from knowledge of the prevailing situation (relevant literature, oil spill surveillance and modelling data) by converting a state of knowledge to a probability assignment.

In the Baltic Sea region the Seatrack Web on-line oil drift forecasting system is used to support the NEBA decision making in oil spill related emergency situations [7]. System covers the entire Baltic Sea area and the eastern North Sea. PISCES (Potential Incident Simulation Control and Evaluation System) is used to simulate development of an oil spill incident with human response and it calculates both the changes to the spill mass due to dynamically varying environmental parameters (e.g. currents, wind, sea state etc.), and also due to the deployment of oil spill response resources, such as booms, skimmers and chemical dispersants [8]. Integrated Seatrack Web and PISCES modelling suite is used to generate the necessary values of input variables for BBNs root nodes.

In situ measurements of flow fields along the Estonian coast of the Gulf of Finland (Baltic Sea) were performed using an oceanographic measuring complex called RDCP-600 from AADI Aanderaa. It applies the Doppler Effect to measure vertical distribution of velocity. Atmospheric forcing conditions are provided by the Estonian Meteorological and Hydrological Institute (EMHI).

The sensitivity maps were based on three different ecosystem elements: the EU Habitat Directive Annex 1 habitats and associated habitat forming species, birds and seals. In each raster cell the maximum value of different layers was calculated to give the final assessment of ecosystem sensitivity by coastal water bodies and the seasons (Figure 1).

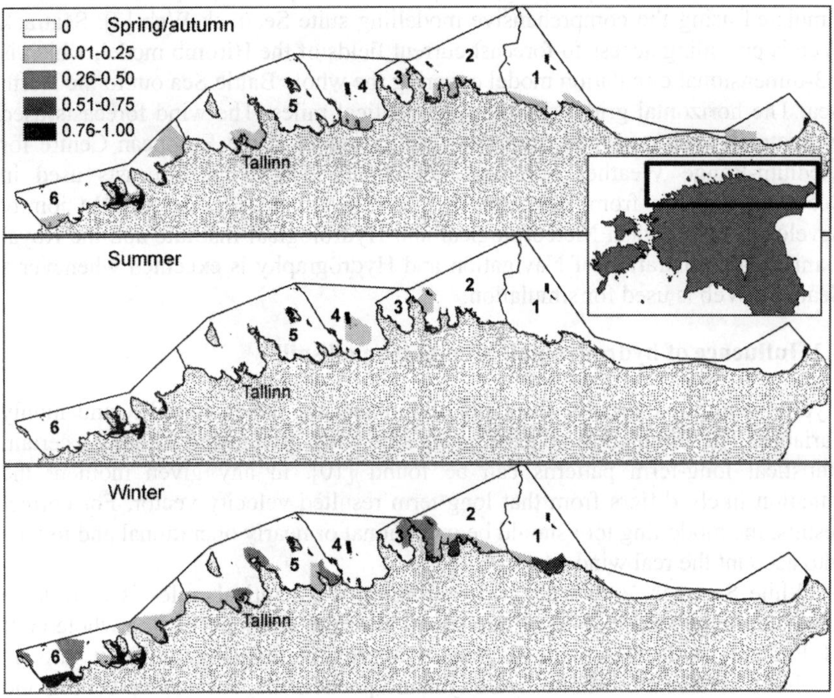

Figure 1: Ecological sensitivities by coastal water bodies of the southern Gulf of Finland (1 – 6) and seasons (spring/autumn, summer and winter). Sensitivity scale according to sensitivity criteria applied: (0) – no sensitivity, (0-0.25) – low sensitivity, (0.26-0.50) – medium sensitivity, (0.51-0.75) – high sensitivity, and (0.76-1.00) – very high sensitivity [9].

3 Results and discussion

3.1 What are the expected drift, behaviour and fate of the spilled oil?

As soon as coastal authorities are notified of an oil pollution incident, they need to gather information on the oil spill (size, location, and the type of oil) and the environmental (weather) conditions in order to evaluate the threat or probable overall environmental impact of the spill [2]. Immediately after notification of a pollution incident at sea, the NEBA is to be performed, and a quick decision is to be taken on the most appropriate response option(s). This decision is based on the following information: 1) what are the expected drift, behaviour and fate of the spilled oil, 2) can the oil spill be combated at sea, and 3) is the oil threatening a sensitive resource?

In a case of oil spill the spatio-temporal fate of spilled oil (transport by currents and wind, spreading, evaporation, dispersion, emulsification) is simulated using the comprehensive modelling suite Seatrack Web [7]. Seatrack Web is providing access to forecast current fields of the Hiromb model, which is a 3-dimensional circulation model covering the whole Baltic Sea out to the North Sea. The horizontal grid resolution is 3 nautical miles. The wind forecasts used in Seatrack Web originate from the weather model at the European Centre for Medium-Range Weather Forecasts (ECMWF). The wind forecasts used in Seatrack Web are from 10 meters height. The oil drift model PADM jointly developed by Swedish Meteorological and Hydrological Institute and the Royal Danish Administration of Navigation and Hydrography is executed whenever a Seatrack Web is used for simulation.

3.2 Influence of hydrodynamic situation on oil spill

Hydrodynamic patterns in the Gulf of Finland are rather complex and highly variable. The circulation scheme is mostly wind-driven and although certain statistical long-term patterns can be found [10], in any given moment the situation likely differs from that long-term resulted velocity vector. For correct results, the modelling tool should be operational or nearly operational and to take into account the real wind situation.

While Seatrack Web with grid resolution of 3 nautical miles is capable to simulate the general hydrodynamic situation of the sea area (i.e. the whole Gulf of Finland) with reasonable degree of approximation (Figure 2), it fails in resolving certain meso-scale hydrodynamic phenomena, such as upwelling and the related baroclinic coastal jets. The typical width (cross-section extension) of coastal upwelling is mainly determined by the barcoclinic (internal) Rossby radius, which in this part of the Baltic Sea ought to be (also depending on stratification conditions) around 1-5 km [11, 12]. The width of the upwelling can be larger due to lateral spread of upwelled water and formation of filaments, though. Still, to resolve the process itself and its accompanying hydrodynamic features (rise in pycnocline, Ekman drift, baroclinic jet), the grid-size should be, according to the published data, about 1/2...1/4 of the Rossby radius [10].

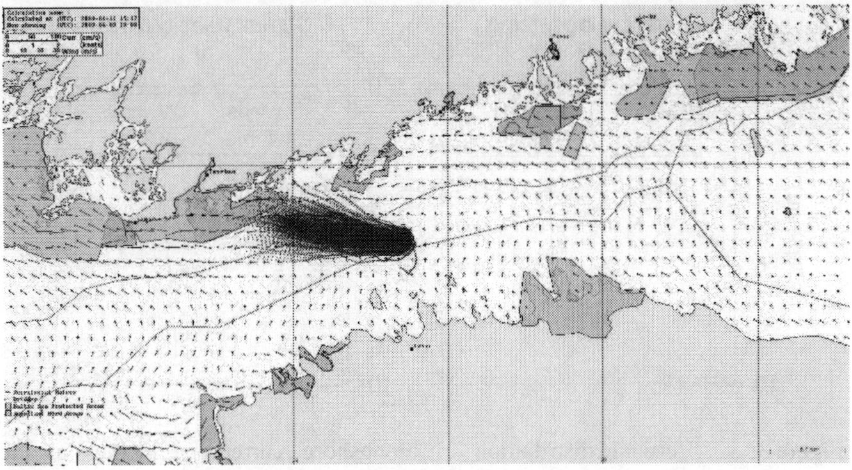

Figure 2: Seatrack Web calculated scenario case: accidental spill of 100 t of fresh medium oil in the Gulf of Finland between Tallinn and Helsinki. Start calculations on 09.04.2009 at 10.00 UTC, and the end of calculations on 06.04.2009 at 15.00 UTC. Arrows are showing the direction and speed of the surface current. Locations of Important Bird Areas and the Baltic Sea Protected Areas are shown along the coasts.

Altogether, 5 measuring sets of multi-layer current dynamics with RDCP has obtained, 2 of them include measurements during extensive upwelling. The first one occurred on July-August 2006 [13, 14], and the second one on August-September 2008. According to MODIS satellite sea surface temperature (SST) images, the upwelling events are more frequent along this relatively straight section of the coast [14]. However, they are rarely covered by direct measurements of hydrodynamics. The frequency of occurrence of upwelling (or downwelling) can be as much as 20-30% in some suitable coastal sections of the Gulf of Finland [15]. As a general rule, persistent westerlies may evoke upwelling along the Finnish coast of the Gulf of Finland, while easterlies and north-easterlies along the North Estonian coast.

Usually, upwelling along one side of the gulf is paired with downwelling along the other coast. Coastal jet appears due to rise in pycnocline during upwelling and evolution of thermohaline stratification. The alongshore current is vertically stratified as well: strong downwind alongshore current in upper layer and relatively weak current (or undertow) in deeper layers. The reason for large surface velocities is simple: a relatively small momentum input is required for driving the relatively thin upper layer. Thus, despite modest wind speeds during upwelling-favour conditions, the velocity can reach 0,6-1 m/s (Figure 3), while much stronger westerly storm winds are capable to yield nearshore velocities up to 0,3-0,6 m/s, which are vertically rather homogeneous, though.

In conclusion, in upwelling-favourable summertime the spilled oil fate may be considerably influenced by changed hydrographical conditions (Figures 3–4).

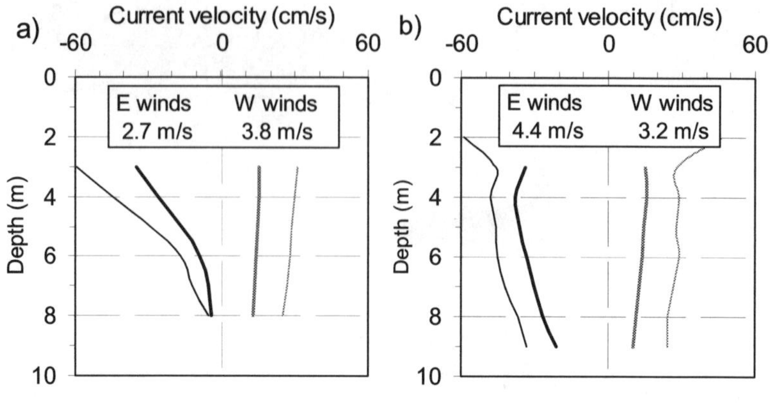

Figure 3: Vertical distribution of alongshore currents velocities at the instrument deployment site (59.56°N, 26.67°E) in the Gulf of Finland (Baltic Sea) in August-September 2006 (a), and August-September 2008 (b) under upwelling conditions (negative values which correspond to westward motions and E-wind forcing) and ordinary conditions (positive current velocity values). Thin lines represent fastest momentary currents and bold lines averages for 5 days with maximum velocities. Possible higher values in upper layers are discarded due to uncertainties in near-surface measuring procedure. Corresponding 5-days average wind forcing is given in the text box.

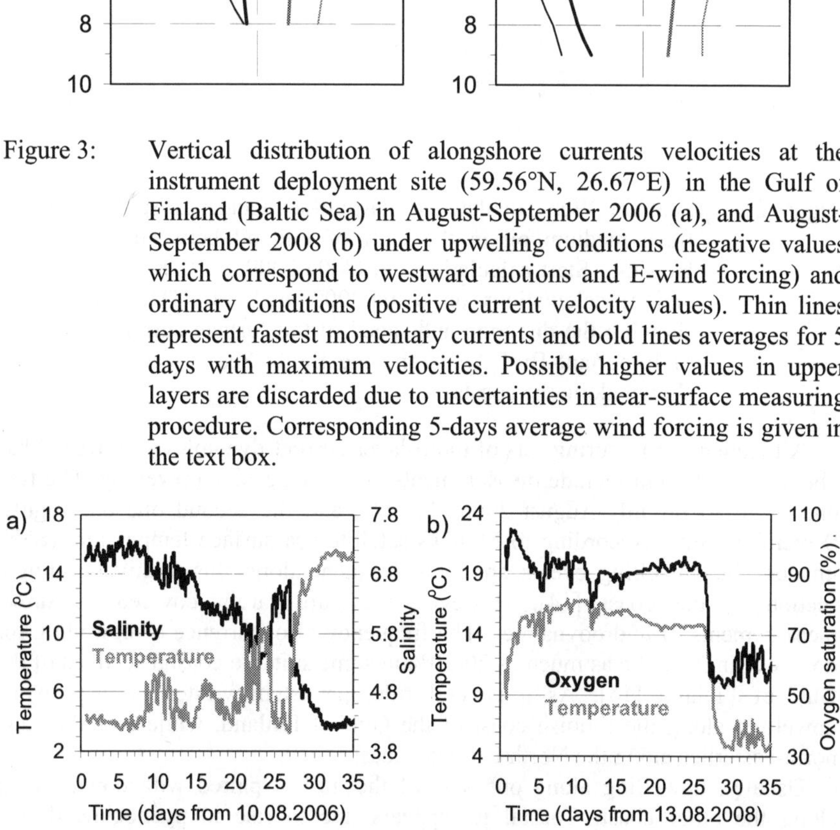

Figure 4: Variations in salinity and water temperature (a) and oxygen saturation and water temperature (b) at the depth of about 10 m at the instrument deployment site (59.56°N, 26.67°E) in the Gulf of Finland (Baltic Sea) during upwelling events in 2006 and 2008. Upwelling conditions apply to low temperature and oxygen content, and high salinity.

Depending on the wind conditions the spilled oil may travel alongshore with a speed up to 50 km per day. The upwelling event itself, however, may last for some weeks. In addition to that, due to low water temperatures (frequently as low as 4....10°C as opposed to normal summertime water temperatures around 20°C in adjacent waters), and lowered oxygen content in the water the spreading, evaporation, dispersion, and emulsification of the spilled oil may also be affected.

3.3 Can the oil spill be combated at sea?

There are obvious advantages if an oil slick that threatens the sensitive coastal sea area can be removed while it is still at sea [16]. Usually, booms and skimmers are the first technique employed to remove oil from marine environments but this technique can usually recover relatively small proportion of the spilled oil. No boom is capable of containing oil against water velocities much in excess of 0.58 m/s acting at right angles to it. The critical current velocity for many crude oils and refined products ranges from 0.7 (0.34 m/s) to 1.2 knots (0.58 m/s). Generally 0.7 knots (0.34 m/s) is accepted as a conservative estimate. Skimmers are used to remove oil from water and put it into storage tanks but how well a skimmer works depends on the type of oil spilled, the thickness of the slick and, the weather conditions.

Salinity, water temperature and depth are problems in the Baltic Sea if dispersants are used [17]. Due to the sensitive ecological conditions in the Baltic Sea area, response to oil should take place by the use of mechanical means as far as possible while response by using dispersants should be limited [18]. Therefore, the option of dispersants use is not considered within this study.

In situ burning has not been usually considered as oil combating response option for the Baltic probably because of very limited window of opportunity and accompanied environmental concerns. This option is also not analyzed in this study.

According to HELCOM Recommendation [19] the Contracting Parties should be able to respond to spillages of oil and 1) to reach within six hours from start any place of a spillage that may occur in the response region of the respective country, 2) to ensure well organized adequate and substantial response actions on the site of the spill as soon as possible, normally within a time not exceeding 12 hours. Decision on deployment of booms and skimmers can be made if mobilization time is less than the calculated time for oil to wash ashore.

The BBN is constructed to assess general situation when answering the question: can the particular oil spill be combated at sea using booms and skimmers? Current and wind speed in the oil incident sea area, predicted time interval of oil coming ashore are imported from Seatrack Web simulations. Mobilization time – the time for a ship/aircraft to get on oil incident scene depends on the time to be ready to go and the time to reach the location of the spill – is imported from the PISCES modelling suite simulation results.

When a Bayesian model is actually used the new information is inserted (current speed, wind speed, oil type, time from spill event, and mobilization time) to bring a variable (alternative: use or no use of booms and skimmers) to a state that is consistent with the new information. For example, the BBN modelling outcome for the favourable weather conditions is presented in the Figure 5.

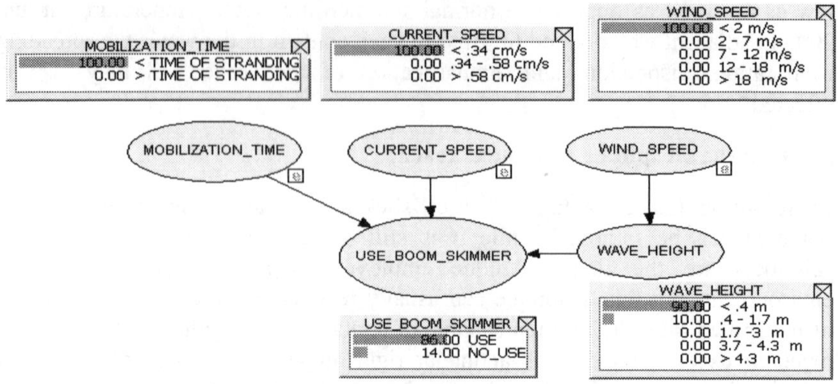

Figure 5: Low current < 0.34 m/s, and a calm wind < 2 m/s. Mobilization time is less than time for oil to wash ashore. Use of booms and skimmers is efficient with probability of 0.86, and inefficient with probability of 0.14.

However, according to results of the BBN simulations the efficiency of booms and skimmers use is rather low under unfavorable weather conditions (Figure 6).

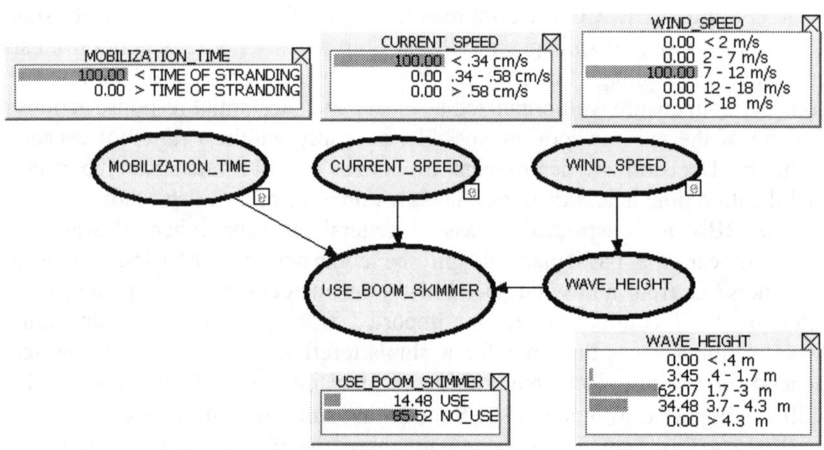

Figure 6: Low current < 0.34 m/s, and a strong breeze 7-12 m/s. Mobilization time is less than time for oil washing ashore. Use of booms and skimmers will be efficient with probability of 0.15, and inefficient with probability of 0.85.

3.4 Is the oil threatening a sensitive resource?

If BBN simulations show that the use of the booms and skimmers is expected to be inefficient, it is almost impossible to prevent the oil from reaching ashore. In this case the advice on sensitive ecological resources likely to be impacted by the oil washing ashore is of critical importance in order to support decisions whether or not a response is necessary or what kind and extent of response is appropriate.

A simple BBN described in more detail in [9] was constructed with an aim to perform the potential oil pollution related predictive ecological risk assessment for the southern part of the Gulf of Finland. A BBN is primarily used to update the ecological risk probability distribution over the states of a hypothesis variable, which is not directly observable. Ecological risk distribution then helps a decision maker in deciding upon an appropriate course of action. According to the requirements of the EU Water Framework Directive the Estonian coastal waters of the Gulf of Finland are divided into 6 water bodies (sea areas) and each water body represents the smallest assessment unit of e.g. water quality and risk analyses (Figure 1). Based on BBN scenario modelling results it is possible to conclude that the western water body of Estonian coastal waters in the Gulf of Finland could be considered as an area of the highest ecological risk for the all seasons.

For example, Figure 7 shows ecological risk distribution calculated for variable "Season" in a state equal to "Winter" and the variable "Ecological Sensitivity" in a state equal to "Low".

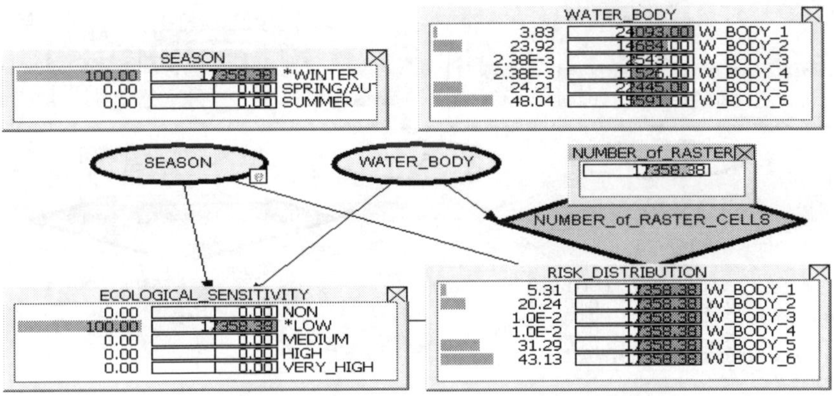

Figure 7: Ecological risk assessment BBN for the southern Gulf of Finland (Baltic Sea) coastal sea area (winter, low ecological sensitivity)

In this case the highest ecological risk is associated with the water bodies 6, 5 and 2 (43.13%, 31.29% and 20.24% respectively). If the variable "Ecological Sensitivity" state is changed to "Medium" (Figure 8) then the highest ecological risk is associated with the water bodies 5, 6 and 2 (30,54%, 26,77%, and 21,83% respectively).

Figure 9 shows ecological risk distribution calculated for variable “Season” in a state equal to “Winter” and the variable “Ecological Sensitivity” in a state equal to “Very High”.

Now, the highest ecological risk is associated with the water bodies 6 and 5 (61,91% and 12,99% respectively) while the ecological risk distribution over the rest of the water bodies is rather uniform and on a low level.

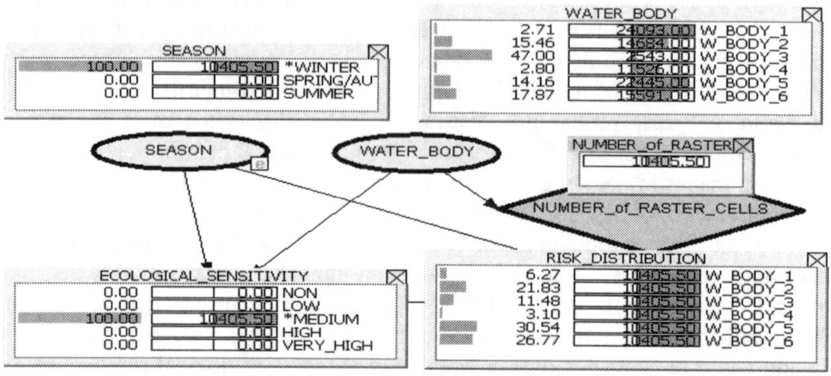

Figure 8: Ecological risk assessment BBN for the southern Gulf of Finland (Baltic Sea) coastal sea area (winter, very high ecological sensitivity).

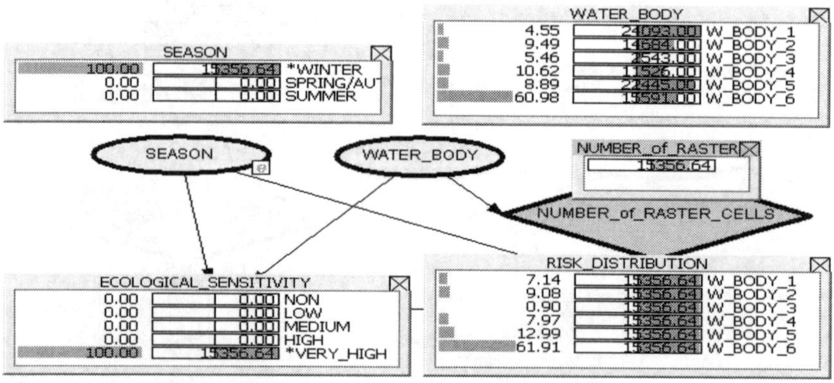

Figure 9: BBN for ecological risk assessment in the southern Gulf of Finland (Baltic Sea) coastal sea area (winter, high ecological sensitivity).

Throughout the paper, the issue of integrating BBN with other simulation tools proved to be an efficient technique in performing the potential oil pollution related predictive ecological risk assessment for the southern part of the Gulf of Finland. Furthermore, it is believed that the combined modelling approach presented in this paper would also be applicable with some modifications to a wide range of oil spill related ecological risk assessment problems.

4 Conclusions

In a case of oil spill the spatio-temporal fate of spilled oil (transport by currents and wind, spreading, evaporation, dispersion, emulsification) can be efficiently simulated using the comprehensive modelling suite Seatrack Web. However, it fails in resolving certain meso-scale hydrodynamic phenomena, such as upwelling and the related baroclinic coastal jets. In reality, depending on the wind conditions the spilled oil may travel alongshore with a speed up to 50 km per day.

Usually, booms and skimmers are the first technique employed to remove oil from marine environments but this technique can usually recover relatively small proportion of the spilled oil because of quite narrow window of opportunity depending on the actual weather conditions.

BBN integrated with other simulation tools proved to be an efficient modelling approach in performing the potential oil pollution related predictive ecological risk assessment for the southern part of the Gulf of Finland. A BBN is primarily used to update the ecological risk distribution over the states of a hypothesis variable, which is not directly observable. Ecological risk assessment is used then to support a decision maker in deciding upon an appropriate course of action.

Acknowledgements

The study was supported by the Estonian target financing programmes SF0180104s08 and SF0180013s08.

References

[1] IPIECA. Choosing spill response options to minimize damage – Net Environmental Benefit Analysis. *IPIECA Report Series Volume 10, International Petroleum Industry Environmental Conservation Association,* London, 20 p. 2000.

[2] Schallier, R., DiMarcantonio, M., Roose, P., Scory, S., Jacques, T.G., Merlin, F. X., Guyomarch, J., Le Guerroué, P., Duboscq, K., Melbye, A., Resby, J.L.M., Singsaas, I., Leirvik, F. NEBAJEX Pilot Project – Final Report. *Royal Belgian Institute of Natural Sciences.* 100 p. 2004.

[3] Borsuk M.E., Stow C.A., Reckhow K.H. A Bayesian network of eutrophication models for synthesis, prediction, and uncertainty analysis. *Ecological Modeling* **173,** pp. 219–239, 2004.

[4] Ellison A.M. An introduction to Bayesian inference for ecological research and environmental decision making. *Ecological Applications* **6,** pp. 41036–1046, 1996.

[5] Ellison A.M. Bayesian inference in ecology. *Ecological Letters* **7,** pp. 509–520, 2004.

[6] Hayes, K. R. Bayesian statistical inference in ecological risk assessment. Crimp Technical Report (No.17), Centre for Research on Introduced Marine Pests, CSIRO, Hobart, Australia. 104 p. 1998.

[7] Manual Seatrack Web. *A user-friendly program for forecasts and presentation of the spreading of oil, chemicals and substances in water, version 2.4.1.*, 31 p. 2008.

[8] Delgado, L., Kumzerova, E., Martynov, M. Simiulation of oil spill behavior and response operations in PISCES. *Environmental Problems in Coastal Regions VI including Oil Spill Studies. Ed. C.A.Brebbia, Wessex Insitute of Technology*, pp. 279-292, 2006.

[9] Aps, R., Fetissov, M., Herkül, K., Kotta, J., Leiger, R., Mander, Ü., Suursaar, Ü. Bayesian inference for predicting potential oil spill related ecological risk. *SAFE 2009. WIT Transactions*. In press, 2009.

[10] Andrejev O., Myrberg K., Alenius P., Lundberg P. Mean circulation and water exchange in the Gulf of Finland – a study based on three-dimensional modelling, *Boreal Env. Res.,* **9**, pp. 1–16, 2004.

[11] Alenius P., Nekrasov A., Myrberg K. Variability of the baroclinic Rossby radius in the Gulf of Finland, *Cont. Shelf Res.,* **23**, pp. 563–573, 2003.

[12] Lehmann, A., Krauss, W. & Hinrichsen, H.-H., Effects of remote and local atmospheric forcing on circulation and upwelling in the Baltic Sea. *Tellus*, 54A, pp. 299–316, 2002.

[13] Suursaar, Ü. & Aps, R., Spatio-temporal variations of hydrophysical and – chemical parameters during a major upwelling event in the southern coast of the Gulf of Finland in the summer of 2006. *Oceanologia*, 49, pp. 209–228, 2007.

[14] Suursaar, Ü., Aps, R., Martin, G., Põllumäe, A. & Kaljurand, K., Monitoring of the pulp mill effluents in the coastal waters of North Estonia. *Water Pollution IX.* (Series: WIT Transactions on Ecology and the Environment), D. Prats Rico & C.A. Brebbia (Eds.), 111, pp. 217–226, 2008.

[15] Myrberg K., Andrejev O. Main upwelling regions in the Baltic Sea – a statistical analysis based on three-dimensional modelling, *Boreal Env. Res.,* **8**, pp. 97–112, 2003.

[16] Clark, R.B., Frid, C., Addrill, M. Marine pollution. *Oxford University Press, NY.* 237 p. 2001.

[17] Lindgren, C., Lager, H., Fejes, J. Oil spill dispersants: risk assessment for Swedish waters. *IVL Swedish Environmental Research Institute, IVL Rapport/report* **B 1439**. 25 p. 2001.

[18] HELCOM. Restricted use of chemical agents and other non-mechanical means in oil combating operations in the Baltic Sea area. *HELCOM Recommendation 22/2.* 2002.

[19] HELCOM. Development of national ability to respond to spillages of oil and other harmful substances. HELCOM Recommendation 11/13. 1990.

Oil accident response simulation: allocation of potential places of refuge

R. Leiger[1,2], R. Aps[1], M. Fetissov[1,2], K. Herkül[1], M. Kopti[1,2], J. Kotta[1], Ü. Mander[3] & Ü. Suursaar[1]
[1]*Estonian Marine Institute, University of Tartu, Estonia*
[2]*Estonian Maritime Academy, Estonia*
[3]*Institute of Geography, University of Tartu, Estonia*

Abstract

This paper explores the problem of allocation of potential places of refuge for a ship in distress along the Estonian coast of the Gulf of Finland balancing the advantage for the affected ship and for the environment resulting from bringing the ship into a place of refuge. The integrated oil accident response simulation environment (PISCES II, ArcGis spatial modeling tools and Bayesian Belief Networks) proved to be instrumental for operational decision support in the case of a hypothetical oil accident. The knowledge of the ecological sensitivity of the Estonian coastal sea in the Gulf of Finland is used to estimate the probability of expected ecological damage associated with different towing directions of the vessel in distress. It is shown that the choice of the most favorable towing direction of the vessel in distress depends on the season, the position of the accident, and the ecological sensitivity level of the coastal sea area concerned.
Keywords: places of refuge, vessel in distress, oil incident, ecological risk assessment.

1 Introduction

The Gulf of Finland is a sensitive brackish water area with a unique nature and environment. At the same time the Gulf of Finland has some of the busiest oil shipping routes in the world. According to Kuronen et al. [1] a total of 263 million tons of cargoes were transported in the Gulf of Finland and the transportation of petroleum products formed 56% of all cargo traffic in 2007. The authors estimate that in the case of slow economic growth the ship transport

WIT Transactions on Ecology and the Environment, Vol 126, © 2009 WIT Press
www.witpress.com, ISSN 1743-3541 (on-line)
doi:10.2495/CP090221

in the Gulf of Finland would reach 322.4 million tons in 2015 (growth 23%), while average growth would yield some 431.6 million tons (growth 64%), and strong growth some 507.2 million tons (growth 93%) of cargo. Despite improved navigation measures there is a growing risk of incidental oil spills and associated oil pollution. Current oil incident emergency response efforts in the Gulf of Finland are concentrated mainly on the deployment of equipment for containing and skimming spilled oil. However, the issue of towing the ship in distress away from exposed coastlines or bringing that ship into a place of refuge has attracted much less attention so far.

In November 2003, the IMO Assembly adopted the Guidelines on places of refuge for ships in need of assistance in a manner that retains a proper and equitable balance between the rights and interests of coastal States and the need to render assistance to ships that are distress at sea [2]. The Guidelines recognize that, when a ship has suffered an incident, the best way of preventing damage or pollution from its progressive deterioration is best carried out in a place of refuge. However, bringing such a ship into a place of refuge may endanger the coastal State, both economically and from the environmental point of view, and local authorities and the population may strongly object to the operation.

According to IMO Guidelines, in the case of an accident, when permission to access a place of refuge is requested, there is no obligation for the coastal authority to grant it, but the coastal authority is going to weigh all the factors and risks in a balanced manner and give shelter whenever reasonably possible. Political decisions on the possible towing destination for a ship in distress are negotiated between coastal authorities, with aim the of selecting the best towing destination alternative through weighting of the advantages and disadvantages of the different towing destinations and of their expected net benefit towards, or net reduction of the overall environmental impact. A comprehensive review of the state of the play as regards the allocation of places of refuge in the Baltic Sea is given by Ohlson [3].

Bayesian inference and the Bayesian Belief Networks (BBNs) are increasingly used in ecological studies [4-5] because BBNs are able to employ subjective interpretations of probability, and they immediately direct the analyst to the full distributional qualities of parameter uncertainty, through the posterior distribution function [6].

Based on the probabilistic modeling the problem of ecological risk assessment related to allocation of potential places of refuge for a ship in distress along the Estonian coast of the Gulf of Finland is analyzed.

2 Material and methods

Ecological sensitivity maps used in this study are based on three different ecosystem elements: the EU Habitat Directive Annex 1 habitats and associated habitat forming species, birds and seals. In each raster cell the maximum value of different layers is calculated to give the final assessment of ecosystem sensitivity by coastal water bodies and the seasons (Figure 1).

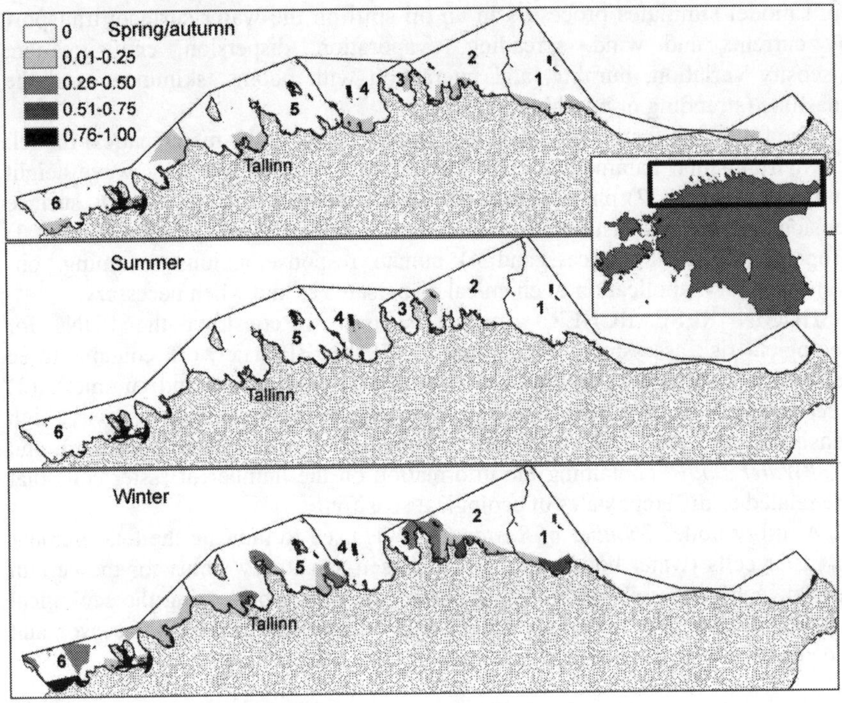

Figure 1: Ecological sensitivities by coastal water bodies of the southern Gulf of Finland (1 – 6) and seasons (spring/autumn, summer and winter). Sensitivity scale according to sensitivity criteria applied: (0) – no sensitivity, (0-0.25) – low sensitivity, (0.26-0.50) – medium sensitivity, (0.51-0.75) – high sensitivity, and (0.76-1.00) – very high sensitivity [7].

According to technical documentation [8] Seatrack Web's modeling suite calculates the spreading of oil that has come out in the Gulf of Bothnia, the Gulf of Finland, the Baltic Sea, the Sounds, the Kattegat, the Skagerack and part of the North Sea. Seatrack Web has access to forecasted current fields of the Hiromb model (HIgh Resolution Operational Model for the Baltic), which is a 3-dimensional circulation model covering the whole Baltic Sea and part of the North Sea. Every 3 hour new current fields are used in Seatrack Web with the horizontal grid resolution of 3 nautical miles. Hiromb gives the currents at 24 different depth levels and those influence the drift and spreading of the substance.

The wind forecasts used in Seatrack Web originate from the weather model at the European Centre for Medium-Range Weather Forecasts (ECMWF) 5 days ahead and High Resoluted Limited Area weather Model (HIRLAM) 2 days ahead. The wind forecasts used in Seatrack Web are from 10 meters height.

PISCES II (Potential Incident Simulation Control and Evaluation System) is used to simulate development of an oil spill incident scenarios [9]. The PISCES II spill model simulates processes in an oil spill on the water surface: transport by currents and wind, spreading, evaporation, dispersion, emulsification, viscosity variation, burning, and interaction with booms, skimmers, and the coastline (stranding or beaching).

The following factors are taken into consideration in the mathematical model: 1) environmental parameters: coastline, field of currents, weather, wave height and water density, 2) physical properties of spilled oil: specific gravity, surface tension, viscosity, distillation curve and emulsification characteristics, 3) properties of spill sources, and 4) human response actions: booming, on-waterrecovery, application of chemical dispersants as and when necessary.

HUGIN RESEARCHER software is used to construct the BBNs for ecological risk assessment. According to [7] this BBN network contains three information variables: (1) "*Season*" (winter, spring/autumn and summer), (2) "*Ecological Sensitivity*" (no sensitivity, low sensitivity, medium sensitivity, high sensitivity and very high sensitivity to potential oil spill related pollution), and (3) "*Water Body*" containing the information on the number of raster cells that are related to different states of ecological sensitivity.

A utility node "*Number of Raster Cells*" is used to indicate the total number of raster cells (water bodies 1 to 6) of chosen sensitivity status for the certain season. A hypothesis variable "*Risk Distribution*" is representing the ecological risk distribution over the all six water bodies concerned for the given season and the chosen ecological sensitivity status.

3 Results and discussion

Assessment of ecological risk related to 1) accidental instantaneous oil spill in the Western Gulf of Finland, and 2) towing of continuously spilling vessel in distress to a closest allocated place of refuge in Western and the Eastern Gulf of Finland respectively is exemplified below by the three following hypothetical scenarios.

3.1 Scenario I

Oil tanker carrying the medium oil gets damaged in the Western Gulf of Finland and this accident resulted in the instantaneous spill of 100 tons of medium oil. Immediately after notification of an oil accident at sea, the quick decision is to be taken on the most appropriate option(s) on handling the ship in distress. This decision is based on the following information: 1) what are the expected drift, behavior and fate of the spilled oil, and 2) is the oil threatening a sensitive resource?

What would be the ecological risk in a case if the decision will not be made to tow the damaged tanker to the nearest designated port of refuge? Seatrack Web modeling suite was used to calculate the hypothetical trajectory of the spilled oil (Figure 2).

Trace of spilled oil is calculated taking into account the weather related uncertainty (Figure 3).

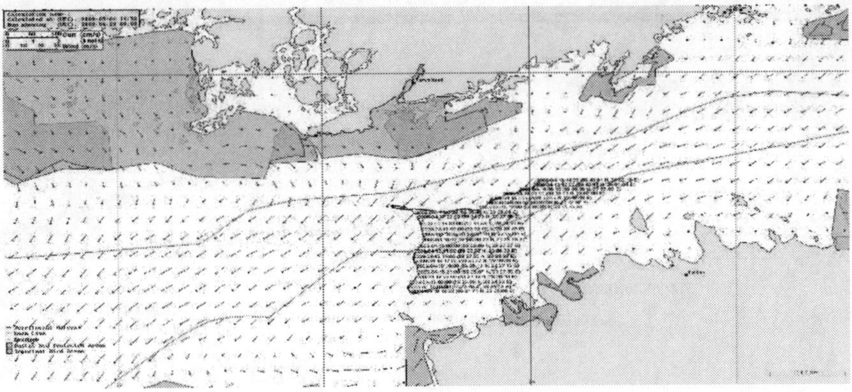

Figure 2: Trajectory of the accidentally spilled medium oil in the Western Gulf of Finland calculated using the actual weather conditions for the period starting on 13 April 2009 (12:00 UTC, Coordinated Universal Time) and ending on 20 April 2009 (00:00 UTC). Arrows show the direction and speed of the surface current. The locations of Important Bird Areas and the Baltic Sea Protected Areas are shown along the coasts.

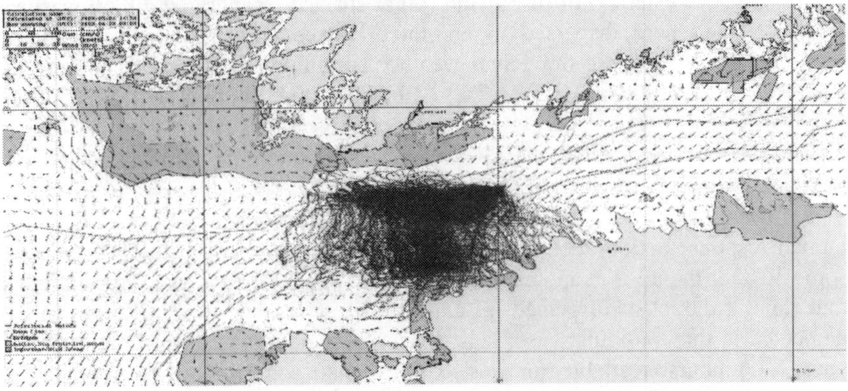

Figure 3: Trace (including uncertainty) of the accidentally spilled medium oil in the Western Gulf of Finland calculated using the actual weather conditions for the period starting on 13 April 2009 (12:00 UTC) and ending on 20 April 2009 (00:00 UTC). Arrows show the direction and speed of the surface current. The locations of Important Bird Areas and the Baltic Sea Protected Areas are shown along the coasts.

In Seatrack Web modeling suite the uncertainty in a drift simulation, so called uncertainty spreading is calculated in a way that each particle is given an additional random velocity whose magnitude is a function of the expected uncertainty in the wind forecast. The idea is to mimic an ensemble of simulations with slightly different forcing while only particles on the surface are affected [8].

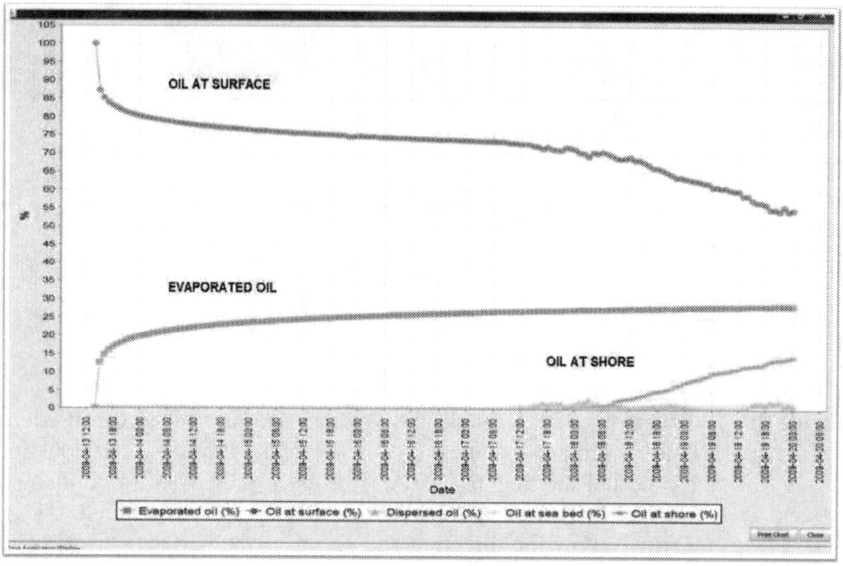

Figure 4: The fate of the 100 tons of medium oil accidentally spilled in the Western Gulf of Finland (oil at surface, evaporated oil, oil washed ashore, dispersed oil, and the oil at sea bed) simulated by Seatrack Web using the actual weather conditions for the time period: 13 April 2009 (12:00 UTC) - 20 April 2009 (00:00 UTC).

The fate of the spilled oil in the sea environment is also calculated by Seatrack Web and presented in Figure 4.

Figure 4 shows that after 180 hours from the accident time and with no human response action about 55% of the spilled oil are still at the sea surface, some 30% of the oil is evaporated, and some 15% of the oil is washed ashore by that time. Amount of dispersed oil and the oil at sea bed is small and can be neglected in that particular case. It is important to note, that in this case there is about 100 hours available for possible response actions (use of booms and skimmers depending on actual weather conditions) before the oil starts increasingly come ashore. According to HELCOM Recommendation 22/2 "Restricted use of chemical agents and other non-mechanical means in oil combating operations in the Baltic Sea area" [10], it was recommended that, due to the sensitive ecological conditions in the Baltic Sea area, response to oil should take place by the use of mechanical means as far as possible while response by using dispersants should be limited. Therefore, the option of dispersants use is not considered within this study. Usually, booms and skimmers are the first technique employed to remove oil from marine environments but this technique can usually recover relatively small proportion of the spilled oil because of quite narrow window of opportunity depending on the actual weather conditions.

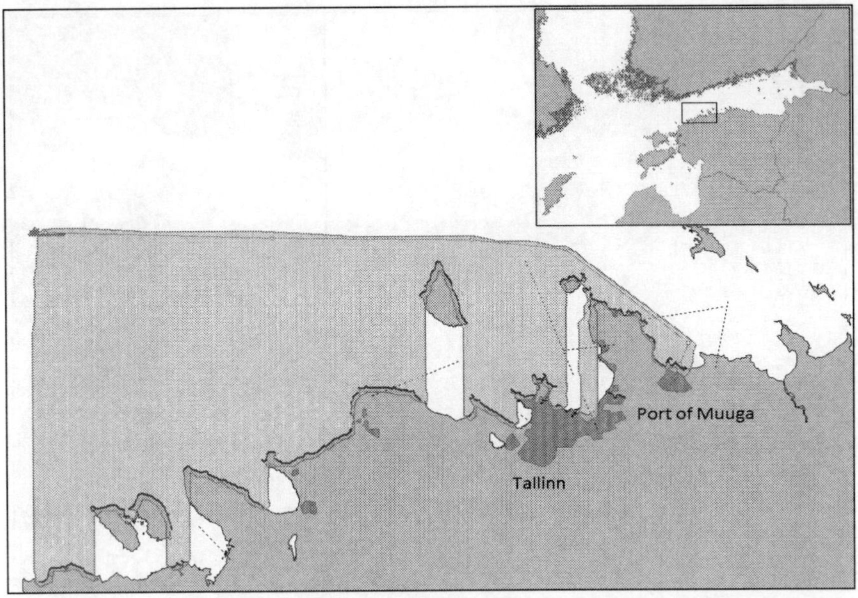

Figure 5: PISCES II simulation of the oil spilling vessel in distress taken from an accident site to the designated port of refuge – Port of Muuga. Distance 74,1 km, time of response action 04:56 hours, spill rate – 60 tons per hour.

3.2 Scenario II

Oil tanker carrying the medium oil gets damaged in the Western Gulf of Finland and this accident resulted in the continuous spill with a rate of 60 tons per hour of medium oil. Decision was taken to bring the vessel in distress to the closest designated port of refuge – the Port of Muuga (Figure 5).

This is one of the worst case scenarios (sea surface current is coming straight towards the Estonian coast of the Gulf of Finland and the situation is not changing during the action) that is representing roughly the severe environmental consequences of the accident despite of the human response action taken. Simulation results suggest that in similar cases everything possible should be done to prevent oil to wash ashore because there are obvious advantages if the spilled oil that threatens ecologically sensitive coastal sea area can be removed while it is still at sea.

3.3 Scenario III

Oil tanker carrying the medium oil gets damaged in the Eastern Gulf of Finland and this accident resulted in the continuous spill with a rate of 60 tons per hour of medium oil. Decision was taken to bring the vessel in distress to the closest designated port of refuge – the Port of Sillamäe (Figure 6).

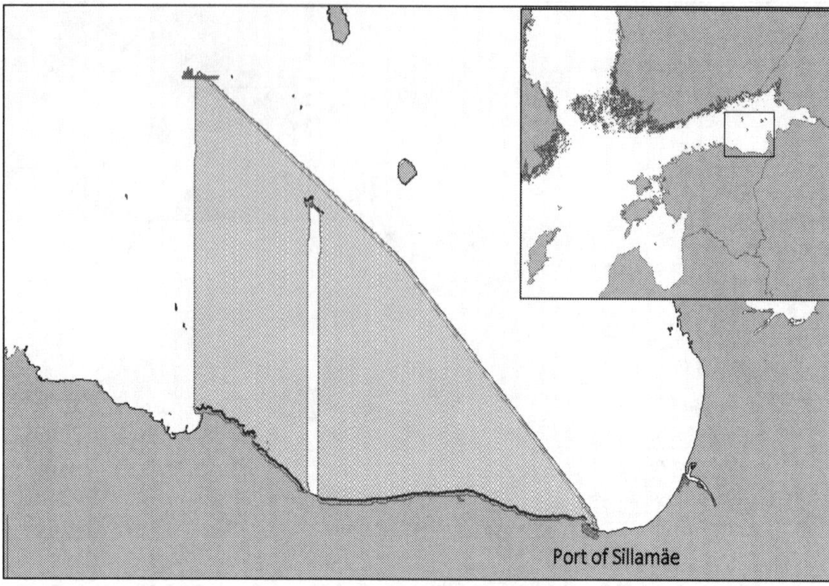

Figure 6: PISCES II simulation of the oil spilling vessel in distress taken from an accident site to the designated port of refuge – Port of Sillamäe. Distance 92,9 km, time of response action 06:12 hours, spill rate – 60 tons per hour.

Conclusions that can be drawn based on the results of the Scenario III simulation are very similar to those related to Scenario II results. In these worst case scenario situations everything possible should be done to prevent the oil to wash ashore.

4 Ecological risk assessment

A BBN constructed with aim to assess the accidental oil spill related ecological risk is primarily used to update the ecological risk probability distribution over the states of a hypothesis variable, which is not directly observable. Ecological risk distribution then helps a decision maker in deciding upon an appropriate course of action. For example, in a case of Scenarios I and II the EU Water Directive water quality assessment water bodies 5 and 6 will be affected (compare Figures 1, 3 and 5). If the ecological risk distribution is calculated for variable "Season" in a state equal to "Winter" and the variable "Ecological Sensitivity" in a state equal to "Very High" then the risk that the habitats of very high ecological sensitivity will be damaged is 61.91% and the 12.99% respectively for the water bodies 5 and 6 (Figure 7).

At the same time, in a case of Scenario III the EU Water Directive water quality assessment water body 1 will only be affected (compare Figures 1 and 6). In this case the risk that the habitats of very high ecological sensitivity will be damaged is only 7.14% for the affected water body 1 (Figure 7).

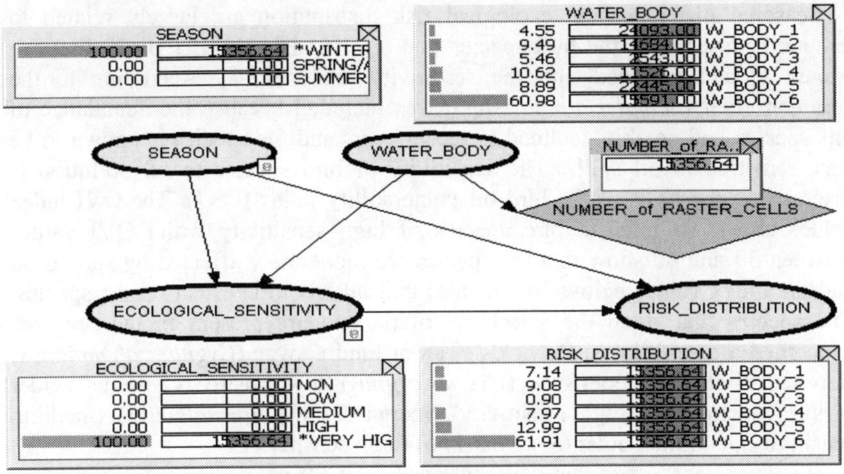

Figure 7: Ecological risk assessment BBN for the southern Gulf of Finland (Baltic Sea) by the EU Water Directive water quality assessment water bodies 1-6 (winter, high sensitivity) [7].

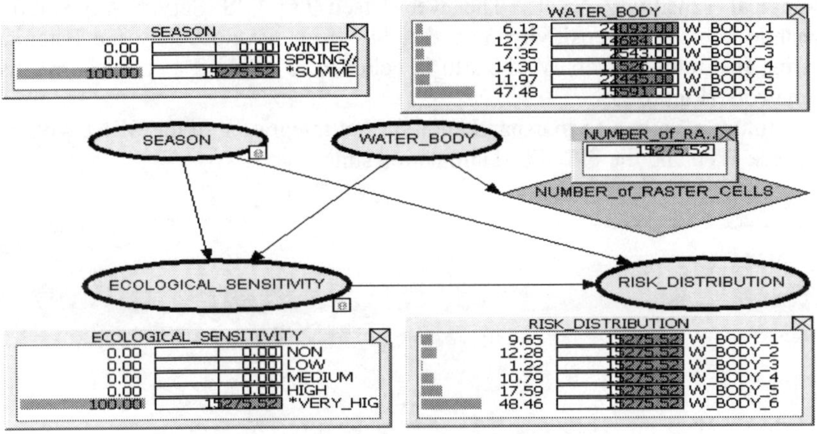

Figure 8: Ecological risk assessment BBN for the southern Gulf of Finland (Baltic Sea) by the EU Water Directive water quality assessment water bodies 1-6 (summer, high sensitivity) [7].

If the ecological risk distribution is calculated for variable “Season” in a state equal to “Summer” (summer time accident) then the risk that the habitats of very high ecological sensitivity will be damaged is in a case of scenarios I and II 48.46% and the 17.46% respectively for the water bodies 5 and 6 while in a case of scenario III the risk that the habitats of very high ecological sensitivity will be damaged is only 6.12% (Figure 8).

Seasonal differences in ecological risk distribution are largely related to temporal presence of the bird species that are listed in the EU Birds directive Annex 1, and included into the sensitivity analyses [7]. According to the common eider (*Somateria mollissima*) was included because the abundance of this species had notably declined in recent years and the species is known to be very sensitive to oil spills. The sensitivity of bird species to oil pollution is assessed on the basis of the bird oil vulnerability index (OVI). The OVI index values above 60 refer to bird species of high sensitivity while OVI values between 30 and 60 show that the species are moderately affected by spilled oil and the index values below 30 indicates that oil has little effect on the species. The species that fitted the selection criteria and are present in the sea area concerned are as follows: Bewick's swan or tundra swan (*Cygnus columbianus*) (low sensitivity), whooper swan (*Cygnus cygnus*) (low sensitivity), Steller's eider (*Polysticta stelleri*) (high sensitivity), merganser (*Mergus albellus*) (medium sensitivity), common eider (*S. mollissima*) (high sensitivity).

It is important to add that the scenarios I – III are presenting the rough worst cases in a sense that the whole area of the water quality assessment water body is damaged by the hypothetical accidental oil pollution. The objective of our future work is to develop the probabilistic risk assessment framework for the Estonian coastal sea areas of any spatial configuration broken down by seasons and the ecological sensitivity levels. The web based (ArcGIS Server Application) dynamic ecological sensitivity map that has been developed by the Estonian Marine Institute, University of Tartu is believed to be used as a basis for that framework (Figure 9).

In future we aim also to expand upon further integration of the BBNs with the Seatrack Web and the PISCES II simulation suits.

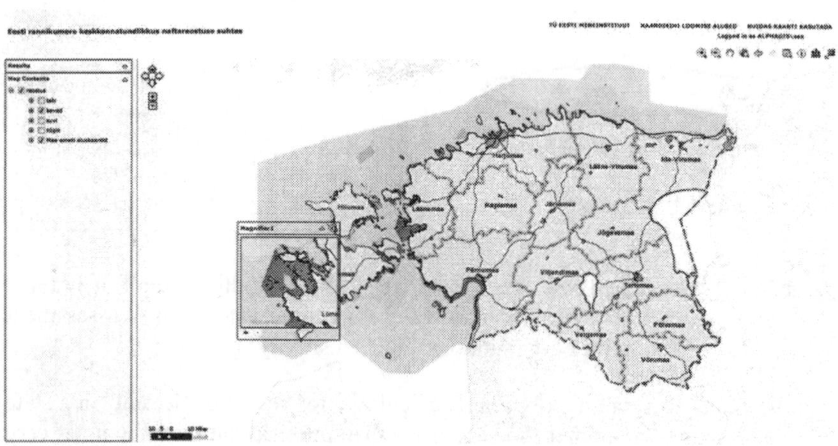

Figure 9: Estonian Baltic coastal sea ecological sensitivity web based ArcGIS map application (four seasons and five ecological sensitivity levels).

5 Conclusions

1. Comparison of the assessed ecological risk in the Western Gulf of Finland related to 1) accidental instantaneous oil spill, and 2) the towing of continuously spilling vessel in distress to a closest allocated place of refuge shows that the level of ecological risk related to the instantaneous medium size oil spill with no human response action is similar to the level of ecological risk related to taking the spilling vessel in distress into the designated port of refuge.
2. Simulation results suggest that in a case of the accident similar to the simulated scenario everything possible should be done to prevent the oil of washing ashore because there are obvious advantages if the spilled oil that threatens ecologically sensitive coastal sea area can be removed while it is still at sea. Use of dispersants is not recommended due to sensitive ecological conditions in the Baltic Sea area.
3. In a case of accident booms and skimmers should be urgently employed 1) to surround a slick as much as possible and reduce its spread, 2) to protect to extent possible the biologically sensitive areas, and 3) when and as possible to divert oil to an area where it can be recovered despite of fact that this technique can usually recover relatively small proportion of the spilled oil due to quite narrow window of opportunity that is depending on the actual weather conditions and the mobilization time of the response equipment.

Acknowledgements

The study was supported by the Estonian target financing programs SF0180104s08 and SF0180013s08.

References

[1] Kuronen, J., Helminen R., Lehikoinen A., & Tapaninen U. Maritime transportation in the Gulf of Finland in 2007 and in 2015. *University of Turku.* 114 p. 2008.

[2] IMO Resolution A.949 (23) – Guidelines on Places of Refuge for Ships in need of Assistance. *Assembly 23rd session.* 14 p. 2003.

[3] Ohlson, J.H. The state of play as regards the allocation of places of refuge in the Baltic Sea. Dissertation, M.Sc. in Maritime Administration. World Maritime University, Malmö, Sweden. 86 p. 2006.

[4] Borsuk M.E., Stow C.A., Reckhow K.H. A Bayesian network of eutrophication models for synthesis, prediction, and uncertainty analysis. *Ecological Modeling 173*:219–239. 2004.

[5] Ellison A.M. An introduction to Bayesian inference for ecological research and environmental decision making. *Ecological Applications 6*, pp. 41036–1046, 1996.

[6] Ellison A.M. Bayesian inference in ecology. *Ecological Letters 7*, pp. 509–520. 2004.

[7] Aps, R., Fetissov, M., Herkül, K., Kotta, J., Leiger, R., Mander, Ü., Suursaar, Ü. Bayesian inference for predicting potential oil spill related ecological risk. *SAFE 2009. WIT Transactions*. In press. 2009.

[8] Technical documentation Seatrack Web. *Physical processes, numerics,* algorithms and references. Version 2.4.0. 27 p. 2008.

[9] Delgado, L., Kumzerova, E., Martynov, M. Simulation of oil spill behavior and response operations in PISCES. Environmental Problems in Coastal Regions VI including Oil Spill Studies. Ed. C.A. Brebbia, Wessex Institute of Technology, pp. 279-292. 2006.

[10] HELCOM Recommendation 22/2. Restricted use of chemical agents and other non-mechanical means in oil combating operations in the Baltic Sea area. 1 p. 2002.

Effects of simulated acid rain on tropical trees of the coastal zone of Campeche, Mexico

R. M. Cerón, J. G. Cerón, J. J. Guerra, E. López, E. Endañu, M. Ramírez, M. García, R. Sánchez & S. Mendoza
Universidad Autónoma del Carmen, México

Abstract

Native trees species of the coastal zone of Campeche Mexico belonging to different families (Blood wood tree (*Haematoxylum campechianum* L.), White mangrove (*Laguncularia racemosa* (L.) Gaertn), Red mangrove (*Rhizophora mangle* L.), Button mangrove (*Conocarpus erectus* L.), Pink flower tree (*Tabebuia rosea Bertol*), Mahogany (*Switenia macrophyla King*) and Red cedar (*Cedrela odorata* L.)) were exposed to simulated acid rain for six weeks to assess visible foliar damage and effects on sulphur, nutrients, and photosynthetic pigment contents. A total of 245 seedlings were exposed four times a week from June to July under controlled conditions by using a simulated rainwater distribution system designed to reproduce rain events of 10 mm. Six treatments with five repetitions each were established at drought and irrigation conditions at different pH values. The results suggest that the water stress condition is related to the severity of the effects shown by individuals exposed to acid rain at pH values of 2.0 and 3.0. A significant increase in sulphur contents was observed on mature leaf tissues, being greater in the Mahogany, Red Cedar and Blood wood trees at pH 2 treatment under drought conditions. The chlorophyll a/b ratio showed a significant decrease in the Mahogany, White Mangrove and Blood wood trees, and nutrient levels were sensitive to the lowest pH values. According to the results, it could be observed that the Red mangrove, Button mangrove, Red cedar and Blood wood trees were more sensitive to acid rain. The Horsfall-Barratt method was applied to develop a severity scale; however, it is necessary to conduct a field survey for long-term exposure in parcels focused on sensible species to obtain a more precise scale.

Keywords: tropical trees, simulated acid rain, nutrients, visible damage, photosynthetic pigments.

WIT Transactions on Ecology and the Environment, Vol 126, © 2009 WIT Press
www.witpress.com, ISSN 1743-3541 (on-line)
doi:10.2495/CP090231

1 Introduction

Tree growth results from multiple interacting physiological processes influenced by an inherited genetic constitution and the environment. Air pollutants that limit the carbon gain or nutrient availability may, however, suppress the growth rate and total biomass production, and thus affect the allocation pattern (McLaughlin and Shriner [1]; Troiano *et al.* [2]).

An increase in precipitation acidity predisposes tree seedlings to a number of environmental stresses, which are reflected in seedling germination, growth and survival (Percy [3]; Jacobson *et al.* [4]; Shepard *et al.* [5]). However, considerable differences have been reported in the response of individual species to acid rain (Percy [3]; Abouguendia and Baschak [6]; Billen *et al.* [7]). Such variability may result from genetic differences in susceptibility, nutrient imbalances in the soil or foliage, or from the direct influence of acid precipitation on the foliage. There has been considerable speculation about forest decline related to acid deposition due to the main physiological mechanisms for observed responses remain unclear (Bäck *et al.* [8]). The reasons for this uncertainty include some factors as the variability in tree growth due to site, competition, tree age and genotype, and the fact that nitrogen compounds of the acid rain can act as fertilizer to improve tree growth.

Experiments of simulated acid rain have been usually considered and designed by many researchers as an effective way for determining the relative importance of the effects attributed to acid rain, and great progress has been made so far (Houbao and Chuanrong [9]). Effects of simulated acid rain (SAR) have been quantified by measurements of tissue damage, physiological response and even by observed changes in host-parasite relationships (Evans *et al.* [10]; Ferenbaugh [11]; Jacobson and Van Leuken [12]; Shriner [13]; Wood and Bormann [14]). Extremely low pH level such as pH 2.0 seems to induce severe necrosis in plants (Haines and Carlson [15]; Shriner *et al.* [16]; Kohno *et al.* [17]). According to Kohno *et al.* [17] some of broad leaf trees exposed to SAR with a pH 3.0 showed acute visible injuries. At this time, it appears that rain with pH values above 4.0 applied in a routine manner to the vegetation, does not cause detrimental effects, however, the frequency of occurrence of rain events below pH 4.0 is important.

Tamm and Popovic [18] and Tveite *et al.* [19] reported long-term exposure experiments of SAR in the field grown *Pinus sylvestris*. However, such field experiments have a lot of limitations to be carried out. In contrast, many short-term exposure experiments of SAR (Izuta and Miwa [20]; Miwa *et al.* [21]; Matsumoto and Maruyama [22]; Shriner *et al.* [16]) are available in the literature. Several works have demonstrated that sensitive plants response quickly to acid rain episodes within few of days after the exposure, showing typical symptoms characteristics, so this property can be used by scientists to define some species as biological indicators. Biological indicators are a common tool for toxicologists. Because of instrument networks used to monitoring of air quality need constant calibration procedures and repairs, less expensive but reliable methods are required, and the use of plants for studying air quality is quite appropriate. For this reason, it is necessary to know which plants react in what ways to one or several pollutants, and which conditions govern the susceptibility of a particular plant to a specific pollutant or a mix of them. The purpose of this paper was to assess the visible damage and the effects on sulphur, nutrients and photosynthetic pigments levels as a consequence of acid rain on endemic and protected vegetation species of Campeche, Mexico, as well as identify the most sensitive species to acid rain to be proposed as regional bioindicators of atmospheric pollution.

2 Methodology

A total of 7 species of tropical trees were selected for this research, considering their regional importance in the coastal zone of Campeche State (species protected and endemic were included). Red mangrove, Button mangrove, White mangrove, Mahogany, Pink Flower Tree, and Blood Wood Tree are protected species. Red Cedar, White Mangrove and Blood Wood Tree are endemic species of the studied region, Red Cedar being a threatened specie according to NOM-059-SEMARNAT-2001 [23].

During propagation stage of species, periodically, sanitation inspections of seedlings were carried out to assure the good health of the individuals. For each of the species, 35 seedlings were potted into 2 kg pots in a mixture of soil and compost and placed in a greenhouse to assure uniform conditions in their growth. Each Potted seedling was labelled with the name of the specie and the number of individual. Individuals labelled from 1 to 10, were assigned to treatments under drought conditions, and the remaining individuals were assigned to treatments under irrigation conditions. A general diagnostic about the health conditions was carried out by measuring physiological characteristics as height, stem diameter, width treetop, weight, leaf size, number and colour of leaves, and the presence of previous diseases, injuries from insects or mechanical damage.

Irrigation of seedlings was done with well water whose quality was verified by continuous monitoring. Seedlings were irrigated daily using an automatic system in a greenhouse specially designed to avoid the natural rainwater.

Trees were treated four times a week during six weeks from June to July 2008 with SAR at pH values of 2.0, 3.0, 4.0 and 5.6, produced by adding H_2SO_4 to a base solution. pH values and the total amount of daily rainfall (10 mm) were selected with base on regional data recorded the last ten years. Two sets of experiments were done; one of them was carried out under drought conditions: the SAR exposure regime was the same as irrigation conditions but individuals were not irrigated during all the study to simulate the predisposing water stress condition. On the other hand, the second set of experiments was carried out under irrigation conditions to simulate acid rain events during rainy season; in this case, individuals were subjected only to automatic irrigation system avoiding natural rainwater.

Six treatments with five replications for each specie were established: 1) SAR at pH 2.0 under drought conditions (individuals labelled from 1 to 5), 2) SAR at pH 3.0 under drought conditions (individuals labelled from 6 to 10), 3) SAR at pH 5.6 under irrigation conditions (individuals labelled from 11 to 15), 4) SAR at pH 4.0 under irrigation conditions (individuals labelled from 16 to 20), 5) SAR at pH 3.0 under irrigation conditions (individuals labelled from 21 to 25), 6) SAR at pH 2.0 under irrigation conditions (individuals labelled as 26 to 30). Individuals labelled from 31 to 35 were used as control individuals so they were not exposed to SAR.

Tissue samples of young and mature leaves were collected at the beginning, in the middle and at the end of the exposures. Each sample was numbered, identified, digitalized to be processed by Photoshop and Image Tools software, and temporally stored under refrigeration in airtight plastic bags until subsequent analysis. Vegetal tissue samples were weighed in fresh, and were divided in two halves, one of them was grinded and used for chlorophyll determination on acetone extracts reading the absorbances at 665, 663.2, 646.8, 470 and 430 nm. Chlorophyll alpha, chlorophyll beta, charotenoids and total chlorophyll content (mg fresh weigh l^{-1}) were estimated from Lichtenthaler equations.

The second half was subdivided into two halves, each one of them was dried at 80 °C during 24 hours and weighed for sulphur and nutrients determination. The first half dried was grinded and digested with nitric and hydrochloric acids in Teflon® closed flasks (Cole-Parmer) of 100 ml, using autoclave equipment as an energy source. Subsequently, digested samples were filtered to determine sulphur as sulphate by turbidimetric method reading the absorbance of barium sulphate at 420 nm. The second half dried was grinded and digested with nitric, sulphuric and hydrochloric acids in Teflon® closed flasks (Cole-Parmer) of 100 ml, using autoclave equipment as an energy source. Subsequently, digested samples were filtered to determine Ca, K, Mg and Mn by atomic absorption spectrophotometry.

Data statistical analysis was carried out by using SAS [24], for each treatment (total chlorophyll, chlorophyll alpha, chlorophyll beta, charotenoids, chlorophyll alpha/beta ratio, nutrients and sulphur) for each collect at $p \leq 0.05$. Variance analysis was carried out by using Duncan Method and a correlation was applied to data using Pearson method. To process digitalized images *Adobe Photoshop Cs* was applied, and healthy and injured areas were calculated by using *Image Tool for Windows V. 1.28* (Osada and Mora [25]). Subsequently, *Horsfall-Barrat* method and *2 LOG V1.0* software were used to generate severity scales and charts to know the damage in field.

3 Results and discussion

3.1 Acute visible symptoms and severity scale

In the case of the three species of mangrove (red, white and button), exposure of SAR at pH 2 and 3 induced reddish-brown necrotic blots and chlorosis symptoms between veins and in some cases, injuries appear as brown blots in the tips of the leaves. Individuals of Mahogany, red cedar, blood wood tree and pink flower tree showed visible foliar damage (necrosis and chlorosis symptoms) when they were exposed to SAR at pH 2 and 3 both dry conditions as irrigation conditions. A diffuse green coloration was observed in some individuals, the affected areas showed a typical bleached of decomposition of pigments (chlorosis) reported for damage induced by atmospheric pollutants. These visible injuries were more evident on mature leaves with treatments under drought conditions.

A severity scale was developed by using digitalized images, and injured areas were obtained applying *Image Tool for windows v. 1.28*. In spite of, severity charts and scales were obtained by applying *Horsfall-Barrat* method and *2 LOG v. 1.0* software, it is necessary to conduct field survey for long-term exposures in parcels focused on sensible species to obtain a more accurate scale.

3.2 SAR effects on photosynthetic pigments levels

Decomposition of photosynthetic pigments has been related to the effects of atmospheric pollutants on vegetation. It has been demonstrated that pollutants as SO_2, O_3 and acid rain induce the formation of free radicals in the cell. Therefore, when accumulation of radicals exceeds the capacity of detoxification of the plant, photo-oxidative injuries can be generated in the pigment-protein system.

Besides the content of pigments, their relative ratios also can be affected. A decrease in the chlorophyll a/b ratio is a good indicator of damage by photo-oxidation process in leaves. From protective function given by charotenoids on

Table 1: Decreasing percentages of photosynthetic pigments for species of mangrove.

Specie		Treatment					
		pH 2 DC.	pH 3 DC.	pH 2 IC	pH 3 IC	pH 4 IC.	pH 5.6 IC
Red Mangrove	C-A	47.69	-	-	21.52	-	-
	C-B	57.65	-	-	12.69	15.37	-
	C-T	51.79	-	-	18.05	8.74	-
	A/B	-	-	-	-	-	-
	T.CH.	-	-	-	-	-	-
White Mangrove	C-A	29.98	-	-	-	-	-
	C-B	17.77	-	-	-	-	-
	C-T	25.9	-	-	-	-	-
	A/B	33.72	27.49	24.32	-	23.38	23.26
	T.CH.	60.87	35.33	17.71	-	41.20	41.08
Button Mangrove	C-A	-	-	-	-	-	-
	C-B	-	39.23	-	2.48	2.18	40.51
	C-T	-	14.58	-	-	-	-
	A/B	-	-	-	-	-	-
	T.CH.	-	-	-	-	-	-

DC: Dry conditions; IC: Irrigation conditions; C-A: Chlorophyll-A; C-B: Chlorophyll-B; C-T: Total Chlorophyll, A/B: Chlorophyll A/B ratio; T.CH: Total Charotenoids.
- Significant differences were not found.

chlorophyll, a decrease in the content of them can be used as an indicator of direct photo-oxidation of pigments.

Photosynthetic pigments were quantified before and after the controlled exposure to SAR. In Table 1, decreasing percentages in photosynthetic pigments induced by SAR treatments for three species of mangrove are showed. It can be observed that chlorophyll-a, chlorophyll-b, and total chlorophyll were influenced by SAR treatments in the case of red and white mangrove at pH 2 under dry conditions, being this decrease more evident in red mangrove. On the other hand, the chlorophyll a/b ratio and the total content of charotenoids decreased in all treatments applied to white mangrove.

From results reported in Table 2, it can be observed that chlorophyll-a, chlorophyll-b, and total chlorophyll were influenced by SAR treatments in the case of Red Cedar and Pink Flower Tree at pH 2 and 3 such as under dry conditions as irrigation conditions, being this decrease more evident in pink

Table 2: Decreasing percentages of photosynthetic pigments for forestry species.

Specie		Treatment					
		pH 2 DC	pH 3 DC	pH 2 IC	pH 3 IC	pH 4 IC	pH 5.6 IC
Mahogany	C-A	-	-	7.86	-	-	-
	C-B	-	-	-	-	-	-
	C-T	-	-	2.91	-	-	-
	A/B	7.23	21.26	10.91	-	35.55	20.01
	T.CH.	-	-	18.64	-	-	-
Red Cedar	C-A	-	-	29.25	22.76	42.28	-
	C-B	-	36.54	37.78	55.42	56.84	13.75
	C-T	-	14.72	32.59	36.62	51.40	-
	A/B	-	-	-	-	-	-
	T.CH.	-	-	64.62	36.42	-	23.74
Blood Wood Tree	C-A	19.43	-	-	-	-	-
	C-B	56.36	-	-	58.26	-	-
	C-T	41.39	-	-	36.29	-	-
	A/B	31.44	-	-	-	-	-
	T.CH.	12.10	5.91	1.31	-	-	-
Pink Flower Tree	C-A	66.55	45.09	52.98	41.47	-	-
	C-B	75.51	52.62	30.74	43.91	20.84	10.81
	C-T	66.87	48.47	42.72	42.49	-	-
	A/B	-	-	52.54	-	-	-
	T.CH.	48.72	30.95	77.11	24.06	-	-

DC: Dry Conditions; IC: Irrigation Conditions; C-A: Chlorophyll-A; C-B: Chlorophyll-B; C-T: Total Chlorophyll, A/B: Chlorophyll A/B ratio; T.CH: Total Charotenoids.
-Significant differences were not found.

flower tree under dry conditions. On the other hand, Mahogany and Blood Wood Tree individuals showed more resistance to effects on photosynthetic pigments induced by SAR. Results suggest that SAR does not predispose Mahogany to drought injuries related to decreasing of photosynthetic pigments. However, in spite of chlorophyll levels in Mahogany individuals were not affected directly by SAR, the chlorophyll a/b ratio decreased in the most of treatments applied. Finally, it can be observed that the total content of charotenoids was affected by SAR treatments in the case of Red Cedar, Blood Wood Tree and Pink Flower Tree.

3.3 SAR effects on sulphur levels

In some regions of Europe Central, characterized by soils with a low buffer capacity, foliar concentration of sulphur is a good indicator of atmospheric pollution and is related to accumulation of sulphates as a result of acid deposition. Sulphur is an essential component of proteins and other organic compounds that play an important role in the nutrition process of the plants, however, sulphur in excess from acid rain cannot be eliminated by detoxification processes and it is accumulated as sulphate in cell vacuole of mesophyll, with a subsequent cation-anion imbalance. Bäck *et al.* [8] reported that sulphur accumulation in the needles and roots of irrigated pine and spruce with SAR treatments was greater compared with the control treatment. Sulphur levels were quantified before and after the controlled exposure to SAR. In Table 3, increasing percentages in sulphur levels induced by SAR treatments for three species of mangrove can be observed.

From results reported in Table 3, it can be observed that sulphur levels were influenced by SAR treatments for the three species of mangrove at pH 2 and 3 under dry conditions, being this decrease more evident in red and white mangrove. On the other hand, forestry species showed an evident influence on sulphur levels induced by SAR treatments. In all treatments applied, forestry species showed increasing percentages greater than 75%, being this increase more evident at dry conditions, in special for mature leaves tissues of Mahogany and Blood Wood Tree, suggesting a senescence acceleration effect induced by SAR.

Table 3: Increasing percentages of sulphur levels for mangrove species.

Specie	Treatment					
	pH 2 DC	pH 3 DC	pH 2 IC	pH 3 IC	pH 4 IC	pH 5.6 IC
Red Mangrove	79.33	198.97	-	-	-	-
White Mangrove	87.41	323.43	-	-	297.26	1691.11
Button Mangrove	73.72	28.35	-	30.56	0.32	-

DC: Dry Conditions; IC: Irrigation Conditions
- Significant differences were not found.

3.4 SAR effects on elemental concentration in tissues

Deposition of acids from the atmosphere onto forest floors may gradually increase soil acidity. Soil acidification leads to an increase in the rates of leaching of base cations such as Ca, Mg and K from the rhizosphere soil, which may cause nutrients imbalance in forest tree species (Izuta *et al.* [26]). However, Hogan [27] did not find a clear indication that foliar levels of calcium, potassium, magnesium, manganese, phosphorous, iron and aluminium had been substantially altered over the short term by SAR applied to sugar maple. On the other hand, some authors (Bäck *et al.* [8]; Izuta *et al.* [28]; Shan *et al.* [29]) reported that Mn concentrations in needles and roots of seedlings increased in SAR treatments in comparison with control treatment; whereas K and Mg

Table 4: Potassium levels (%) in plant tissues before and after SAR.

Specie	Treatments											
	pH 2 DC		pH 3 DC		pH 2 IC		pH 3 IC		pH 4 IC		pH 5.6 IC	
	B	A	B	A	B	A	B	A	B	A	B	A
Red Mangrove	4.89	1.97	6.02	4.54	6.19	3.95	3.5	3.8	3.14	4.09	3.8	4.1
White Mangrove	6.52	2.52	7.89	2.35	8.98	2.42	10.1	0.51	6.47	2.98	4.76	1.09
Button Mangrove	1.85	0.67	1.53	0.99	2.46	0.99	0.96	0.64	0.92	1.12	1.75	0.64
Mahogany	4.62	2.35	5.14	1.66	4.58	1.24	3.98	2.37	3.77	2.42	2.04	1.09
Red Cedar	15.4	4.21	11.3	4.4	12.2	2.48	11.5	2.42	15.2	1.36	13.2	3.66
Blood wood tree	12.8	1.83	13.7	1.89	12.4	1.12	12.2	4.83	16.08	1.17	11.9	2.28
Pink flower tree	5.21	1.64	6.16	1.85	7.3	1.28	6.0	1.88	6.58	0.78	5.0	1.73

DC: Dry Conditions; IC: Irrigation Conditions; B=Before SAR treatments (at the beginning of the experiment), A=After SAR treatments (at the end of the experiment).

Table 5: Magnesium levels (%) in plant tissues before and after SAR.

Specie	Treatment											
	pH 2 DC		pH 3 DC		pH 2 IC		pH 3 IC		pH 4 IC		pH 5.6 IC	
	B	A	B	A	B	A	B	A	B	A	B	A
Red Mangrove	2.65	3.83	2.8	3.47	1.72	2.89	2.48	4.09	2.36	3.18	2.19	2.86
White Mangrove	3.07	5.06	3.45	4.44	3.74	3.71	3.97	3.49	3.79	2.37	3.09	3.44
Button Mangrove	8.91	10.2	8.56	11.9	7.66	14.8	7.94	12.9	9.28	11.8	6.14	12.2
Mahogany	2.73	1.7	2.29	1.22	2.07	1.5	1.81	1.42	2.72	1.66	1.69	1.79
Red Cedar	3.84	5.99	3.48	5.17	2.88	4.8	4.09	3.24	3.18	3.31	2.86	3.6
Blood wood tree	5.75	6.43	1.65	7.26	0.4	4.59	0.44	3.55	0.41	6.53	0.82	5.86
Pink flower tree	5.92	2.68	4.66	2.89	2.41	1.57	2.61	2.22	2.92	4.2	2.88	2.78

DC: Dry Conditions; IC: Irrigation Conditions; B=Before SAR treatments (at the beginning of the experiment), A=After SAR treatments (at the end of the experiment).

Table 6: Manganese levels (%) in plant tissues before and after SAR.

Specie	Treatment											
	pH 2 DC		pH 3 DC		pH 2 IC		pH 3 IC		pH 4 IC		pH 5.6 IC	
	B	A	B	A	B	A	B	A	B	A	B	A
Red Mangrove	0.03	0.03	0.01	0.01	0.006	0.007	0.009	0.02	0.01	0.02	0.009	0.02
White Mangrove	0.03	0.02	0.02	0.03	0.02	0.02	0.02	0.04	0.03	0.03	0.03	0.02
Button Mangrove	0.03	0.05	0.03	0.06	0.02	0.05	0.02	0.04	0.033	0.05	0.04	0.06
Mahogany	0.05	0.16	0.03	0.02	0.06	0.03	0.02	0.03	0.033	0.02	0.02	0.03
Red Cedar	0.14	0.16	0.09	0.12	0.13	0.4	0.11	0.08	0.17	0.11	0.13	0.09
Blood wood tree	0.56	0.38	0.24	0.37	0.19	0.22	0.19	0.77	0.18	0.21	0.14	0.32
Pink flower tree	0.1	0.04	0.06	0.04	0.05	0.02	0.03	0.03	0.024	0.1	0.03	0.05

DC: Dry Conditions; IC: Irrigation Conditions; B=Before SAR treatments (at the beginning of the experiment), A=After SAR treatments (at the end of the experiment).

Table 7: Calcium levels (%) in plant tissues before and after SAR.

Specie	Treatment											
	pH 2 DC		pH 3 DC		pH 2 IC		pH 3 IC		pH 4 IC		pH 5.6 IC	
	B	A	B	A	B	A	B	A	B	A	B	A
Red Mangrove	3.14	4.73	2.91	4.02	2.75	4.85	4.28	11.1	3.2	4.5	2.93	4.57
White Mangrove	4.81	5.3	4.99	6.4	7.05	11.4	5.0	7.77	5.17	14.6	4.66	5.64
Button Mangrove	1.73	4.87	1.59	8.41	0.98	5.22	1.26	4.88	3.53	9.17	1.8	8.5
Mahogany	2.7	1.31	1.6	1.3	3.09	1.52	2.89	1.08	2.19	1.36	1.57	1.61
Red Cedar	3.24	6.76	2.42	3.5	5.13	5.73	2.89	2.66	5.33	3.19	3.55	2.87
Blood wood tree	14.4	8.76	6.74	7.16	2.99	4.92	3.23	18.3	4.7	5.07	4.03	6.0
Pink flower tree	3.12	1.73	2.72	1.59	1.79	0.98	1.69	1.26	1.63	3.53	1.42	1.8

DC: Dry Conditions; IC: Irrigation Conditions; B=Before SAR treatments (at the beginning of the experiment), A=After SAR treatments (at the end of the experiment).

concentrations decreased after SAR treatments were applied. In this study, nutrients levels were quantified before and after the controlled exposure to SAR. From the results of Table 4, it can be observed that potassium levels in vegetal tissue decreased in all cases, being this decrease greater in button mangle, mahogany, and red mangrove at pH 2 and 3 under irrigation and drought conditions. On the other hand, it was not observed a clear tendency in magnesium data (Table 5).

From Table 6, it can be observed that Mn levels in vegetal tissue increased in blood wood tree, red cedar, button mangrove, and red mangrove, possibly due to acid-induced increased solubility of the cations in the soil. In addition, it can be observed (From Table 7) that calcium levels in leaves of red cedar, button mangrove, blood wood tree, red mangrove and white mangrove increased as a result of SAR treatments, especially at pH 2 and 3 under dry and irrigation conditions.

3.5 SAR effects on the water stress condition

Our results suggest that water stress condition of studied individuals predispose them to suffer damage more severe. In all cases, individuals exposed to acid rain under dry conditions were more susceptible to visible injuries, showing a decreasing in photosynthetic pigments, increasing in sulphur levels and nutrients imbalance. From these results it can be inferred that the occurrence of acid rain events among dry periods can have severe consequences on sensitive vegetation species. In places with a monsoon climate like Southeast of Mexico, mid-summer drought is a common meteorological phenomenon (Mid-summer drought constitutes a relative minimum between two peaks of maximal precipitation during rainy season, and usually it takes place in August). On the other hand, at the beginning of the rainy season, acid rain episodes after long dry periods can take place in May. Therefore, we could identify two critical periods along year in the region of Campeche, in which plants can be more sensitive to acid rain episodes.

3.6 Identification of sensitive species with potential to be used as bioindicators

In general, all studied species showed sensitivity to SAR, but, considering the observed effects on photosynthetic pigments, sulphur and nutrient levels, we can propose the three mangrove species, red cedar and blood wood tree as sensitive species with potential to be used as bioindicators in the region of Campeche.

4 Conclusions

Severe visible damage observed was similar to symptoms reported for chlorosis and necrosis induced by acid rain. Nutrient levels were sensitive to lowest pH values. After SAR treatments, potassium levels decreased and calcium and manganese levels increased. According to the results, it could be observed that Red mangrove, Button mangrove, White mangrove, Red cedar and Blood wood tree were more sensitive to acid rain, and they can be suggested as regional bioindicators of atmospheric pollution. The Horsfall-Barratt method was applied to develop a severity scale; however, it is necessary to conduct field survey for long-term exposures in parcels focused on sensible species to obtain a more accurate scale.

References

[1] McLaughlin, S.B; and Shriner, D.S. (1980). Allocation of resources to defence and repair. In *Plant diseases*, ed. J.B. Horsfall and E.B. Cowling. Vol. 5. Academic Press. New York. 407-431.

[2] Troiano, J; Colavito, L; Heller, L; McCune, D.C; and Jacobson, J.S. (1983). Effects of acidity of simulated rain and its joint action with ambient ozone

on measures of biomass and yield in soybean. Environ. Exp. Bot. 23: 113-119.

[3] Percy, K. (1986). The effects of simulated acid rain on germinative capacity, growth and morphology of forest tree seedlings. New. Phytol. 104: 473-484.

[4] Jacobson, J.S; Heller, L.I; Yamada, K.E; Osmeloski, J.F; Bethard, T; and Lassoie, J.P. (1990). Foliar injury and growth response of red spruce to sulfate and nitrate acidic mist. Can. J. For. Res. 20: 58-65.

[5] Shepard, L.J; Cape, J.N; and Leith, I.D. (1993). Influence of acidic mist on frost hardiness and nutrient concentrations in red spruce seedlings. 1. Exposure of the foliage and the rooting environment. New Phytol. 124: 595-605.

[6] Abouguendia, Z.M; and Baschak, L.A. (1987). Responses of two western Canadian conifers to simulated acid precipitation. Water, Air & Soil Pollution. 33: 15-22.

[7] Billen, N; Schätzle, H; Seufert, G; and Arndt, U. (1990). Performance of some growth variables. Environ. Pollut. 68: 419-434.

[8] Bäck, J; Huttunen, S; Turunen, M; and Lamppu, J. (1995). Effects of acid rain on growth and nutrient concentrations in Scots pine and Norway spruce seedlings grown in a nutrient-rich soil. Environ. Pollut. 89. 2: 177-187.

[9] Houbao, F; and Chuanrong, L. (1999). Effects of simulated acid rain on seedling emergence and growth of five broad-leaved species. J. For. Res. 10. 2: 83-86. Evans, L.S; Gmur, N.F; and Da Costa, F. (1977). Leaf surface and historical perturbations of leaves of *Phaseolus vulgaris* and *Helianthus annu*us after exposure to simulated acid rain. Am. J. Bot. 64: 903.

[10] Evans, L.S; Gmur, N.F; and Da Costa, F. (1977). Leaf surface and historical perturbations of leaves of *Phaseolus vulgaris* and *Helianthus annu*us after exposure to simulated acid rain. Am. J. Bot. 64: 903.

[11] Ferenbaugh, R.W. (1976). Effects of simulated acid rain on *Phaseolus vulgaris* L. (Fabaceae). Am. J. Bot. 63: 283-288.

[12] Jacobson, J.S. and Van Leuken, P. (1977). Effects of acidic precipitation on vegetation. In: Proc. Of the fourth International Clean Air Congress, Tokyo. May 16-20. 124-127.

[13] Shriner, D.S. (1976). Effects of simulated rain acidified with sulfuric acid on host-parasite interactions. In: Proc. First Intern. Symp. On Acid Precip. And the Forest Ecosystem, (L.S. Dochinger and T.A. Seliga eds.). U.S. For. Serv., Gen. Tech. Rep. NE-23. 919-925.

[14] Wood, T. and Bormann, F.H. (1974). The effects of an artificial acid mist upon the growth of *Betula alleghaniensis* Britt. Environ. Pollut. 7: 259-268.

[15] Haines, B.L; and Carlson, C.L. (1989). Effects of acidic precipitation on trees. In: Acidic precipitation 2. Springer-Verlag.

[16] Shriner, D.S; Heck, W.W; and McLaughlin, S.B. (1990). Response of vegetation to atmospheric deposition and air pollution. NAPAP SOS/T Report 18.

[17] Kohno, Y; Matsumura, H; and Kobayashi, T. (1994). Effect of simulated acid rain on the development of visible injuries in tree seedlings. J. Japan Soc. Air. Poll. 29: 206-219.
[18] Tamm, C.O. and Popovic, B. (1989). Acidification experiments in pine forests. National Swedish Environmental Protection Board. Report 3589.
[19] Tveite, B; Abrahamsen, G; and Stuanes, A.O. (1990/1991). Liming and wet acid deposition effects on tree growth and nutrition: Experimental results. Water, Air and Soil Pollution. 54: 409-422.
[20] Izuta, T. and Miwa, M. (1990). Growth response of *Cryptomeria* seedlings to simulated acid rain. Man & Environ. 16: 44-53.
[21] Miwa, M; Izuta, T; and Totsuka, T. (1993). Effects of simulated acid rain and/or ozone on the growth of Japanese cedar seedlings. J. Japan. Soc. Air. Pollut. 28: 279-287.
[22] Matsumoto, Y. and Maruyama, Y. (1992). Some negative results of simulative acid mist and ozone treatments to *Cryptomeria japonica* seedlings in explanation of mature *C. japonica* decline in the Kanto plains in Japan. Jpn. J. For. Environ. 34: 85-97.
[23] NOM-059-SEMARNAT-2001. Protección Ambiente. Especies Nativas de México de Flora y Fauna Silvestres-categorías de riesgo y especificaciones para su inclusión, exclusión o cambio-Lista de especies en Riesgo.
[24] SAS Institute Inc. SAS/STAT Guide for personal computers. The SAS System for Windows V. 8. United States of America.
[25] Osada, V.H. and Mora, G. (1997). DosLog V.1. Colegio de Postgraduados, Montecillo. Texcoco, Estado de México. Free Software.
[26] Izuta, T; Yamaoka, T; Nakaji, T; Yonekura, T; Yokoyama, M; Funada, R; Koike, T; and Totsuka, T. (2004). Growth, net photosynthesis and leaf nutrient status of *Fagus crenata* seedlings grown in brown forest soil acidified with H_2SO_4 or HNO_3 solution. Trees. 18: 677-685.
[27] Hogan, G.D. (1998). Effect of simulated acid rain on physiology, growth and foliar nutrient concentrations of sugar maple. Chemosphere. 36. 4: 633-638.
[28] Izuta, T; Yamaoka, T; Nakaji, T; Yonekura, T; Yokoyama, M; Matsumura, H; Ishida, S; Yakazi, K; Funada, R; and Koike, T. (2001). Growth, net photosynthetic rate, nutrient status and secondary xylem anatomical characteristics of *Fagus crenata* seedlings grown in brown forest soil acidified with H2SO4 solution. Water, Air & Soil Pollution. 130: 1007-1012.
[29] Shan, Y; Izuta, T; Aoki, M; and Totsuka, T. (1997). Effects of O_3 and soil acidification, alone and in combination, on growth, gas exchange rate and chlorophyll content of red pine seedlings. Water, Air and Soil Pollution. 97: 355-366.

Section 7
Planning and beach design

New requirements on beach design: limiting states condition

J. C. Santás & J. M. de la Peña
Centro de Estudios de Puertos y Costas (Harbours and Coasts Studies Centre), CEDEX, Spain

Abstract

The natural evolution of a beach follows a sequence of different situations: accretion, erosion, annual oscillations and stable phases, but the beach is not an isolated coastal unit and interacts with its surrounding areas. Problems appear when its evolution affects other entities on the coastal zone related to infrastructures or leisure use or to the territorial development of the coastal region.

Problems also brought about by urban pressure and the natural tendency to invade and use the coastal zone has led to designing the coastal edge without taking into account the effects of bad weather states or storms, as it is for other coastal structures. When this happens and damage occurs, public opinion establishes a relation of damage with other circumstances or factors different to the real question, which is that the beach design has not taken into account extreme conditions: the land development has considered the beach only in its best or medium weather conditions.

This paper presents new criteria to designing beaches taking into account its state of maximum sea impact, that is, its *limit state*.

Keywords: beach design, stormy boundary conditions for beach design.

1 Introduction

Often beaches, which have not had problems throughout their history, suddenly appear to hatch an unusual situation, sometimes due to coastal erosion, but at other times because the beach is considered as a static unit in its best state while the worst climate conditions were considered for other coastal systems. This led to designing coast without taking into account the evolution that may happen in

WIT Transactions on Ecology and the Environment, Vol 126, © 2009 WIT Press
www.witpress.com, ISSN 1743-3541 (on-line)
doi:10.2495/CP090241

storm conditions. When these climatic periods appear, public opinion tends to be based around a belief that the beach has always been in a good condition, without considering that it was designed based on those good conditions. Comments like 'The beach has disappeared' or 'These are the consequences of the climate change that is coming!' are regularly expressed. It does not mean that the mean level variation does not exist, but there are other factors to be taken into account. The most important factor is due to the coastal land uses development or Coastal Zone Development (CZD), which has changed for new uses and citizen demands. It has conditioned the beach, its evolution and its own state. Figure 1 shows the coastal historic evolution of the Spanish area between Gerona to Blanes on the Spanish Mediterranean coast in the Catalunia Region.

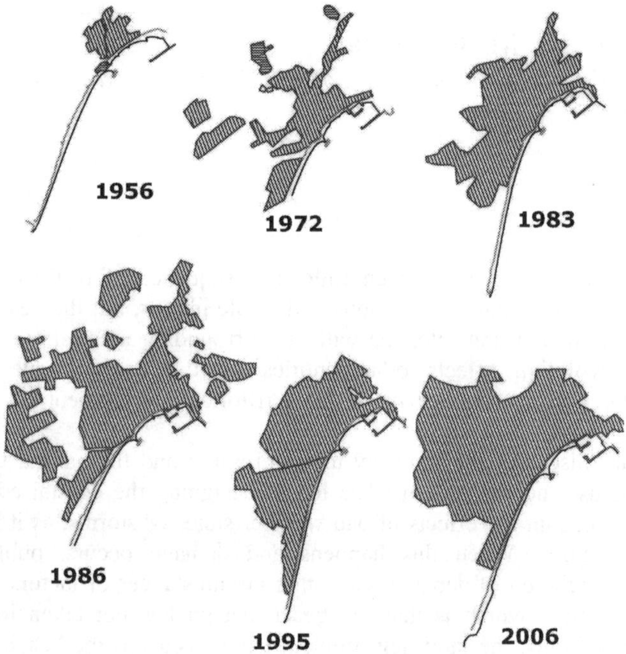

Figure 1: Coastal evolution of Gerona to Blanes coast.

2 Current concept of 'beach limit'

The basics to designing a beach, if it is in an urban area, are managed by the following aspects:

-The beach is in equilibrium when it is in its summer state, summer profile.

-The beach width has to be designed in view of the public use. Some designers consider that this may be the line where the vegetation begins or ought to begin.

These considerations have created problems when the beach is included in an urban domain. In some cases these assumptions have produced breaks on promenades and loss of foundations of the nearest buildings, figure 2.

Figure 2: Loss of foundation of temporal stands located on a beach due to a storm.

This last concept creates problems because there are three functions of a beach: as coastal defence, as a place for public use and recreation, and as a biological habitat. On the other hand, it is focussed on a static beach without external actions that are inevitably going to modify it. The littoral dynamic is continuously acting on it and the beach width has to be compatible with the evolution forced by these external factors.

The beach is a dynamic element as real as the forces that acts upon it and it has a very high energetic dissipation capacity. It seems that the design should considerer all the external forces in its 'maximum state' as it does for other civil coastal structures, and that consideration should be made for its future in 50 or 100 years time. These concepts dealing with the *natural states of a beach* are going to be introduced on the Spanish Coasts Authority Recommendations (Peña and Sánchez [1]).

Other questions that need to be included are about the uncertainties that aspects like beach equilibrium profile and sea waves climate have (Sánchez-Arcilla [2]).

In this line the idea of 'beach extreme states' would accept also that these are *bands* instead of *lines*. The equilibrium profile would be the centre line that represents the 'medium conditions' and there were a fork, whose width ought to be related to return period of waves and meteorological actions, including sea level change. All these sea actions and their effect need to be well defined from field data.

3 Minimum width of a beach

The dry beach width has to be enough in all the circumstances in the way of defence and protect such infrastructures and the environmental values located behind the beach. The first question to define is the time for which this beach width has to be maintained, the *beach width return period A_1*.

This *minimum value of beach width* has to be a value to ensure the minimum needed to protect structures and goods, and values of the beach, taking into consideration the total amount of all the predicted losses of width, permanent, progressive or time dependent.

These individual losses are:

- A1: erosion, defined as the loss of beach width for the return period
- A2: annual oscillation defined by the tilt summer–winter on the profile, evaluated from the maximum distance waited into the return period.
- A3: width variation by sea level increase due to the maximum annual increase due to other factors (run-up, meteorological and tidal elevations) to be waited into the return period
- A4: backup width to ensure the roll of the beach as defender of goods.
- A5: width variation caused by sea level increase due to the climatic change along the estimated return period.

So the total width change could be estimated as $A1 = \Sigma A_i$ (i= 1 to 5).

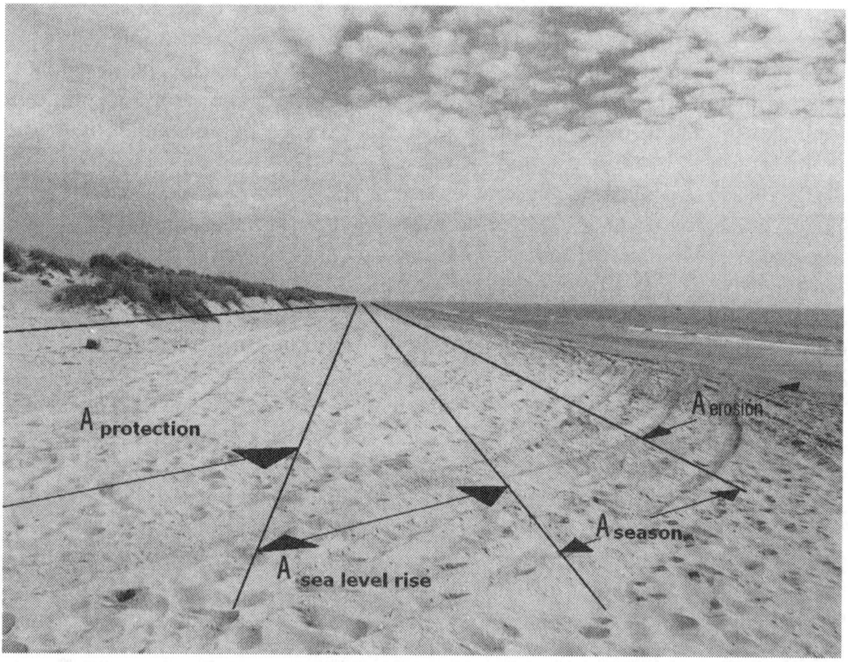

Figure 3: Different components for the beach width.

www.witpress.com, ISSN 1743-3541 (on-line)

Only two of the upper individual considered aspects, A1 (annual oscillation) and A3 (tilting due to run-up, meteorological, astronomical and waves actions) have to be estimated from extreme climate conditions into the return period. These are those that configure the limiting states of the beach.

4 Limiting states of the beach

The limiting states of the beach are two virtual lines: the upper dry beach line and the real closure depth line. The beach is defined between these two limits but they are not fixed lines. The deepest line is the real limit for all kind of movement of the sedimentary layer that is defined by the sea states (wave actions and tide) and the characteristics of sediment. For the dry limit the definition is very complex as we have shown before. To prevent extra actions coming from outside the beach that affect it, a backup distance for both sides has to be added: backup for dry beach and backup for submerged limit.

The closure depth, or limit for sediment motion, is defined by 2 ways: 'active depth', d_i, as the limit for transversal and longitudinal motion of sediment obtained from sea waves and tide characteristics (Birkemeier [3]), formula or other similar expressions, and the deeper limit, d_l, for all kind of sediment motions, given, as an example, by the CERC's [4] formula as:

$$d_i = 3{,}5.H_{s,0.137} \tag{1}$$

that give us the deepest limit obtained on the bibliography, function of the significant wave height that shows the p=1-0.137 probability (approximately: 12 hours in a year) on average distribution of waves. For this limit all the profiles obtained coincide. Figure 4 shows the different parts of profile, as Hallermeier [5].

The maximum depth and the active depth are defined in terms of Hs, but it would be in relation to the return period. It is: by the storm regime instead of medium regime of Hs. In the same way the upper limit has to be defined by the maximum level reached by waves in storm conditions and the profile of the dry beach that is related to the initial state of the beach, the grain size and the slope.

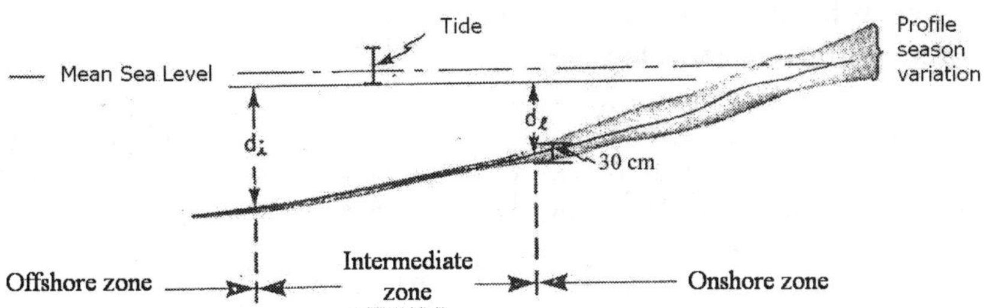

Figure 4: Layout of the beach profile and summer-winter oscillation.

Starting from an equilibrium profile the beach answers to the energy of an incident storm by a change in its profile as well as moving material from the dry part to the submerged part of the profile. Usually submerged bars appear to diminish the waves' energy by turbulence and movement of sediment. This action also modifies the submerged part of the profile with less slope and fine sediment in comparison to the shore line where the beach increased the grain size and slope.

Before the surf zone, waves coming by very shallow water show an elevation of the mean sea level, it is the *set-up* that is a function of the significant height H_{s0} on depth water. After the breaking zone the waves go up by the beach slope until its final stop by losing all the energy, returning the water layer to the sea. The maximum point reached by water gives the *run-up* level.

The study of what is the maximum level reached by the water has to be done after knowing the sea waves climate. Maximum levels for the set-up and the run-up are determined in function of the characteristics of the incident sea waves (direction, height and period) as well as the beach slope.

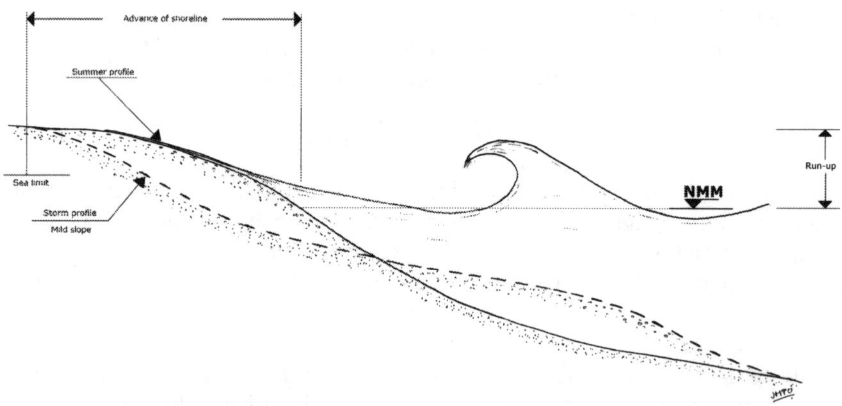

Figure 5: Summer and winter profile, run-up and maximum elongation over the beach for storm.

5 Beach limit from the point of view of Coastal Zone Development

Following the previous described aspects, the design of an urban beach has to ensure an appropriate width as well as the assumption of the natural dynamic conditions and the natural evolution of the beach, for erosion and sea level displacement due to the climatic change.

Figure 6 shows a clear case where the beach could have not been designed for citizen uses. It is the 'La Concha' beach, San Sebastián, Vasc Country, Spain, located 200m westward from Zurriola beach. The city invaded the beach area three centuries ago and now it is not possible to fix the situation.

The equilibrium between these points affects the Coastal Zone Development, mainly to the space located between the urban area and the beach. In the Spanish case, the Coastal Authorities, *Dirección General de Sostenibilidad de la Costa y del Mar*, has given a regulation that analyse the CZ from three points of view:

a. Transformation of marine urban facades.
b. Treatment of walking ways
c. Access, transit and public use of the coast.

Figure 6: 'La Concha' beach. High spring tide and low spring tide.

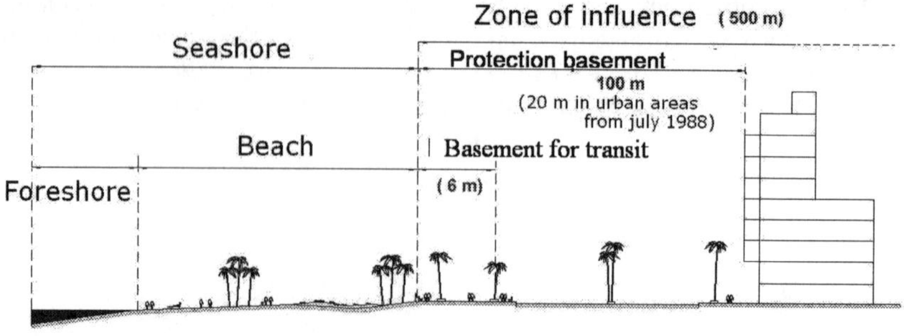

Figure 7: Coastal Zone regulation, Law 22/1988, Environment Spanish Ministry (Peña and Sánchez [1]).

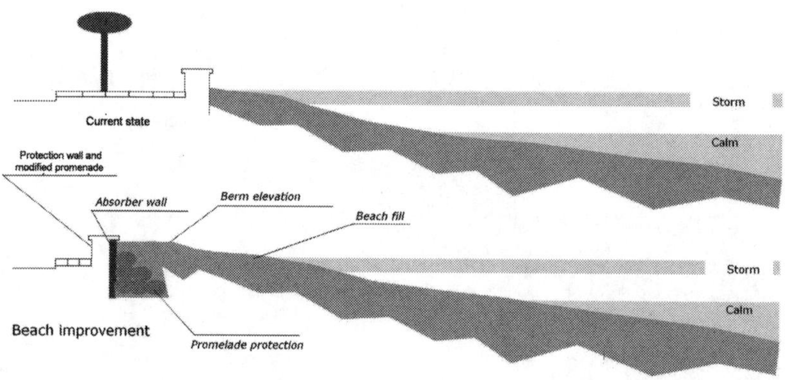

Figure 8: Transformation of a maritime front to allow free movement of the beach.

The limits that have to be considered for the regulation of a beach are the maximum elevation of sea level for a period of time equal or greater than the return period of a littoral urban structure, a walking promenade for example, after the addition of a backup margin due to the uncertainties of the data. This margin could be in the order of 30%.

Also, it needs to be taken into account that the beaches are going to be an area for citizens' use, with growing use in the summer time owing to more visits by its population.

6 Conclusions

1) It is not possible to maintain the current parameters used for the beach design overall for urban beaches, as well as their relationship to the Coastal Zone Development. These parameters are: the equilibrium profile adding a supplementary width that is defined for the citizen uses, because this concept has drive us to big problems and failures on the coastal zone of coastal cities.

2) The state of maxima sea actions (waves, tides, winds and currents) has to be considered for the beach limits and beach design in relation to the coastal border. For that a return period has to be used never less than the life period of the adjacent urban developments, typically between 50 and 100 years.

3) Due to the uncertainties of data and especially for the maximum elevation values of sea level, an error band ought to be taken into account. If not able to be accurately predicted based on available data then a confidence band of 30% could be included as a safety parameter.

References

[1] Peña, J.M. and Sánchez, V., *Asistencia técnica para la redacción de dos instrucciones para la Dirección General de Costas*;(in Spanish) Centro de Estudios de Puertos y Costas, CEDEX(report nº 22-408-5-001 for D.G. de Costas), 2008

[2] Sánchez-Arcilla, A.; Personal communication (20th-April-2009), Head of Engineering Laboratotium, LIM; Civil Engineering School, Politechnical University of Barcelona. Spain.

[3] W.A. Birkemeier, Field data on seaward limit of profile change, *Journal of W.,P.,C. and Oc. Eng., A.S.C.E.*, vol 111, 3; pp. 598 to 602. , 1985.

[4] CERC, 1998, Coastal Engineering Research Centre, *Engineering and design- Coastal Eng. Manual, Part III, Cross Shore sediment transport processes*; Dep. of the Army, US Corps of Eng.. Pub.: 1110-292.

[5] Hallermeier, R.J. (1978). /Uses for Calculated Limit Depth to Beach Erosion/; Proceedings of 16th International Conference on Coastal Engineering (pp. 1493 a 1512)

Geographic information systems for integrated coastal management and development of sustainability indicators

J. L. Almazán Gárate & the Maritime and Portuary Engineering Investigation Group
Polytechnic University of Madrid, Spain

Abstract

The anthropic pressure on the coastal zone in countries such as Spain make it necessary to carry out adequate performances in planning, as well as management and administration, to achieve an optimal sustainable development. In such a fragile zone with adequate complex information strategies, to achieve that objective requires full data integration from different areas of knowledge, which is the main purpose of the research project titled "Development of new technologies of physical modelization and tools for its use for Integral Coastal Management in Spain" developed by the authors.

The main goals set for this research project are the development and optimization of an integrated system to carry out successive high precision bathymetries; integration of results measured through a period of time on a coastal stretch using computerized coastal dynamics mathematical models; design of a geographic information system to integrate information needed to back the decisions for an optimal Integrated Coastal Management; determination of criteria to identify and quantify the alternatives proposed as coastal interventions when needed and development of sustainability indicators comprising socioeconomic, geographic and environmental aspects, to evaluate and follow up consequences of decisions adopted.

The research project is divided into two phases. The first of these is focused on the description of an application for satellite positioning and computerized methods for high precision bathymetries and its applications to coastal dynamics mathematical models and the incorporation of all data to the Geographic Information System. The second phase is focused on setting up the criteria for the identification and quantification of indicators for the interventions considered

WIT Transactions on Ecology and the Environment, Vol 126, © 2009 WIT Press
www.witpress.com, ISSN 1743-3541 (on-line)
doi:10.2495/CP090251

for a particular coastal zone. The sustainability indicators comprise socioeconomic, geographic and environmental aspects, making it possible to evaluate jointly the interventions and their effects on coastal dynamics in an integrated way.

Keywords: geographic information system, survey, bathymetry, indicators.

1 Introduction

As far as economy is concerned, the tourism industry has a significant weight, and this weight is even more significant concerning Spanish coasts and all the leisure activities associated with them. Internal migrations to the coast in Spain have created a strong anthropic pressure on the coastal zone, aggravating those problems associated with seasonal tourism.

Despite efforts from public service organizations to develop non coastal tourism, which is already having pretty good results, the increase in population and tourism keeps creating pressure on the rather fragile coastal zone. Beaches are the main destination for tourism, and are defined as coastal features formed by sediments that are deposited and that were conducted by rivers to the sea.

The lack of water and main rivers water regulation, along with desirable measures against flooding, have resulted in a lack of sediments in the Mediterranean and South Atlantic coasts (Cádiz and Huelva), and that has produced erosions in many popular beaches in Spain. Considering concepts, a coastal feature formed by sediments in erosion is transformed into an erosion landform.

Global warming processes that are being experimented on our planet, associated with a slow water level increase, have contributed to the erosion processes in zones with low tectonic activity.

These processes can be regarded as slow processes, but if we consider a four year spell in between elections in Spain, those processes are significant in terms of usual business and are vertiginously fast using a geological scale [1, 2].

Erosion processes can be modified by civil engineering actions that require inversions, and resources assignments should be related to objective parameters to ease decision making processes, having currently enough technical tools to analyze this problem from three different points of view:

A. Litoral behaviour knowledge against environmental actions, mainly waves and currents.
B. Quantification for costs of execution of those civil engineering actions necessary to modify undesirable behaviours from natural elements appointed to be preserved.
C. Financial evaluation and estimation for profitability both in financial and economic terms, considering all the aspects regarding environmental accounting as well as social aspects associated with actions to take place in a coastal zone in a country like Spain.

Spain is well portrayed by its coastal length and also by the importance that the coast has in its economy regarding tourism and how its management can be decisive on the environment and also in eco-systems sustainability in the coastal zone and continental shelf. In this sense it is necessary to analyze such a great coastal system after the natural and human actions. Also, coastal management should consider global changes, e.g. water level increase and climate change [3].

One of the main objectives for coastal engineering is how to modelize a coastal zone and all the processes acting on that particular zone, i.e., from an initial state, make predictions with enough precision and accuracy through a period of time. Nowadays there are many models that fulfil those tasks. On the other hand, computer rising makes it possible to improve those models. However, bathymetry methods are not precise nor accurate enough for the most advanced models. That makes those advanced models useless [4, 5].

The project that is being introduced in these paragraphs is expected to apply the latest satellite positioning techniques and computer methods to obtain high precision bathymetries and use them in beach evolution models. It is also expected to develop a new bathymetric system that is able to get more than 10,000 bathymetric points per hour with centimetric precision [6–8].

The main goals set for this research project are the following:

- Development and optimization of an integrated system to carry out successive high precision bathymetries.
- Integration of results measured through a period of time on a coastal stretch using computerized coastal dynamics mathematical models.
- Design of a geographic information system to integrate information needed to back the decisions for an optimal Integrated Coastal Management.
- Determination of criteria to identify and quantify the alternatives proposed as coastal interventions when needed.
- Development of sustainability indicators comprising socioeconomic, geographic and environmental aspects, to evaluate and follow up consequences of decisions adopted.

2 Description of the location researched

The physiographic region analysed includes the municipalities of Chipiona and Rota, with a total area of 117.2 km^2 and a total population of 45753 [9].

The following beaches were surveyed:

- Regla beach: this is an urban fine sand beach, with moderate waves, a high occupation level and a length of about 1700 meters. It is considered the most emblematic beach in Chipiona.
- La Ballena beach: this is an urban fine sand beach, with light waves, a high occupation level and has a length of about 4500 meters.
- La Costilla beach: this is an urban fine sand beach, with light waves, a high occupation level and has a length of about 2200 meters. It is considered as one of the best beaches in Spain, and it is located to the west of Rota Port.

- El Rompidillo beach: this is an urban fine sand beach, with light waves, a high occupation level and has a length of about 1500 meters. It is located in a sheltered zone, eastwards of Rota Port.

3 Methodology

Each of the beaches briefly introduced above was surveyed twice in a year in order to calculate sand accumulations and erosions through a direct comparison between the surfaces obtained in the surveys, as well as summer and winter profiles for these beaches due to seasonal alternation between dead calms and storms.

Additionally comparisons with data from surveys made a year before were made to evaluate the long term evolution undergone by the beaches. Those data used for the long term comparisons come from surveys made by the local coastal demarcation of Andalucía-Atlántico.

The methodology used to carry out bathymetries can be explained through the following steps [11]:

1. Data compilation: terrestrial cartography compilation from the Spanish Army Geographic Service and the National Geographic Institute. Data compilation on geodetic vertexes from the National Geodetic Network for the zone surveyed. Nautical charts and maritime climate compilation were also use as a complement for the data already compiled.
2. Reference stations location election for each of the beaches surveyed, as well as the link with the vertexes from the National Geodetic Network and the vertex located next to La Ballena beach.
3. Device calibration for the surveyed zone.
4. Survey execution for summer and winter campaigns.
5. Data integration.
6. Map designing and drawing using suitable software.
7. Database design.
8. Product generation.

4 Geographic information system design

All the data aforementioned were broken down in order to include them in a geographic information system. In order to achieve this it was necessary to validate and transform all data in a compatible version for the software used. The activities were carried out following the steps described as follows:

1. **Base map display showing all four beaches surveyed:** Base map conversion from AutoCad to a format compatible with ArcView. All reference bases were also included in the map.
2. **Data integration for all 140 sample points:** All the data from the samples analysis were integrated in tables associated to maps. Those data included profile number, depth, D_{50}, etc.

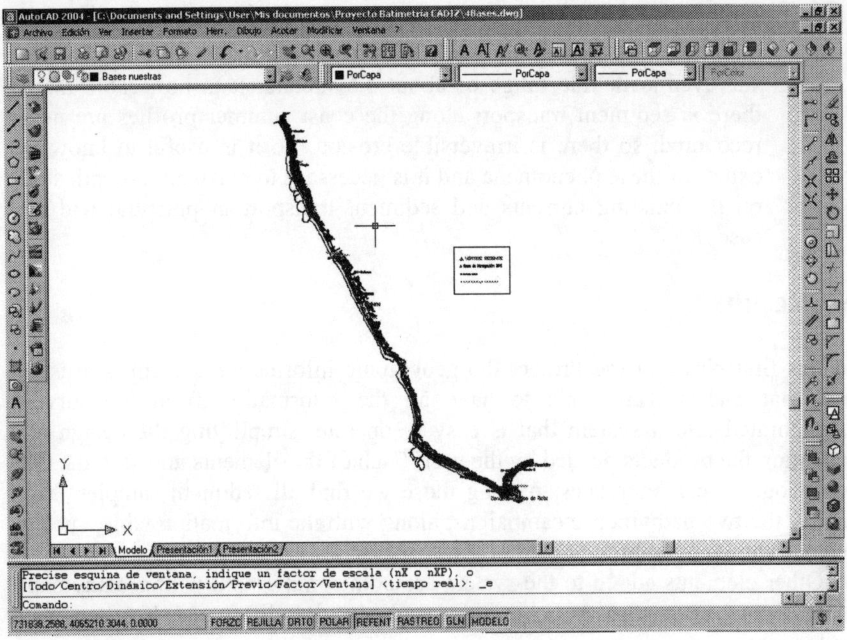

Figure 1: AutoCad map.

3. **Echo sounder data conversion:** All data was converted into an ArcView compatible format. In order to carry out that conversion it was necessary to maintain the coordinate references and so all points are completely arranged with an x, y, z trio.
4. **Bathymetry lines incorporation with their respective labels:** All data obtained in AutoCad format was converted into an ArcView compatible format and it was necessary to process each of the points surveyed.
5. **Comparison between bathymetries corresponding to both winter and summer campaigns.**
6. **Profile drawing for each of the four beaches surveyed:** Each of the profiles were drawn associated to an ArcView profile in another layer.
7. **3D models for each of the beaches surveyed:** Each of the 3D models drawn were added to a layer associated with each of the beaches surveyed. It is important to consider that 3D models were made from data taken following lines perpendicular to the coast line with an equidistance of 100 meters, making the data density different between data parallel to the coast line and those that are perpendicular, making the model much more liable on the zones near the profile lines.
8. **Integration of bathymetry comparisons:** As there are seasonal variations between beach profiles in winter and summer, which are even more noticeable after severe storms, beach profiles adopt characteristic shapes recognizable in the bathymetries, experiencing severe erosion in the emerged zone and those sediments are accommodated in the zone

affected by coastal dynamics. Those zones are not well defined when there are remarkable tides and that is the particular case of the zone surveyed with tide range of about 3.5 meters. It is noticeable that if there is sediment transport along the coast summer profiles are never recovered, so there is irreversible erosion. So it is useful to know the extent of these phenomena and it is necessary to make an in-depth study on the existing currents and sediment transport in potential and real cases.

5 Results

In this first phase of the Project the geographic information system is a useful tool that makes it possible to have all the information from the surveys incorporated into a system that is easy to operate, simplifying the design and obtaining the products needed by the user. Each of the elements added to the GIS has geographical references. Among those we find all sediment samples taken during the two bathymetric campaigns, along with the information table, such as the one shown in table 1.

Other elements added to the system are both summer and winter bathymetry maps, which can also be overlaid to obtain the difference between both bathymetries.

The profiles for each beach were also drawn, geo-referenced and associated to an image that represents the profile.

Table 1: Table showing D50 data from all samples.

Sample Location		Depth						
Beach	Profile No.	4	2	0	-1	-2	-3	-5
REGLA	2	0.32	0.36	0.72	Rock	Rock	Rock	*
	8	0.27	0.21	0.82	0.29	0.10	0.37	Rock
	13	0.32	0.25	0.25	0.14	0.18	Rock	Rock
	16	0.17	0.17	0.18	0.17	Rock	Rock	Rock
	18	0.27	0.17	0.18	Rock	Rock	Rock	Rock
BALLENA	8	0.20	0.21	Rock	Rock	0.42	0.17	Rock
	16	0.35	0.22	0.24	0.28	0.16	0.15	0.16
	22	0.32	0.22	0.28	0.40	0.15	0.16	0.09
	30	0.36	0.51	0.32	0.34	Rock	0.14	Rock
	36	0.16	0.36	0.28	0.22	Rock	0.32	0.31
COSTILLA	2	0.18	0.34	0.29	0.56	Rock	Rock	Rock
	6	0.32	0.33	0.54	0.36	0.45	Rock	Rock
	12	0.22	0.32	0.31	0.37	0.31	Rock	Rock
	16	0.23	0.51	0.33	0.20	Rock	Rock	Rock
	24	0.30	0.27	0.90	0.23	Rock	Rock	Rock
ROMPIDILLO	2	0.25	0.10	0.25	0.38	0.30	0.25	Rock
	4	0.32	0.16	0.13	0.28	Rock	Rock	Rock
	6	Rock	0.23	0.17	0.29	Rock	Rock	Rock
	9	0.23	0.76	Rock	0.12	0.18	Rock	Rock
	12	0.25	0.30	0.22	0.13	0.18	0.15	Rock

* Not enough depth

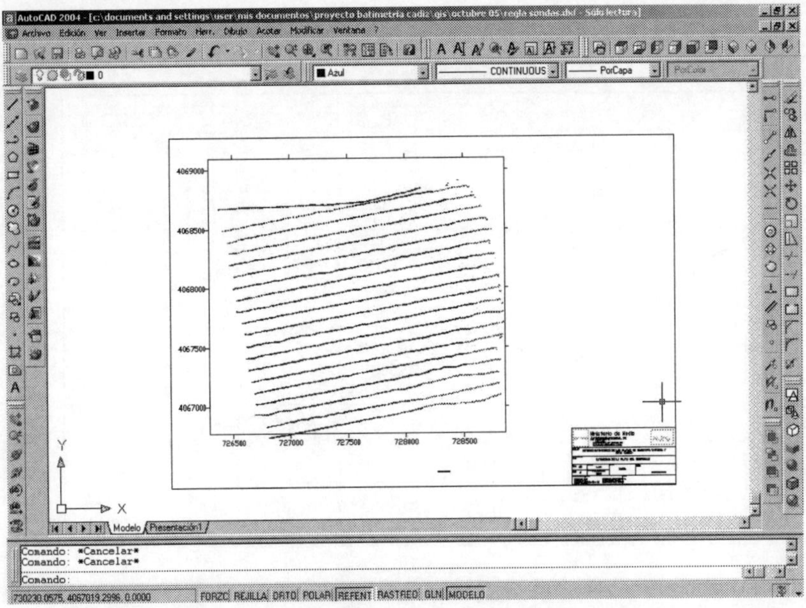

Figure 2: Bathymetry example using AutoCad.

On the other hand, more products can also be obtained depending on the information needed by the user. In this particular case two of those multiple products are shown, and these are the layer layout selected by the user and the printed report for all data previously selected.

6 Development of sustainability indicators

Given the fragility of the ecosystems present in the coastal zone surveyed, the importance derived from its legal nature justifies the diverse attempts to provide instruments and orientations for the integral coastal management in the European Union [11].

The interest and the necessity for sustainable development, the current concerns being faced that threaten the environment and the defective management carried out lately have also led to a re-examination of the means at our disposal to evaluate and monitor the evolution and trends of the environment, as well as those of the use of natural resources and development processes (Winograd [12]; EU [13]; Backhaus [14]).

The decision making process along with the analysis and monitoring of policies and development strategies often uses data, statistical economic and social indicators at regional and national level (UNEP [15]; UNDP [16]). However, regarding the environment, the equivalent information is not available to the user and frequently it does not even exist, preventing decision making regarding the environment from being carried out unless it considers components and characteristics from real processes. So it is necessary to provide the

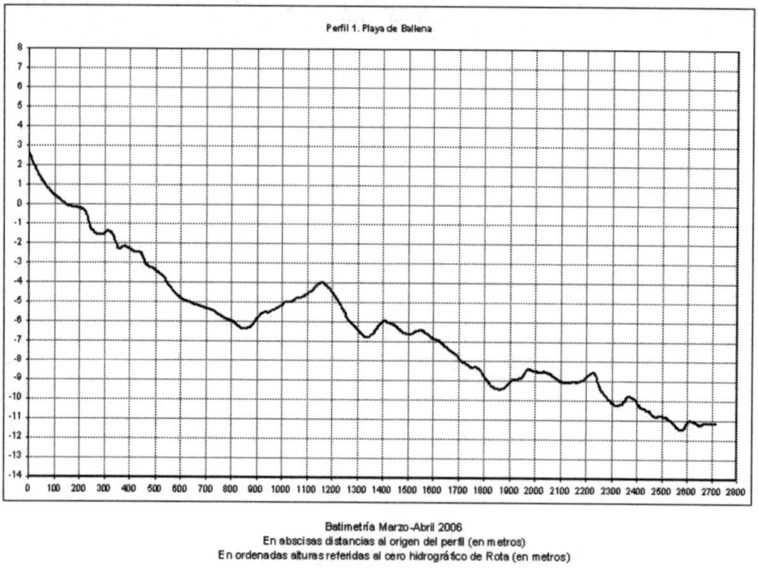

Figure 3: Profile example.

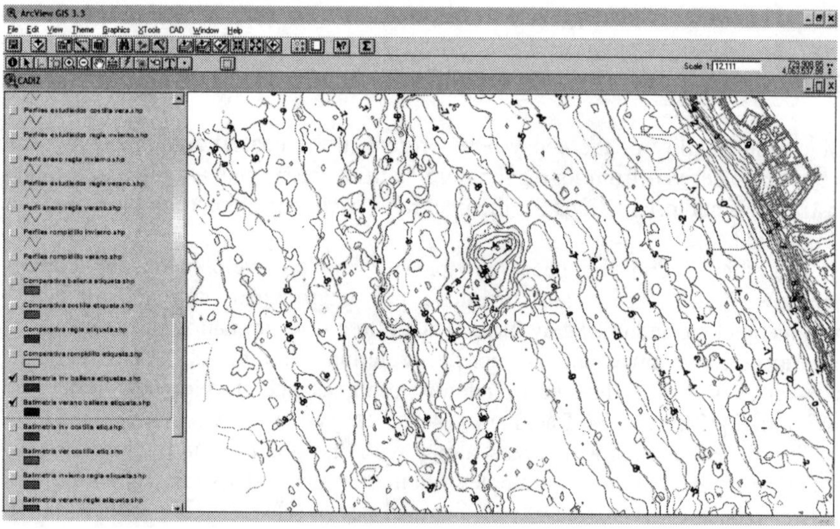

Figure 4: Comparison between winter and summer bathymetries.

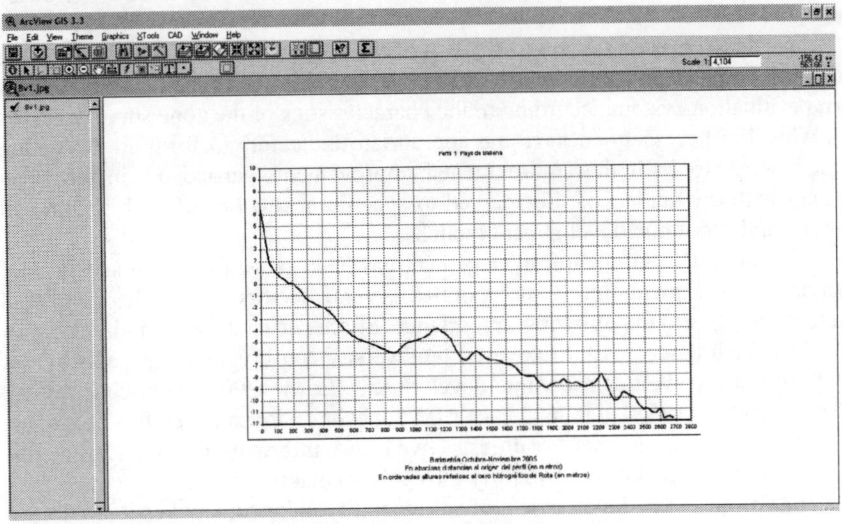

Figure 5: Example of profile.

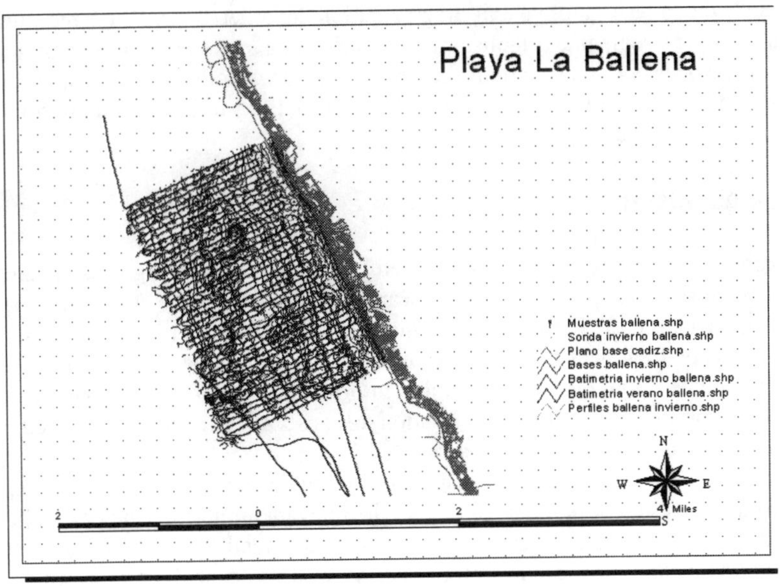

Figure 6: Example of layer layout.

indicators needed to create an important tool in terms of communication, making scientific and technical information available to many groups of users (Lindsey et al. 1977).

In order to make a correct evaluation of the actions proposed and to check the consequences of those actions it is necessary to have a coherent measurement and evaluation system according to the characteristics of the zone surveyed. That is why it is necessary to have the appropriate indicators to form an evaluation and control tool for the environmental improvements carried out in the zone, along with the quality of life adapted to specific necessities of each region and social and economic development model.

The coastal zone integral management has been recognized as a sub-field of environmental planning. It incorporates several sectors and gathers regional relevance, given the growing importance of the coastal section in terms of healthy feeding, decrease of poverty, biodiversity conservation, and reduction of natural risks as well as economic development (Lowry 2002). Consequently, all those actions carried out in the coastal zone have to consider the economic, social and environmental realities. Given the information complexity it is necessary to have tools that simplify the decision making.

In this sense, the development of sustainability indicators makes it possible to consider the criteria that integrate the three spheres of action (economic, social and environmental), which are also integrated into the geographic information system, making it possible to update and look up any information on the indicators in a quantity and quality way.

From a methodological point of view it is useful to make a diagnosis to achieve the physiographic unit fragmentation in order to evaluate the current state of every segment, establishing each of the characteristics in common and the differences and interrelations that they have between them. This analysis is to establish possible guidelines to follow from the actions to be carried out for each coastal zone considered.

7 Conclusion

Geographic information systems are useful tools when managing great quantities of spatial information as well as cartographic data. The Integral Coastal Management and the performance of the corresponding water law necessitate the simultaneous treatment of a mass of information, a great percentage of which is geo-referenced, making the geographic information system an adequate tool for the appropriate use of that information because all data is transformed into cartographic or alphanumeric data.

With the integration of sustainability indicators into the geographic information system it is intended to give the relevant authorities the possibility to use a tool that is useful to evaluate the actions intended to be carried out or just for the analysis of restoration proposals, as well as to analyze and justify the decisions adopted or not adopted to the group of stake holders from the coastal field.

References

[1] Almazán, J., (2008), "El Régimen jurídico de los Puertos del Estado. Agenda de Legislación" Edit. E.T.S. Ing. Caminos, C. y P.

[2] Almazán, J., (2006), "El Régimen jurídico de los Puertos del Estado." Edit. E.T.S. Ing. Caminos, C. y P.

[3] Almazán, J., (2000), "Introducción al diseño de obras de defensa de formas costeras de depósito". Edit. E.T.S. Ing. Caminos, C. y P.

[4] Caturla, J.L. (1988), "Sistema de posicionamiento global (GPS)".MOPU.

[5] Caturla, J.L. (1978), "Determinación de puntos de Laplace". Madrid. Inst. Geográfico Nacional.

[6] Hein, G.W.(1986), "Integrated geodesy: State-of-the art 1986 reference text" in: Sünkel, H.(ed.), Lectures Series in Earth Sciences, Vol. 7, Mathematical and Numerical Techniques in Physical Geodesy, New York.

[7] Hein, G., B. Eissfeller: "The basic equations of carrier phase measurements to the Global Positioning System including general orbit modelling, Schrifttenreihe University Studiengang Munchen, num. 19, 1986.

[8] Almazán, J., (2002), "Posicionamiento y navegación de precisión en 2D y 3D: batimetrías de alta precisión". Edit. E.T.S. Ing. Caminos, C. y P.

[9] Gavala Laborde, J. (1992), "Geología de la Costa y Bahía de Cádiz". Diputación de Cádiz.

[10] Bosque, J. (2006), "Sistemas de información geográfica y localización de instalaciones y equipamientos", Edit. Ra-Ma.

[11] De la Peña, J. (2007), "Guía técnica de estudios litorales: Manual de costas", Edit. Colegio de Ingenieros de Caminos, Canales y Puertos.

[12] Winograd M., 1995, Environmental Indicators for Latin America and the Caribbean: Towards Land Use Sustainability, GASE in collaboration with IICA- GTZ Project, Organization of American States and World Resources Institute, Washington, D.C., 85 pp.

[13] "EU Member State experiences with sustainable development indicators". European Comission, 2004.

[14] Backhaus, R., 2002, "*The spatial dimension of landscape sustainability".*

[15] *"Selected Satellite Images of Our Changing Environment". UNEP 1993.*

[16] *UNDP, 1994.* "Statements and Recommendations from Major International Meetings on Water Resources, Water Supply and Sanitation". UNDP Science, Technology and Private Sector Division.

Author Index

Environmental Problems in Coastal Regions VII